Principles and Design
of Production
Control Systems

EVAN D. SCHEELE *Vice President*
Central Management Division
Rapid American Corporation

WILLIAM L. WESTERMAN
Executive Vice President
The Rapid Electrotype Company

ROBERT J. WIMMERT, Ph. D.
Associate Professor
Department of Industrial Engineering
University of Florida

Principles and
Design

34-36 Beech St., London, E. C. 1

# of Production
Control
Systems

Prentice-Hall International, Inc. *1960*

PRINCIPLES AND DESIGN
OF PRODUCTION CONTROL SYSTEMS

Evan D. Scheele
William L. Westerman
Robert J. Wimmert

© 1960 by PRENTICE-HALL, INC.
Englewood Cliffs, N. J.

Library of Congress Catalog Card Number:
60-15161

PRINTED IN THE UNITED STATES OF AMERICA
70114-C

Preface

The purpose of Principles and Design of Production Control Systems is to develop a scientific approach to the solution of planning and control problems in *any* type of production or management activity. It describes specific steps, in chronological order, for the design and use of Production Control Systems. With emphasis on design, the text presents the fundamental guides for organizing sound systems for Production Control including the following essentials:

1. Applying the principles of Production Control to any activity.
2. Making use of existing equipment and techniques as aids to Production Control.
3. Demonstrating methods of analyzing and evaluating production control problems.
4. Developing the ability to adapt systems for continued effectiveness.
5. Setting up the relationship of the system to other management functions.

All of the vital elements of a Production Control System are discussed in the text with regard to the principles for their effective operation and their relationship to the total system.

Each chapter is devoted to an individual major function of Production Control or a related management and industrial engineering technique. This enables the reader to proceed in a systematic manner through the succession of Production Control fundamentals, functions and tools of design and analysis. Summaries and chapter questions and problems are also provided.

The reader will also find descriptions of the latest mathematical techniques useful in Production Control. He will also find tested and proven methods of solving Production Control problems as well as new ideas for continuously evaluating and improving his Production Control systems.

We wish to express our deep appreciation for the help and encouragement obtained from many friends and associates. We are indebted particularly to the staff of the Ordnance Management Engineering Training Agency and its Director, Mr. A. Lynn Bryant, for the contributions and interest in the manuscript; to Dr. M. E. Mundel and Dr. I. P. Lazarus for their advice and assistance; to Mr. M. Riklis, President of Rapid-American Corporation for his cooperation; to Mrs. Marianne Pirrung for her typing assistance; and to our wives, Elizabeth Ann Scheele, Rosemary Wester-man, and Jane Wimmert for their patience and encouragement throughout the preparation of the manuscript.

Evan D. Scheele
William L. Westerman
Robert J. Wimmert

Contents

ONE

page 1

What is Production Control?

Need for production control, *Comparison of military history with production control*, Objectives and benefits of production control, *Directing, Coordinating, Controlling, Innovating*, Organization of production control systems, Fundamental approach vs. case study, Purposes of book, *Summary*

TWO

page 12

Human Relations in Production Control

Summary

THREE

page 16

The Functions and Principles of Production Control

The planning phase, *Prior-planning, Action-planning phase*, Action phase, Follow-up or control phase, *Progress reporting, Corrective action*, Principles of a sound production control system, *Summary*

vii

FOUR

page **27**   **Types of Production Activities and Processes**

Continuous or mass production, Similar process, Job-order or intermittent, Custom work, *Summary*

FIVE

page **35**   **Basic Designs of Production Control Systems**

Flow or continuous process production control, Similar-process control, Job-order control, Special-project control, *Summary*

SIX

page **40**   **Forecasting Techniques**

Need for forecasting, and its uses, Types of forecasting, Forecasting techniques, *The historical estimate, The sales force estimate, Trend lines, Market survey, Correlation, Prior knowledge as a substitute for forecasting,* Use of judgment in forecasting, When the forecast should be made, Analyzing the forecasting function, *Summary*

SEVEN

page **55**   **Forecasting Control**

When the forecast should be revised, Control-limit application, Example, Limitations of control-limit application, *Summary*

EIGHT

page **66**   **Work Authorization**

Basic requirement, Relationship between work authorization and other management controls, Externally-generated work, Internally-generated work, Analyzing the work authorization function, *Summary*

NINE

page **72**   **Product Planning**

Important considerations in extending the original product

information, Value analysis, Problems created by lack of product planning, *Summary*

TEN

page **79**   **Process Planning and Routing**

Prerequisite information needed for process planning, Steps in process planning, Information generated as a result of process planning and its related activities, Breakeven analysis, Relationship between process planning and methods improvement, Advantages of thorough process planning, Analyzing the process planning function, *Summary*

ELEVEN

page **90**   **Material Control**

Use of the bill of materials, types of material control situations, Physical control, Record keeping, Two-bin material control system, The supermarket concept, *Summary*

TWELVE

page **109**   **Material Management**

Economic order quantity, *Procurement or set-up costs, Carrying costs, Objective of determining the economic order quantity, Formula determination, Simplification of calculations, Modifications of the basic formula,* Classification of inventory by dollar usage, *Modifications for limited quantities,* Modifications for quantity discount and variable usage, *Quantity discount, Variable usage,* Introduction to reorder point determination, "Scientific" reorder point formula, *Procurement lead time, Usage during lead time, Cost considerations, Reorder point formula, Classification of inventory for reorder point determination, Summary*

THIRTEEN

page **131**   **Tool Control**

Procurement of tools, Control of tools, Analyzing the tool control function, *Summary*

FOURTEEN

page **138**   **Work Measurement**

Stop watch time standards, Other forms of work measurement, Analyzing the work measurement function, *Summary*

FIFTEEN

page **150**   **Loading and Scheduling**

Information required for loading and scheduling, Developing a loading and scheduling system, *Selection of the unit of measure, Selection of the facilities or operations to be controlled, Selection of the techniques to be used,* Types of loading and scheduling systems, *Master scheduling, Perpetual loading, Order scheduling, Loading by schedule periods,* Determination of the available standard hours, *Summary*

SIXTEEN

page **168**   **Mathematical Loading and Scheduling**

Statistical load control, Index method of scheduling, Linear programming method of scheduling, *Summary*

SEVENTEEN

page **183**   **Commercial Loading and Scheduling devices**

Sched-U-Graph, Produc-Trol Board, Boardmaster, Visual control panels, Summary

EIGHTEEN

page **192**   **Dispatching**

Plant manager level, Plant superintendent level, Foreman level, Operator level, Centralized and decentralized control of dispatching, *Centralized control of dispatching, Decentralized control of dispatching,* Summary

NINETEEN

page **199**   **Progress Reporting—Principles**

Why "too much"? *Progress reporting at fixed time intervals, Progress reporting by work accomplishment,* Why "too late"? Basic methods of progress reporting, *Written systems of progress reporting, Oral systems of progress reporting, Electronic systems of progress reporting, Summary*

TWENTY

page **209** **Progress Reporting—Systems**

Prewritten papers, Pneumatic tubes, TelAutograph tele-scriber, Teletype equipment, Telephone and intercom equipment, Radio and closed-circuit television, Telecontrol equipment, *Summary*

TWENTY-ONE

page **220** **Data Processing**

Binary numbering system, Marginally punched cards, Punched cards, Digital computers, *Summary*

TWENTY-TWO

page **237** **Corrective Action**

Factors creating the need for corrective action, *External factors, Internal factors,* Methods of corrective action, *Schedule flexibility, Schedule modification, Capacity modification, Summary*

TWENTY-THREE

page **245** **Procedure Analysis**

Standard procedure for procedure analysis, *Step 1—objective, Step 2a—analysis, Step 3a—evaluation (procedures), Step 2b—analysis, Step 3b—evaluation, Step 4—improving, Step 5—test, Step 6—trial, Step 7—application, Summary*

TWENTY-FOUR

page **266** **Work Sampling**

Case study: time distribution of production control dis-patching office, *Summary,* References

TWENTY-FIVE

page **277**   **Evaluating a Production Control System**

Qualitative evaluation, Quantitative evaluation, *Work load analysis, Planning and performance analysis, Cause analysis, Summary*

TWENTY-SIX

page **285**   **The Place of Production Control**

Over-all goals, Range of activities, Reporting level, Specific authority, Relationships to other management functions, *Summary*

TWENTY-SEVEN

page **296**   **Designing the Production Control Organization**

Define the specific production control functions, Define the type of personnel and determine where they are to be obtained, Determining the size of the production control staff, Fitting men to functions, Training personnel to perform production control functions, Planning and controlling the production control activities, Establishing the methods of evaluating the performance of production control, *Summary*

page **305**   **Appendix I—Statistical Control Charting**

page **308**   **Appendix II—The Least-Squares Method of Linear Regression**

page **311**   **Appendix III—The Modi Method of Linear Programming**

page **319**   **Appendix IV — Data Processing Equipment**

page **329**   **Questions and Problems**

1

What Is Production Control?

Production Control is defined as the design and use of a systematic procedure for establishing plans and controlling all the elements of an activity.

The key words are "design" and "use," as will be explained in considerable detail in later chapters. At this time let us fix in mind that the main problems in Production Control are involved with:

1. *Designing* a sound and systematic procedure.
2. Properly *using* the system that has been designed.

In order to understand fully the meaning of our definition, it may be worthwhile to consider some of the other key words:

Systematic Procedure: A complete set of facts which have been arranged in a rational dependence or connection and are carried on with thoroughness and regularity.

Plans: Methods or schemes of action, procedure, or arrangement.

Control: To exercise restraint or direction over; to check or regulate.

We can see from these definitions that Production Control includes:

1. A complete plan.
2. A follow-up procedure for determining how closely the plan is followed.
3. A means of regulating execution to meet the plan's requirements.

1

The function of Production Control is to investigate the various means of setting up plans to perform an activity so that all the facts pertinent to the activity will be available *before* the activity has started. These facts must be combined in such a manner that all the elements of the activity take place in a harmonious relationship. Then a means must be provided for determining whether or not the plans are being properly followed and, if not, for taking action to regulate the performance of the activity to meet the requirements of the plans.

Generally speaking, the term *"Production* Control" implies the manufacturing activity, but we would avoid such references and implications for a very good reason. In the following chapters the intention is to develop a *scientific approach* to the solution of the problem of planning and controlling *any* type of activity. This may include activities as diverse as Manufacturing, Maintenance, Research and Development, Engineering, Warehousing, Purchasing, Quality Control, Time Study, Methods Engineering, Production Control, and General Office Activities. Production Control is planning and control in its broadest sense; the end result of whatever work is done is the object of the production.

NEED FOR PRODUCTION CONTROL

Now that we have the definition of Production Control, the logical questions that might be asked are:

Why is Production Control needed?
What will it do for an activity?

These questions can be answered simultaneously. The answers date far back into history.

The fundamental concepts of Production Control have been in use many centuries, and yet they have been scarcely considered in many industrial plants of today. Let's take a brief look into the history of Production Control practices and then consider what their status is now.

Comparison of Military History   The real beginning of Production
With Production Control   Control may be traced to the military;[1] in fact one of the first good examples of sound practice is found among the Roman legions. The strength of the Romans depended upon three things:

---

[1] It can be shown also, that many other present day concepts of "scientific management" have been modelled after military practices. These include Organization Analysis and Statistical Quality Control. Even the latest "scientific management" techniques such as mathematical programming have mainly been

1. Careful selection of men.
2. Strict discipline.
3. Prompt improvement in areas of deficiencies as suggested by experience.

It was mainly by following these principles rigidly that the military arm of Roman empire became unbeatable.

Defeat came to Roman armies after conditions made it necessary to violate one of the principles—that of careful selection of men—by placing heterogeneous classes together and hiring professional fighters to replace the civilian troops. This destroyed the pride for Rome and the respect for valor impressed in Romans. In its place was only the lust for wealth. The loss of morale and moral nerve in its legions combined with other factors to doom the Roman empire as a tightly-knit political hegemony.

Now compare this with the defined concepts of Production Control. Obviously, the smooth functioning of any activity—its strength—is based on the same principles:

1. Competent, qualified personnel must be selected to design and use any Production Control system.
2. The system itself is the discipline. Unless it is adhered to rigidly, it is useless.
3. If any deficiencies are found either in the system or in the activity which the system is designed to control, they must be corrected promptly if the activity is to continue functioning smoothly.

After the decline of Rome, the center of the world's military might was transferred from Rome to Constantinople, the seat of Byzantine power. The military leaders of this empire made war a game of clever plans and finesse. They never plunged recklessly into any battle, but always first assured themselves of every possible advantage. Only after the destruction of its patriotic middle classes and the robbery of buffer-states protection did the empire break into fragments.

Parallel application to industrial activities can be readily drawn. Plans and finesse are vitally important to assure every possible advantage in accomplishing the objectives of an activity. To plunge recklessly into the execution of the work more often than not spells failure or certainly minimal results.

After the Roman and Byzantine empires had fallen, political rule in Europe belonged to relatively small feudal lords. Changes in weapons made armor and fortification of towns necessary. Defense took precedence over offense until increased raiding demanded combination into larger units. Combining was hastened by the introduction of firearms. Soon also the fortifications of castles and cities were useless.

---

the result of efforts of the military to provide a more sound approach in supplying its vast needs. These techniques, once proven, have been eagerly grasped and further developed by commercial industry.

The greatest example of combining forces and obtaining absolute rule is Frederick the Great. He concentrated all controlling powers in his own hands, reducing everything else to clerical duties and mechanical obedience. No subordinate dared use his own initiative and individuality. Frederick's calculations were accurate and his plans were limited to what was possible. His greatest fault, however, was that he trained no one to carry on his work, apparently being able to do things only through himself. He handed down a system to his successor which required a Frederick to administer.

After his death, his prestige was so great that it was considered blasphemy to suggest that any change could improve the Prussian army. The result was a complete rout of this army by Napoleon twenty years after Frederick's death.

Although the aims of Frederick and Napoleon differed in that Frederick was devoted wholly to the advancement of Prussia while Napoleon desired personal power and was selfish and vain, they possessed the same type of defect. They failed to provide for an adequate staff to relieve themselves of details. During Napoleon's early years he made up for this lack by remarkable physical activity. In early middle age his energy began to decline. In his thirst for personal honor he would not permit able military leaders under him to share his honors. Without this advice and assistance he plunged to his Waterloo.

This pattern of complete individual control can be readily found in industry. Systems are all too often built around personalities rather than on sound principles. This is particularly true where strong-minded managers are concerned. Their way is the only way, with no concern about the necessity of carrying on in the event that they should suddenly be absent from the picture. A sound system of Production Control includes thoroughly training subordinates in all of the minute details of administering the system and the maintaining of continuous audits for the purpose of constantly evaluating the system for improvements.

In modern military tactics great emphasis is placed on preliminary planning and careful assignment of duties to separate individuals without overloading any one person. Great flexibility is obtained by centralizing only for the development of the general features of the plan of operations to assure harmony in the execution of the plan. The details of execution are entrusted largely to the officers in the field, and those in direct command of the minor divisions of troops. Discipline is now broadly interpreted to mean that each individual shall apply sound principles in every emergency.

This calls for a superior class of executives of all ranks, adequately prepared for their duties. Each executive, at whatever level, is required to use the principles of sound planning and control to assure the satisfactory accomplishment of the work to which he is assigned.

Throughout history we can find example after example of how a great power arose as the result of thorough planning and coordination based on sound principles of planning, operations, and administration. In nearly every case its subsequent downfall was the result of overconfidence or the over-reaching of power and capabilities which in effect was the violation of those principles upon which their power rested.

In surveying a few points of military history in comparison with industrial organization we do not wish to imply that business should be run exactly as an army is run or that business is a kind of warfare. The purpose of comparison is to show:

1. Production Control has been in use for many centuries in the military.
2. General principles are the same and form the working basis of both.

It would be rather difficult to conceive of a successful field commander who did not follow a well-recognized and orderly procedure in the planning for and during the conduct of an operation. Having been assigned a mission, his first step is to *estimate the situation*. Then follows *preparation of the plan of action* with computation of the necessary support, time schedules, and phase lines of intermediate objectives leading to the main objective. The conduct of the directed operations is the *scheduled execution of the plan*. The status of execution is then reflected by complete and up-to-date situation maps and reports which show at a glance the *progress made* as well as the *spots where trouble is developing*. Complete *review and analysis* of all areas of action is *continuously* being made.

Considering the tremendous success of planning and control of military operations, it is little wonder that industry so often looks to the military for the latest techniques of management. Prime examples of these are (1) Production Control, (2) Statistical Quality Control, and (3) Operations Research.

The first record of a complete production and control system applied to a manufacturing plant was at Watertown Arsenal in the 1880's. This was followed by Taylor's and Gilbreth's concepts of scientific management which brought greater emphasis to the need for better planning and control and the tools and techniques which would bring it about. The continuing work in this area has resulted in the refinement of these tools and techniques to the point where it would be almost unthinkable for a modern industrial plant to be without them; certainly the plant that does not have them is very inefficient and usually fighting for survival.

Sound Production Control is basic to good management. It is the only means by which all levels of management can accurately and effectively forecast, plan, and control the operations of the organization. An industrial plant might have the most efficient production facilities and methods in the world, but they are of little value without controls to insure the accomplishment of what is intended.

OBJECTIVES AND BENEFITS OF PRODUCTION CONTROL

Sound Production Control may result in many tangible and intangible benefits if properly installed and operated. In order to obtain these benefits, however, it is essential that adequate auditing of the system be made continuously. Methods of auditing Production Control systems will be discussed in Chapter 25.

Following are some of the objectives and results of a sound Production Control system, categorized in terms of the common management functions of Directing, Coordinating, Controlling and Innovating:

Directing

1. Efforts can be directed and channeled into those production areas that will contribute most towards accomplishing a given objective or task.
2. Programs can be closely geared to the wants and needs of the company.
3. The risk is reduced that higher management will become absorbed in minor details that appear important at the moment to the subsequent neglect of the over-all goals of the basic program.
4. Manufacturing or production cycles are shortened, which in turn reduces in-process inventory costs and provides better customer service.
5. Work must be performed according to the preplanned schedules or priority lists; easy and hard jobs are distributed according to the objective needs of the scheduler and cannot be selected or assigned out of sequence.
6. Supervisors are forced to take corrective action when it is necessary.

Coordinating

1. The employees can be made more aware of the part they play in the business or company.
2. Information is provided quickly to customers concerning the status of their orders.
3. Over-all expenses are reduced because of systematizing and reducing the amount of paperwork involved.
4. "Production" is maximized by making greater use of facilities, equipment, and manpower through sound scheduling and loading.

Controlling

1. Less time will be spent by management and supervision reading and analyzing reports on activities.
2. Necesary information can be provided for determining where and when preventive or corrective action is necessary.
3. A "yardstick" is provided by which management can measure both the progress and the effectiveness of the activities in which the company engages.
4. Administration of the activity is put on a factual basis rather than one of experienced guesswork.
5. Tool and material costs are minimized by placing responsibilities where

they belong and maintaining up-to-date records of all activity in these areas.

6. Reports are more timely, adequate, and accurate.
7. Continuous evaluation of the effectiveness of the planning and control system and of other functions is made possible.
8. Graphical or visual presentation of data is facilitated.

Innovating

1. Time becomes available to work out details that would otherwise be left to improvisations.
2. Time-phasing of all elements of the activity becomes a necessity. This includes the preparatory functions as well as the execution functions.
3. More flexibility is obtained to accommodate necessary changes that occur in schedules or orders.

ORGANIZATION OF PRODUCTION CONTROL SYSTEMS

Recognizing that Production Control exists in any kind of activity because of the fact that the functions have to be performed by someone at some time, we can classify the organization of Work Planning and Control systems.

Generally speaking, Production Control can be organized three ways:

1. Formal
2. Informal
3. Combination of the formal and informal

Formal Production Control should be understood to mean centralizing the coordination of the various functions of Production Control in order to establish and sustain the best method known at any given time. It means that a systematic paperwork procedure takes the place of people physically following each step in the process. It means that the foremen and supervisors are relieved of the responsibilities of doing most of the detail planning and can concentrate on "supervising" and training. It means that management is given "up-to-the-minute" reports on the progress of all activities or that a "line of balance" can be struck at any given time without physically taking inventory of the production position of the activity.

The *informal* system is one in which the first-line supervision, or the manager, does everything—plans, requisitions materials and tools, determines when each job shall be done, determines what shall be done, what equipment to use and who shall do the job, and then personally watches each job to be assured that it is being run exactly as he wishes it run. The only paper involved is usually a "little black book" which the supervisor keeps in his hip pocket.

The *combination* formal and informal system is one which has elements of both. In such a system, for example, a Production Control department may do the overall planning of a department and forward work to the department as soon as the work authorization is written. The department foreman would then do all the planning for materials and tools and scheduling himself. The foreman would then report completed jobs to Production Control. A growing industry or activity will usually work out such a combination of "systems" during the transition period from informal production control to formal production control.

We might also break down the various control systems into another set of classifications:

1. Systems operated by chance
2. Systems operated by custom
3. Systems operated by choice

Either the formal or the informal system may be any one of these three types. In most cases we will find that those systems which are operated either by chance or by custom are lacking many of the requisites of a sound system.

Chance systems usually result in a haphazard assortment of forms, filing equipment, graphs, verbal instructions, and expeditors that have been accumulated over a long period of time. Much of the chance system is due to high-pressure salesmen of forms and equipment, and to changes in management, each new manager throwing out a little something and putting something else in its place or just adding a little more to the existing conglomeration "to get control" of the situation.

Most of the informal systems of planning and control can be readily identified as those operated by custom. When this system is used one often hears the reasoning, "Well, it worked for us for the past 20 years and we made money, so why should we change?" True, the system may have worked quite well in the past under the circumstances existing at that time, but it is usually not very difficult to identify the reasons why it worked. The primary reason is usually the fact that the supervision had the knowledge and experience personally to "control" the existing work. But what would happen to the activity if these experienced people were suddenly no longer available? It doesn't take much imagination to anticipate the inevitable result—complete chaos. The same thing would probably result if the company or the activity were forced to expand the number of different "products" that it is producing. At some point it becomes unfeasible to retain the informal system or the system established and operated by custom. It should be pointed out, that the criterion of profit given above is no defense of any system unless the proper benchmarks are used. The appropriate question is not "How much profit *did* the activity make?" but "How much profit *could* it make if it operated with maximum efficiency?" The former statement gives no objective measure of efficiency.

Systems operated by custom are not limited to the informal type. Indeed, most activities will retain procedures of all types merely because "That is the way we have always done it." For example, in one plant a particular foreman was keeping a rather elaborate chart of his activities. It was placed on the wall of his office for all to see. An assistant spent about one-half hour each day gathering the data and neatly posting it on the chart. When the foreman was asked what the data was, he readily and proudly explained it. After some questioning, he admitted that he did not use the information on the chart for any particular reason and that it was never necessary for him to refer to the chart for the information in any event. However, the chart was there when he took over the job and he was merely keeping it because it had always been done. The foreman had been on this job for about two years. After further investigation, it was learned that the chart was started because someone from higher management had asked for the information at one time and that the foreman in charge was not able to give a quick answer. This resulted in a sound reprimand for the foreman who quickly set up the chart so that he would not be "caught" again. Yet, in the two years that the present foreman was on the job, the information was never requested because a change in the paperwork system gave management the information it needed.

This, of course, is more an indictment of the management for not continuously auditing its systems than it is of the foreman who was just carrying on in a customary way. This is a very common occurrence in many activities.

Obviously, our objective is to present methods of designing the best system of Production Control, everything considered. This would be a system of *choice*. In most cases this would involve the formal system, keeping in mind, however, that the informal system may be the most feasible under certain existing conditions. This leads, then, to the fact that in order to be able to design the "best" system we must have some guides or principles to follow.

FUNDAMENTAL APPROACH vs. CASE STUDY

"There are no two things exactly alike." This single statement perhaps more than anything else will describe why the case-study[2] approach to the design of a system for Production Control will not suffice. In nature and in business it is virtually impossible to find two things which are identical in every respect. Although many things *seem* to be exactly alike upon first observation, upon minute inspection they will be found to differ in some respects.

---

[2] The case-study approach is the study of actual or hypothetical examples of production control systems in given situations.

It may be possible to find two manufacturing plants making the same type of product, using the same materials, tools, and equipment, having the same types of facilities, and being approximately the same size. We may still find, however, that a planning and control system of a certain type may work in one of these plants and be a failure in the other. Why? The answer may well be in the one factor which is always different—people.

It would follow then, that because differences will always be found in people, we will rarely find a complete Production Control system that we can "lift" from one plant or activity and install in another plant or activity. However, the basic principles or guides governing the design of the system will be equally applicable in all cases.

It might be concluded, then, that the problem is to develop a set of principles that will act as guides in the design of any Production Control system regardless of the particular conditions encountered. These principles will act as qualitative measures of the soundness of the system that is in use or that has been designed.

This does not mean that the case-study approach should be entirely discarded—quite the contrary. However, the emphasis should be placed on the fundamentals and principles and then using the case study or practical experience to develop or design the niceties of particular application. A thorough grounding in the basic principles eliminates the risk that is taken in administering an activity by the trial-and-error method and reduces the cost of educating those who become wise only through experience in disasters.

PURPOSES OF BOOK

So far we have considered the background of Production Control and the reasons why it is essential to good management. Although Production Control was one of the first contributions of scientific management, the so-called scientific approach has rarely been utilized in the design of Production Control systems. The following statement from Holden and Shallenberger's text *Industrial Management* sums up very adequately the position that Production Control holds in the management field.

> Of the many contributions of scientific management, none is of greater significance than Production Control. Unfortunately, it has been one of the least understood or fully comprehended features. Perhaps the greatest impetus to a wider appreciation of the subject was provided by World War II. The almost unbelievable performance of American industry during that period could have been achieved only through the application of good Production Control in most of the plants engaged in war work. Every student of industrial management should have a grasp of what Production Control involves, what are the objectives to be realized, how are these to

be attained, what are some of the basic techniques in common usage, where does the function fit in the normal factory organization, what factual information is necessary, how is it obtained, compiled and utilized.[3]

In order to accomplish the objectives of Production Control that have been previously described, it is essential that the designer of the system has knowledge and abilities other than just a knowledge of the steps in the scientific method. This book is intended to provide the necessary information to obtain the required knowledge, and to present the methods and techniques for developing the abilities necessary for designing sound systems for work planning and control. The purposes of the book are as follows:

1. To develop the fundamental functions and principles of Production Control as applied to any type of activity.
2. To present the variety of equipment and techniques presently used and obtainable as Production Control aids.
3. To demonstrate methods of analyzing Production Control problems and designing or evaluating and improving Production Control systems.
4. To describe the factors essential for effectively using a system and for continuously improving and adapting the system to changing conditions.
5. To describe the relationship of Production Control functions to the other management functions and to the work being programmed.

Only when the designer has obtained the knowledges described in 1, 2, and 3 and has developed his ability in 4 and 5 through experience can he hope to obtain the "best" system possible under existing conditions.

Summary   Production Control is essentially a systematic procedure. Effective Production Control is dependent upon a soundly designed system and proper evaluation and auditing of its use after proper installation. "Production" in this book is considered in its broadest sense—as applied to any constructive activity. It is virtually impossible to "case study" every system in advance, and so general principles of design and use are developed.

Although Production Control can be applied both formally and informally, this text will primarily present the formal organization.

People, as human personalities, play a major part in the use of Production Control systems. Because of this, the designer of the system should be cognizant of human relations. This subject is more fully discussed in Chapter 2.

---

[3] Holden and Shallenberger, *Industrial Management: Teacher's Manual* (Englewood Cliffs, N. J.: Prentice-Hall, Inc., 1953), p. 281.

2

Human Relations In Production Control

Human relations has been a problem for a much longer period of time than scientific management has been in use. The Industrial Revolution made crucial the need for consideration of human relations. However, in practical application, there was little or no thought given to the problems of human relations until scientific management was introduced. The advent of scientific management emphasized the necessity for consideration of the human being who performs work.

Certainly we recognize that human relations is a problem in Production Control. Everything that Production Control does affects some person or group of persons within the organization. Sometimes it is the worker who performs the actual operations, sometimes the manager of the worker, but most often it is both.

It was pointed out that differences exist between apparently similar types of activities and that one of the main differences is in the people. Obviously, working with people involves problems of human relations.

In Chapter 1 it was pointed out that the basic problem in establishing a Production Control system is to design a systematic procedure and after designing the procedure, to use it. Problems in human relations exist in both of these phases.

In designing a system for Production Control, the problem may be either to improve a portion of an existing procedure or it may involve the complete redesign of all possible phases of the activity. In either event it is necessary to collect data which will be used as a guide in the improvement or redesign of the procedure. This, of course, involves surveying the activity, talking to people who are involved in the activity, and collecting and collating the information obtained.

Anyone with experience in making such surveys can readily point out the difficulties encountered in trying to obtain factual data. Information may be highly distorted by some people, while other people will deliberately or unconsciously omit some very valuable points. This may be partly due to their natural resistance to any activity that may involve changes in the existing methods of operation or procedures being followed. Failure to get the full picture from supervision or management may be due to the fact that these people have a feeling that they are being criticized in the way they are doing their job. In either event, it requires skill in the art of human relations to sell people on the idea that the survey is being made for the purpose of improving the procedures which will ultimately result in benefits to everyone in the activity.

The design and use of sound Production Control requires cooperation from all levels of the organization, from the top executive down to and including the worker. This also includes full cooperation from salesmen, engineers, accountants and all others who may be directly or indirectly concerned with the new design and improvements. To get this cooperation, it is necessary to conduct either a formal or informal educational program on the benefits that will be obtained by everyone concerned. In some cases, this requires no less than a "super-salesman." It also requires that the designers and installers of the system proceed only as fast as people will accept their ideas.[1]

Many times failure may result due to the fact that the installer of the system attempts to do too much at one time. He would do much better to evolutionize the system instead of trying to revolutionize the system or procedure. Rarely will people accept a complete change very readily. Most people, however, will accept some change as long as it does not dis-

---

[1] It should be pointed out here that in some cases it may be necessary to force the acceptance of improvements on some people who refuse to go along with the majority. This usually requires some action on the part of top management. Except in rare cases, no individual should be allowed to disrupt what is good for the company or activity as a whole.

rupt their way of doing things too much for the present. Once the change has been made, and it is found that the new procedure operates more satisfactorily than the one used in the past, people will accept additional changes very quickly. People will also tend to accept new ways of doing things if they can observe these things being done satisfactorily by someone else. By designing and installing new procedures gradually throughout the plant, people will get an opportunity to observe what the improvements can do for them also.

Many of the differences of opinion that arise in designing new procedures can be resolved by following simple rules of human relations. People in general like recognition, and by giving them recognition in asking for their advice and opinions on matters concerning the procedures that will affect them, they are most likely to accept the ideas that are presented. At the same time, very valuable information often will be obtained from the people who are the most violent objectors.

A common method of obtaining information is to hold conferences with key operating and staff personnel. These individuals generally are the best source of information since they are most intimately connected with day-to-day operations, and should know the problems of their own specific operations. Frequently, however, operating personnel are reluctant to identify problem areas in their own organizational elements. As the designer and surveyor of new procedures gains the confidence and trust of operating personnel, the conference method will assume greater importance in identifying problem areas. By properly conducting conferences, each individual gets the recognition that he desires.

Another successful technique for stimulating operating officials' interest in the objectives for the improvement or redesign of planning and control functions is for the top official or management officer to discuss the plans at a staff meeting with all office and departmental heads present. A dual purpose is achieved by such meetings.

1. The key official can personally emphasize the support of the program.
2. The operating officials will have an opportunity to suggest desirable modifications in the plans.

One of the greatest obstacles to effective Production Control is the lack of full cooperation and support by top management. It is, therefore, necessary that management recognize and avoid the many pitfalls which prevent or discourage sound and effective Production Control. This, of course, means that proof of the value of Production Control must be presented to top management. This proof should not be just some vague opinions of those who are designing the improved Production Control system. It should be facts which have been obtained through survey of the activities involved. In addition, qualitative and quantitative audits may be made of the activity to gather evidence to support the opinions that planning and control is in need of improvement or redesign. Such audits will be presented in detail in Chapter 25.

In addition to the use of the audit and survey, it is desirable to present, in a summary form, all findings and proposals. This summary includes a complete resumé of both the tangible and the intangible savings that will be accrued with the proposed method.

With all the facts presented to top management, it is not likely that the proposals will be rejected. After all, what is more important to management than having all the facts pertaining to the ability of an activity to perform work? If the cost of getting these facts is less than the savings resulting from getting them, no manager worthy of the title could conscientiously refuse to accept the value of a sound Production Control system.

It is also of great importance in the art of human relations for Production Control that everyone connected with the Production Control system has a clear understanding of his duties and the authority that he obtains or maintains under the improvement or redesigned procedures. Of great value in this respect is a good organizational manual describing the duties and authorities of each person involved in the system. These aspects of Production Control will be discussed in Chapters 26 and 27. Another vital aspect is that decision-making for Production Control be placed at the proper level for the problems involved. This will facilitate obtaining management decisions on problems which are vital to improving and redesigning Production Control.

Summary   It is virtually impossible to design and use Production Control systems effectively without properly applying the art of human relations. A good designer must obtain full cooperation and support of all of the people concerned in order to be the most effective.

Backing by top management is another vital need for effective Production Control. Application of principles of human relations can aid in obtaining such support.

After recognizing some of the fundamentals involved in Production Control design and use, the potential designer of the system is now ready to investigate and develop a knowledge of the functions and principles of Production Control. These functions and principles will be defined in Chapter 3 and discussed in detail in succeeding chapters.

3

The Functions and Principles of Production Control

The over-all purpose of this text, it should again be emphasized, is to develop the knowledge necessary to design and improve systems for Production Control and to use them efficiently and effectively once they are installed. No one would dispute the fact that a good designer of anything, whether it is systems, tools, or machines, must have the basic or fundamental knowledge of the area in which he is designing. For instance, the machine designer may spend four or five years in college learning the fundamentals and principles of mechanical drawing, mechanisms, kinematics, strength of materials, and other related subjects. He may then be the understudy of experienced designers for several more years to learn how to apply his basic knowledge. It is certainly reasonable to expect that anyone who has the responsibility for designing systems for Production Control should also have a thorough knowledge of the fundamentals and principles of Production Control. This knowledge should include the basic functions of Production Control and their interrelationships. This is not to imply that the designer of Production Control systems should necessarily have a college education in this area (although it would be to his advantage) before attempting to design. Not all of the better designers in other fields have such a background in formal education, but it is almost certain that they have had the equivalent in apprenticeship or on-the-job training and experience. The Production Control systems designer may have just such a background. On the other hand, systems have been "designed" by those who had no more than their position of authority and responsibility as "background." Although common sense is a big factor in designing sound Production Control systems, just as it is in designing tools and machines, it will usually be found that considerably more than common sense is required to meet the specifications of a sound system.

The functions of Production Control may be divided into three main categories or phases:

1. The planning phase
2. The action phase
3. The follow-up or control phase

These three phases make up the main body of functions of Production Control. There are other secondary functions which are essential contributors to the efficient performance of the production control functions. In addition, there are other functions which are supported by these three phases but which are not generally considered to be direct functions of Production Control. These include quality control, cost control, and so on.

There is a very close interrelationship among the phases and functions of Production Control and they are mutually supporting. For instance, realistic planning is quite dependent upon the data which is compiled during the action phase. Action, in turn, is dependent upon continuous planning of the work to be performed by the activity. Follow-up is the comparison of the work that was originally planned against the work that was actually accomplished and is of concern to those who did the planning as well as to those who actually did the work. It must also be understood, that if plans are lacking, are not stated properly, or are recorded inadequately in terms of objectives, starting and completion dates, and utilization of resources, there will not be a basis for a comprehensive follow-up phase. Production Control must be viewed in its entirety.

In this chapter, we are going to view the complete scope of Production Control as it would generally exist in the formal organization. Table 3-1 shows, on the left side, all the functions that would normally exist in the job order type of manufacturing situation. Obviously, some of these functions would not exist formally in many other types of activities; we have chosen the manufacturing situation for its inclusiveness. The right side of Table 3-1 shows the terminology usually used to describe the activity that performs the functions shown on the left side.

In describing the functions shown in Table 3-1, it should be recognized that the terminologies may not only vary considerably between different organizations, but may also be found to vary considerably even within the same organization. It is not necessarily important that we use any particular terminology to describe a given function; the important thing is that the *function* is clearly understood. It may also be found that in some organizations other functions may be included in the Production Control department that are not even shown in Table 3-1. It is feasible under some conditions (especially considering the people involved) that purchasing, product design, production engineering, and others may be included. However, if the basic concepts of Production Control are kept in mind, the feasibility of the inclusion or exclusion of any other functions will become quite apparent after careful analysis and evaluation of the situation involved.

All of the functions described here must be performed by someone, at some time, in some degree, either formally or informally. They will be described here as applied to formalized Production Control, with a high degree of refinement. In later chapters parallels will be drawn to differ-

TABLE 3-1

FUNCTIONS OF PRODUCTION CONTROL

General Function	Phase		Formal Terminology
1. Estimation of Future Work		Prior Planning	1. Forecasting
2. Preparation of Work Authorization			2. Order Writing
3. Preparation of Specifications			3. Product Design
4. Preparation of Work Detail Plan	PLANNING PHASE		4. Process Planning and Routing
5. Determination of Requirements & Control of Materials			5. Material Control
6. Determination of Requirements & Control of Tools		Action Planning	6. Tool Control
7. Determination of Requirements & Control of Equipment & Manpower			7. Loading
8. Determination of When Work is to be done			8. Scheduling
9. Starting the Work	ACTION PHASE		9. Dispatching
10. Collecting Data	FOLLOW-UP PHASE	Progress Reporting	10. Data Processing
11. Interpreting the Data			11. Data Processing
12. Making Current Work Corrections		Corrective Action	12. Expediting
13. Making Plan Corrections			13. Replanning

ent situations showing varying degrees of refinement and formality. Constant reference will be made to Table 3-1 throughout the text. The reader should mark the location for ease of reference.

THE PLANNING PHASE

The planning phase consists of two parts:

1. Prior Planning.
2. Action Planning.

Each of these parts will be discussed along with the specific functions of which they consist.

Prior-Planning   Since "planning" already implies that a course of action is established in advance, we would expect "prior-planning" to refer to an activity in advance of the normal planning stages. However, to underscore the Production Control axiom that the whole activity must be planned, and exist "on paper," before the very first action takes place, we use "prior-planning" here to refer to whatever goes on in the planning phase prior to the first step initiating the activity.

This is done in order to make a clear distinction between planning on the "as you go" basis and planning as it should be done—well in advance of starting the work. The prior-planning phase in Table 3-1 is separated from the action planning phase by a horizontal dotted line. This line separates those functions which are not generally considered to be part of the Production Control organization from those below which are generally considered to be Production Control functions.

1. *Estimation of Future Work—Forecasting.* Forecasting is defined as the estimation of future activities. It is the basis for the projection of work load into the future, beyond the work which has been definitely assigned to the activity. It includes both long-range and short-range objectives or programs, from several months perhaps to five or ten years. These estimates of the type and quantity of future work assignments provide the basis for establishing the future requirements for men, materials, machines, time, and money.

By its very nature forecasting is subject to possible wide variations in accuracy. However, this error can be reduced to acceptable tolerances. Chapters 6 and 7 are devoted entirely to the consideration of the tools and techniques of forecasting and forecasting control.

2. *Preparation of Work Authorization—Order Writing.* If work is to be controlled, it must begin with a specified document authorizing it. The activities involved may be as varied as the documents themselves,

and some idea of the variety will be given in a later chapter. In an enterprise whose sales requirements are anticipated the document may be a manufacturing order; if the demand is a specific one, the document will be a customer's order, and so on. Equally variable will be the office that is authorized to issue the ordering document. The function itself, however, is invariably present, and in the well-controlled activity it is carefully specified. The order-writing function is described in detail in Chapter 8.

3. *Preparation of Specifications—Product Design.* After the work authorization has been prepared, the next step is to collect all the information necessary to describe the work to be done for all the people who are to do the work. In the manufacturing activity this would include blueprints or drawings, a list of specifications, a bill of materials, and so on. In other types of activities this may involve only a very brief written description of the specifications of the work required. The work authorization may even be sufficient description to accomplish the job. In the formal type of organization this activity would come under the product engineering department or an equivalent. This function is described in detail in Chapter 9.

Action Planning Phase   Under the dotted line in Table 3-1 are the steps which are usually performed by the Production Control department in the formal type of manufacturing activity. In any type of work activity these are the steps required for planning the details of the work to be done. All of them must be done by someone in some manner at some time.

4. *Preparation of Work Detail Plan—Process, Planning and Routing.* The function of preparing the work detail plans consists of two parts:

1. The determination of the most economical methods of performing an activity, all factors being considered. The formal terminology used to describe this function is "process planning."
2. The determination of *where* the work is to be done. For example; in continuous manufacturing (see definition in Chapter 4), as it is determined *how* the work is to be done, it is also determined *where* the work is to be done, building it into the "line." The arrangements of the work stations are determined by the route. This function is called *Routing.* In intermittent or job-order activities, routing generally refers to the route that is established for a specific job lot to follow through existing work stations.

In order to do the job of process planning, it is necessary to have information such as:

1. The volume of work to be done.
2. The quality of work required.
3. The equipment, tools, and facilities *available* to do the work.
4. The personnel *available* to do the work.

5. The schedule to show *when* the equipment, tools, and personnel will become available.

There is a distinction between *available* tools, equipment, personnel, and so forth, and the *availability* of these. Point 5 indicates that if tools and other needs are not immediately available to do the job and that if it is necessary to purchase them, it must be determined how soon they can be acquired. If on the other hand, the facilities are physically available but are presently in use, it is necessary to determine the time of their availability.

Routing is normally dependent on the plant or activity workload. In many situations the routing is done to departments or above, and the loading and routing of work stations is left to the discretion of the foreman or supervisor.

It should be pointed out that only by sound process planning and routing can valid preplanned standards of performance be established. These standards are essential to obtaining a controlled situation. The complete steps in process planning and routing will be discussed in detail in Chapter 10.

The next step in the planning and control of an activity is to determine what is needed to do the work. This includes materials, tools, equipment, and manpower required to do the work. Each one of these requirements can, in the formal activity, require a complete control system in itself.

5. *Determination of Material Requirements and Control of Materials —Material Control.* This function, generally speaking, receives more attention than any other Production Control function. Material or Inventory Control is vital to an activity not only because of the necessity to assure sufficient raw materials to satisfy production needs and finished products to satisfy customer needs, but also because of the money investment required. For these reasons it is desirable to maintain *optimum* inventory levels at all times. Because of the importance of material control in most manufacturing activities, refined, formal systems have been developed. Details of the principles and the design of material control systems are discussed in Chapter 11 and their use or management are discussed in Chapter 12.

6. *Determination of Tool Requirements and Tool Control—Tool Control.* Procuring and maintaining tools is another job in the sequence of completing the planning for an activity. It must, however, be coordinated with other phases of Production Control. Tool control may be subdivided into two categories:

1. Design and procurement of new tools.
2. Control storage and maintenance of tools after procurement.

Tool control, like materials control, is a complete planning and control activity in itself in the most formal activity. It is discussed in detail in Chapter 13.

7. *Determination and Control of Equipment and Manpower Requirements—Loading.* In order to determine equipment and manpower requirements and to control them in the formal activity requires the functions of loading and manpower control. These may be combined under one activity which we will call *loading.* In most activities the loading function is combined with the functions of routing and scheduling. It is very difficult in the formal system to distinguish or to separate these functions. Therefore, these functions are usually considered simultaneously.

Loading may be defined as the assignment of work to a facility. The facility can be equipment, manpower, or both. All production levels of an activity must eventually be loaded, down to and including work stations. Loading is discussed in detail in Chapter 15 along with scheduling.

8. *Determination of When Work Is to be Done—Scheduling.* The function of determining *when* work is to be done in an activity is called *scheduling.* Scheduling consists of time-phasing of the work load; that is, setting both the starting and ending time for the work to be done. There are many different techniques used in scheduling. The selection of the specific technique used depends upon the type of activity and degree of formality required for control. As was pointed out previously, common practice frequently dictates that routing, loading, and scheduling be performed simultaneously.

Scheduling is discussed in detail in Chapters 15, 16, and 17, along with the function of Loading. It will become obvious after reading these chapters that the information required to route and load is also required to schedule.

This completes the functions in the planning phase of Production Control and leads us to the second phase—the *action phase*—in which the work is started.

ACTION PHASE

There is only one Production Control function existing in the action phase—dispatching—as shown in Table 3-1. All of the other functions are either planning or follow-up.

9. *Starting the Work—Dispatching.* Dispatching is the transition from the planning phase to the action phase and consists of the actual release of the detailed work authorization to the work center. In the formalized Production Control system the dispatching function is commonly performed by an individual called the *dispatcher.* Every job that goes through a department or an activity goes through the dispatcher. The

dispatcher is the first step in the line of communication from the work center back to the Production Control function. In the informal Production Control system dispatching may be done by the foreman or supervisor and may even be done verbally. The function of dispatching will be discussed in detail in Chapter 18.

FOLLOW-UP OR CONTROL PHASE

Once the work has been started in an activity, it is necessary to evaluate continuously the progress in terms of the plan so that deviations can be detected and corrected as quickly as possible. The follow-up phase accordingly consists of two parts:

1. Progress Reporting
   A. Data collection
   B. Data interpretation
2. Corrective Action

Progress Reporting   This part of the follow-up phase is primarily a matter of communications. As will be pointed out later, one of the principles of a sound Production Control system is that it must furnish timely, adequate, and accurate information about the performance of an activity. The application of this principle is the responsibility of the progress reporting function and consists of the collecting of the data as well as the interpretation of the data. These two elements are described in the following paragraphs.

10. *Collecting Data—Data Processing.* The first step in Progress Reporting is to collect data for what is actually happening in the activity. This is a problem of communications with "Production," and demands sound design of the media by which data will be gathered and transmitted. In the formal activity where a dispatcher is used, he is the originating point of communication to management. Therefore, the dispatcher is a key link in the management-by-exception chain. Where there is no dispatcher, the foreman or supervisor is the originating point of communication and would be responsible for reporting to management. This originating point is the vital link in the communication system, and where paperwork is involved, it is the nerve center of the Production Control system.

11. *Interpreting the Data—Data Processing.* After data has been collected, then it is necessary to interpret it by comparing the actual performance against the plan. This usually consists of some form of progress reporting. The system must be designed to report progress in a simplified manner. The important point to remember in the design of progress reports is that they must almost automatically evaluate the situation for

management. Management should not be required to interpret the raw data in order to come up with the evaluation. The principles and design of progress reporting systems is discussed in detail in Chapters 19 and 20.

Corrective Action   This part of the follow-up phase is the "pay-off" for the effort and cost of collecting and interpreting data. This is the third vital link of Control. The whole purpose of Production Control obviously would be defeated if corrective action was called for but not taken. Corrective action may consist of one or both of two courses of action—expediting the work being performed, and replanning.

12. *Making Current Work Corrections—Expediting.* A plan is made for only one purpose—to be followed. If the data from the production unit indicates that there is a significant deviation from the plan and the plan cannot be changed, then some action must be taken to get "back on plan." The progress report should indicate the reasons for the deviation. In the formal Production Control system, the function of following-up to eliminate the cause of deviating from the plan is performed by the *expediting* group. In the informal Production Control system, the function of following-up is usually performed by the person directly in charge of the activity such as the foreman or supervisor.

Obviously, no Production Control system is perfect, and therefore some expediting will always be required. However, it should be minimized by continuously improving the Production Control system. This is fully discussed in Chapter 22.

13. *Making Plan Corrections—Replanning.* It should be emphasized that a plan is not made to be changed, but to be followed. However, if after expediting to correct deviations, it is found that it is impossible to perform according to the plan, it would be foolhardy to attempt to continue with the original plan. It may also be found that there were errors made while developing the plan. In all such cases, changes in the plans are mandatory.

The most important point to remember in replanning is that changes in the original plan should never be made *just* because of deviations. Careful analysis is always required. This function is discussed in detail in Chapter 22.

PRINCIPLES OF A SOUND PRODUCTION CONTROL SYSTEM

Having outlined the fundamental functions and principles of Production Control, our next step is to describe the principles of a sound Production Control system.

All of the following principles are rules that should be followed in designing a Production Control system. They may also be considered as qualities or specifications that should be incorporated into any sound Production Control system.

The system must:

1. Furnish timely, adequate, and accurate information.
2. Be flexible to accommodate necessary changes.
3. Be simple and understandable in operation.
4. Be economical in operation.
5. Force prior planning and corrective action by the user of the system.
6. Permit "management by exception."

All of these principles must be applied to all systems of Production Control. They may also be used as qualitative measures for evaluation of the system. In Chapter 25, we will discuss in detail both the qualitative and the quantitative evaluation of a Production Control system and refer back to these principles to provide the complete procedure for the measurement of a sound system.

1. *Furnish timely, adequate, and accurate information.* This is probably the most important of all of the principles. Unless the information is timely, it is useless. This, of course, also applies to adequate and accurate information. All information that is obtained in the system must have these three qualities. As previously pointed out, communication or information flow is the basis of any Production Control system. Without this there is no system.

2. *Be flexible.* When reference is made to the flexibility of a system, it refers not only to the system's ability to adjust for variations in workload, but also, flexibility in terms of modifying the system itself to accommodate changes in the operation or the conditions that exist in the activity. Most activities are dynamic; i.e., they are in a constant state of flux. This is particularly true under conditions of rapid growth. Just because a system that has been designed is adequate for conditions as they exist today does not necessarily mean that it will be adequate for conditions that will exist in the near future. Because of this, it is necessary for the designer to keep in mind at all times that a system must be adjusted to changing conditions within the organization. This must be done without complete disruption of the work that has been done in the past.

3. *Be simple and understandable.* By simplicity we mean, that the system must be understandable to everyone connected with it. This does not mean that the system in all phases must be understandable to everyone. It only means that the part of the system which applies to an individual in a particular function must be understandable to that individual.

4. *Be economical.* Economy, of course, is the basic reason for having a Production Control system. In all cases, we must get back more than a dollar for every dollar that is spent in planning and control. This is

also one of the most difficult steps to evaluate. Many of the benefits ac-crued by Production Control are intangible, and cannot be given a dol-lars-and-cents valuation. Only by comparing cost of operation in times when no formal system of Production Control was used against the cost when a Production Control system is used could economy be measured accurately.

5. *Force prior-planning and corrective action.* The system itself should require prior-planning and corrective action and cannot be effec-tively used unless these things are done. In other words, the system must do its own policing.

6. *Permit management by exception.* A system which permits manage-ment by exception is one which reports to management only those things which require action by management. At the same time, the system must assure management that the unreported events are proceeding according to plan. The important point here is that management must be able to *assume* that unreported things are going according to plan and that it is not necessary for management continually to follow up the details.

With these principles as guides, the systems designer should be able to develop a sound system once he has obtained a full knowledge of the functions of Production Control and has achieved the other objectives described earlier. Constant reference will be made throughout the text to these principles. It is advisable that any person connected with Pro-duction Control have them firmly in his mind so that it will be second nature for him to evaluate the system by checking against these principles.

Summary   Production Control consists of three phases: the planning phase, the action phase, and the follow-up phase. The first requirement for the designer of a Production Control system is a thorough knowledge of the specific functions involved in completing each phase. Most of the remainder of this text is devoted to the detailed discussion of these functions. In addition, the designer must have a com-plete understanding of the basic principles of a sound Production Con-trol system. These principles or guide lines provide a means of continuous evaluation of any control system and should be "second nature" to the designer.

All work activities can be classified into one of four basic types of "Production." Production Control systems in these basic types are classi-fied in the same manner. The designer of the Production Control system should be familiar with these basic types in order to have a basis for prac-tical application of principles. These are described in detail in Chapters 4 and 5.

4

Types of Production Activities and Processes

In Chapter 1 it was pointed out that Production Control can be applied to any type of activity. It was also pointed out that because of the vast number of different types of activities it would be very impracticable to attempt to define the required features for any specific Production Control system. It was emphasized that the objective of the text was to develop the principles or rules which are applicable to all "production" activities. To illustrate the use of these principles in the best manner, the manufacturing type of activity with the formal Production Control system will be used because it incorporates *all* of the elements of Production Control. It is possible, however, to classify the numerous types of activities into basic groups or classifications and in turn describe the basic processes for performing these basic activities. This chapter will be devoted to this task.

"Production" activities may be classified according to the following eight general classifications:

1. *Manufacturing.* The United States Department of

Labor classifies all manufacturing into 19 distinct classifications and one miscellaneous:

1. Food and kindred products
2. Tobacco manufacturers
3. Textile mill products
4. Apparel and other finished textile products
5. Lumber and wood products (except furniture)
6. Furniture and fixtures
7. Paper and allied products
8. Printing, publishing, and allied industries
9. Chemicals and allied products
10. Products of petroleum and coal
11. Rubber products
12. Leather and leather products
13. Stone, clay, and glass products
14. Primary metal industries
15. Fabricated metal products
16. Machinery (except electrical)
17. Electrical machinery, equipment, and supplies
18. Transportation equipment
19. Instruments and related products
20. Miscellaneous manufacturing industries

2. *Service.* This will include the following types of activities:

1. Professional services, such as medical and legal
2. Plant maintenance and engineering
3. Production Control
4. Industrial engineering
5. Warehousing
6. Inspection and quality control
7. Materials handling and plant layout
8. Communications
9. Personal service (barber, laundry, and so on)
10. Business services (advertising, employment)
11. Automobile repair, other repairs
12. Amusements and recreation
13. Hotel and motel

3. *Sales.* This will include activities such as:

1. Retail sales
2. Wholesale sales
3. Direct contact sales

4. *Clerical.* This would include the following types of activities:

1. Banking
2. Mail order business of all types
3. Insurance

5. *Research and development.* This is distinguished from the applied engineering function and would include both pure research and applied research.

6. *Engineering.* This refers to the redesign and preparation of specifications, drawings, bills of material, and so on, for production purposes of products that have been developed through research and development.

7. *Administration.* This is concerned primarily with the management function and consists of decision and policy making.

8. *Construction (or destruction).* This would include the following types of activities:

1. Building construction
2. Bridge construction
3. Road construction
4. Demolition

It would be difficult to find work of any kind in the industrial complex which cannot be placed into one of these eight classifications. Obviously, within these basic activities, there will be sub-functions or sub-levels of activities which will be fundamentally the same as the basic activities listed. A manufacturing activity, for example, will usually have all of the other basic activities included as sub-functions. There may even be a sub-function of construction under the servicing activity which is a sub-function to the manufacturing activity. Therefore, it is important that we understand that there is not only a basic activity but there are different *levels of activity* which must be considered. The relationship of the sub-activities under the main activity is extremely important in developing and designing a Production Control system. Some form of Production Control can be applied to all levels of the activity, and it must be applied to the main activity in order to assure success.

The next step is to define the basic processes of operation to perform the work of the activities listed above. The basic process of production would be dependent, first of all, upon the *type of production* and secondly, upon the *volume of work.* We will subsequently see that the Production Control system depends upon the process of performing the production. These basic processes may be classified into the following types:

1. Continuous or mass production
2. Job order or job shop
3. Similar process
4. Custom work

CONTINUOUS OR MASS PRODUCTION

Examples of the continuous or mass production process are as follows:

1. Automobile manufacturing
2. Refining
3. Household appliance manufacturing
4. Toy manufacturing
5. Mass production housing development
6. Glass manufacturing

The characteristics which distinguish this process are as follows:

1. The end product or work performed is highly standardized.
2. The quantity of work performed or product produced is large.
3. The type of equipment used to perform the work is specially designed for that particular purpose.
4. The equipment is physically arranged within the plant according to the product produced and its arrangement is seldom changed.
5. The materials-handling equipment is usually built in to provide a smooth flow of work.
6. The in-process inventory is usually small in relationship to output.
7. The workers' skill is relatively low. In most activities only semi-skilled workers are required.
8. Supervision is relatively easy.
9. Job instructions are usually given only at the outset of the new job and few instructions are required thereafter.
10. The unit cost of the service or product produced is relatively low.
11. The initial planning must be extremely detailed and complete.
12. The planning and control is easy after the initial installation is complete.
13. There is a small degree of flexibility.
14. The over-all production cycle time is greatly reduced.
15. Generally a higher equipment investment is required.
16. The balancing of the work load between operations is usually relatively difficult.
17. Disruptions in work flow either by equipment failure or personnel difficulties in the line usually results in considerable amount of lost time and production.

SIMILAR PROCESS

In this method of operation, the work being performed is similar in nature from order to order, but not identical. Examples of activities of this type are:

1. Shipping and receiving
2. Packaging and packing in a mail-order house
3. Most clerical operations
4. Auto repair
5. Shoe manufacturing
6. Clothing manufacturing
7. Book printing
8. Medical clinics

The characteristics which distinguish the similar process are as follows:

1. The product or end result of the work is highly standardized.
2. The order or work quantity is usually very large.
3. The type of equipment required, if any, is usually highly specialized.
4. Equipment is laid out by the type of end product.
5. The materials handling equipment may be both mobile and permanent installation conveyor.

6. The end-product inventory is relatively low.
7. The required worker skills are relatively low because of the repetitiveness of the work.
8. It is relatively easy to supervise the workers.
9. Relatively few job instructions are required because of the similarity of the work.
10. Prior-planning is essentially completed at one time and is relatively easy compared to the prior-planning required on custom and job-order work.
11. Control of the process is relatively easy because of the repetitiveness.
12. There is some degree of flexibility but not as great as in the custom and job-order types of process.
13. The cycle time is relatively short.
14. The balancing of the work load is relatively difficult because the work is laid out according to the end product of the work.
15. A relatively high equipment investment is required in manufacturing operations.
16. Disruptions in the flow of work usually result in a considerable amount of lost production.

JOB-ORDER OR INTERMITTENT

The job-order (job-shop or intermittent) type of process is characterized by a wide diversity of end products. In these activities the work centers are arranged around particular types of work or equipment. For example; in a manufacturing plant doing machine shop work, the drilling machines would be in one department, the turning machines in another department, the milling machines in another department, and so on. The work then flows through these departments or work centers in batches of work called lots. Examples of this type of activity are:

1. General machine shop operations
2. Machine tool manufacturing
3. Aircraft manufacturing
4. General engineering
5. Custom boat building

The characteristics which distinguish the job-order, job-shop or intermittent type of process are as follows:

1. The product or end result of this activity is nonstandard.
2. The order or work-unit quantity is generally small.
3. The equipment, if any, is usually of the general-purpose type. This permits its use for any job that comes through the activity. The objective of this is to obtain greater utilization of equipment.
4. The equipment is usually arranged according to the *type of work* that is performed and not according to sequence of operations on specific products.
5. Materials-handling equipment is of the mobile type which can be used in many locations.
6. In-process inventory is relatively high.

7. A much higher level of worker skill is required.
8. It is generally very difficult to supervise the workers.
9. Job instructions usually must be complete and detailed.
10. Prior-planning of work is relatively complex and must be done for each job.
11. Control of the work is relatively complex because every job must be individually controlled.
12. There is a greater degree of flexibility for scheduling the work of the people and equipment than with the similar or continuous types of processes.
13. The cycle time to complete an order is relatively long.
14. Balancing the work load in this type of activity is much easier than in the similar and continuous types of processes.
15. The unit cost for the product or work produced is relatively high compared to a mass produced item.
16. Disruptions due to equipment or people usually do not seriously affect lost production. Because of the high degree of flexibility, people can be shifted from job to job and other equipment can be used to complete orders.

Seldom do we find a pure job-order, or job-shop type of process. Usually there is a combination of the continuous and job-shop, which we term intermittent, type of process. Even in some of the smallest firms one might identify elements of both continuous and job-order type of process. Obviously, wherever advantage can be taken of the continuous process, it will result in over-all lower cost. By properly analyzing all of the work that goes through an activity, it might be possible to identify areas in which the sub-operation might be set up in such a manner as to simulate the continuous process. In some cases it may be cheaper to move a machine than it is to move materials from one station to the next. For example; in a garment manufacturing plant, machines can be moved around quite easily. Instead of setting up the layout on a job-shop basis for lots of garments, it might prove economical to move the machines into assembly-line setup for each product. This reduces all of the handling of the in-process inventory between operations and greatly reduces the cycle time in order to produce the finished product. This, of course, would only be done in the event that the job-orders are of sufficient size to warrant the cost of "rearranging" the machines.

CUSTOM WORK

This is the kind of process in which the end product is unique for each job, so that its processes are distinct from those of any other job. The only similarity would be in principle alone. Examples of custom work are as follows:

1. Bridge construction
2. Ship construction

3. Pure research and development work
4. Administrative activities
5. Model and pattern making
6. Certain types of maintenance work
7. Building construction

Characteristics that distinguish this type of process are as follows:

1. The product or end result of the work is unique.
2. Work-unit quantities ordered are very small, and often no more than one.
3. In most cases where equipment is required it is the general-purpose type so that it can be used for a high variety of different types of work.
4. Equipment arrangement is by process or type of function performed.
5. Where materials-handling equipment is required, it is usually mobile so that it can be used on a variety of jobs.
6. In-process inventories are relatively high for manufacturing activities.
7. Usually very highly skilled workers are required.
8. Supervision is more difficult than in other types of activities, and requires outstanding administrative ability.
9. Job instructions must be given complete and in great detail.
10. In custom manufacturing industries the finished-goods inventories usually are nonexistent. As each product is completed, it is immediately shipped to the customer.
11. Prior-planning is very complex and difficult. Each job is complete in itself and usually little past experience is available to do detailed prior-planning.
12. Control of the work is the most complex. This is usually the result of limited prior-planning procedures.
13. There is usually a high degree of flexibility in both equipment and manpower used to perform the work.
14. The cycle time to complete the work is usually longer than in any of the other processes.
15. Balancing work load is usually easy because of the high degree of flexibility.
16. The unit cost of the work is very high.
17. Disruption in work flow with either people or equipment does not normally result in serious loss of time or output.
18. It is usually found that in this type of process excessive expenses are incurred to expedite work through various stages in order to meet the customer's requirements.

Obviously, the governing factor in this process is the workers' skill level. The highest level of skills in supervisory abilities is also necessary in order to assure success.

Summary   We have established that the type of process is one of the most important factors in determining the type of Production Control system. Following our broad concept of Production Control to any type of activity, we have described four different types of processes.

The reader should be very familiar with these before attempting to design the Production Control system. It is, of course, necessary for the reader to identify his own activity within the types defined.

For convenience, the types of processes are summarized in Table 4-1.

SUMMARY OF CHARACTERISTICS OF TYPES OF PRODUCTION PROCESSES

Characteristics	Custom	Job-Order (Intermittent)	Similar Process	Continuous (Mass)
1. Product or end result	Unique	Non-standard	Standard	Standard
2. Examples	a. Research	a. General machine shop	a. Shipping and receiving	a. Autos
	b. Bridge construction	b. Machine tool manufacturing	b. Clerical	b. Electrical appliances
	c. Model making	c. General engineering	c. Clothing manufacturing	c. Refining
3. Order or work unit quan.	Small, usually one	Small	Large	Large
4. Type of equipment	General purpose	General purpose	Special purpose	Special purpose
5. Equipment arrangement	By process	By process	By product	By product
6. Material handling equip.	Mobile	Mobile	Mobile and Conveyor	Conveyor
7. In-process inventory	Relatively high	High	Relatively low	Low
8. Worker skill level (relative)	High	High	Low	Very low
9. Supervisory difficulty	Highly difficult	Very difficult	Relatively easy	Very easy
10. Job instructions	Great amount and very detailed	Many and detailed	Relatively few	Very few
11. Prior-planning	Complex	Complex	Relatively easy	Very complex, but once for all units
12. Control	Complex	Complex	Easy	Very easy
13. Degree of flexibility	High	High	Some	Very little
14. Cycle time	Long	Relatively long	Relatively short	Very short
15. Balancing of work load	Easy	Easy	Relatively difficult	Very difficult
16. Unit cost	Very high	High	Relatively low	Low

TABLE 4-1

5

Basic Designs of Production Control Systems

In Chapter 4 we described the basic types of activities and production processes. We established that the type of production process is the most important factor in determining the type of Production Control System. The reader, then, should be familiar with these basic types and should be able to translate the process of his own or any activity in terms of the basic types.

In this chapter, we will *define* the four basic Production Control systems in terms of the basic production processes. However, we will discuss these four systems in terms of the manufacturing activity and it is up to the reader to make the translation to any other specific type of activity which he wishes to consider. By doing this, the definitions of the types of Production Control systems can be simplified and the reader should have no difficulty in applying the basic principles to any type of activity and process.

Describing the basic Production Control systems in terms of the basic types of production processes we have the following four types:

1. Flow or continuous Production Control
2. Similar process Production Control
3. Job-order Production Control
4. Special project Production Control

The distinguishing characteristics of the types of Production Control systems follow very closely the characteristics of the different types of production processes described in the preceding chapter. The following paragraphs consider these distinquishing characteristics in detail.

FLOW OR CONTINUOUS PROCESS PRODUCTION CONTROL

Production Control systems for the flow or continuous-process type of activity will be distinguished by the following characteristics:

1. The objective is to maintain a constant optimum rate of output.
   1.1 Materials and parts must flow through production at a constant rate.
2. There is little or no change in the kind of products being made from day to day.
3. Prior-planning is one of the most important elements of this system because part of it is built into the physical arrangement for work flow.
   3.1 Highly accurate work measurement is required.
   3.2 A high degree of engineering skill is needed.
   3.3 Line balance is an extremely important factor and is one of the most difficult problems in prior-planning.
       3.31 Manpower is easier to balance than machine loads.
       3.32 Bottleneck operations must be eliminated.
       3.33 Line and equipment dependability must be maximized.
       3.34 Standby equipment may be used in critical situations.
   3.4 Assignment of work to individuals is practically automatic.
4. Planning and scheduling of individual operations is not necessary. Only the over-all plan for production rate must be set; after that, the arrival of work-units from the previous stage is the cue to begin the next stage.
5. Operators on the continuous production line do not need daily instructions. Instructions for each product that goes through the line need be given only at the start of manufacturing. From then on, the work is highly repetitive and no further instructions are needed.
6. Day-by-day control is highly simplified.
   6.1 The rate of production is fixed.
   6.2 Production reports consist only of hour, shift or daily count of production. In some cases this is done automatically with production recorders on the machines.
7. Control of production costs by lots is unnecessary and no attempt is made to do it. The information in the prior-planned standards is sufficient for control.
8. Production output can be varied only by changing the number of hours that the line is working.
   8.1 Changing the working hours.
   8.2 Addition of extra shifts.

    8.3 Using relief workers to eliminate shut-down of the line during rest breaks.

    8.4 On manual operations such as assembly the output can be changed by increasing or decreasing the number of workers performing the assembly.

9. Input of raw materials and output of finished products must match the production rate. Inventory control must assure that no stoppages occur because of lack of material.

10. Shut-down of lines for any reason is usually costly.

    10.1 Reports of machine breakdown and maintenance information are vital to the control of downtime.

11. It is extremely important that purchasing, receiving, and shipping are closely coordinated and scheduled to permit smooth flow of production.

12. Production Control in continuous flow operations consists in directing the quantity of each product being made. All of the functions of Production Control are present in continuous manufacturing, but they are not always obvious because they are built into the line at the time it is set up.

SIMILAR-PROCESS CONTROL

As pointed out in the previous chapter, this type of process is one in which all the products made are similar enough to pass through essentially the same processes. It requires a little more control than is necessary in flow control, but does not require as much detail planning as is needed in job-order operations. The review of the characteristics of the similar-process operation will indicate that this type is most common in the non-metalworking industries. Examples of similar-process industries are: printing, garment manufacturing, book printing, and shoe manufacturing.[1]

The characteristics which distinguish the similar-process control are as follows:

1. All products go through the same processes. It is similar to flow control in that there is rarely any variation in the parts or operations required. Detailed instructions for the operator are unnecessary.

2. The instructions for quantities to be made are in lots or batches.

    2.1 The lots are assigned lot numbers for control purposes.

    2.2 Every part or item within the lot is identified with the lot number.

3. Permanent instructions, specifications, and drawings are provided for the worker.

4. Each time a new lot or batch goes through the process new instructions concerning the kind and quantity of products to be produced must be sent to all points affected by the change.

---

[1] The control systems used in these industries are sometimes called *block* or *load* controls. This text will use only the term *similar-process control*.

5. Materials handling becomes somewhat more difficult than in flow control because of the additional control required by the numbered lots.

6. In-process inventory is usually higher than in flow control because it is used to balance the work load between operations.

JOB-ORDER CONTROL

The previous chapter described the characteristics of the job-order production process. These job orders may come directly from the customer or may come from an internal department in the plant, such as inventory control. If the product is being sold from stock, the order comes through the finished-stock control department. Usually when a company produces to the customers' orders, it is called a job-shop.

Following are some of the characteristics which distinguish the job-order control system:

1. Usually each order is different from the preceding order in several respects:
    1.1 In quantity
    1.2 In specifications and quality
    1.3 In materials

2. Prior-planning in advance of receipt of the order is usually difficult.
    2.1 Producing sub-assemblies and parts for stock in large quantities is very expensive.
    2.2 There may not be a high degree of standardization between customers orders.

3. Each "job" or "shop order" is assigned a number by which the job is controlled.
    3.1 Production control maintains control of the individual order by use of the number.
    3.2 Cost control collects and analyzes information for each order by the order number.

4. All of the basic functions of Production Control are involved in job-order control systems.

5. A high degree of control is required. A detailed production control system must be used because the control problems are so difficult.

6. Job-order control is completely dependent upon process and product planning.

Because of the complex nature of job-order control, practically all of the text will be devoted to the development of principles and techniques used in this system. The transition from this to other forms of Production Control must be made in most cases by the reader.

SPECIAL-PROJECT CONTROL

The special-project control system is associated with the custom work type of production process. Following are some of the characteristics which distinguish the special-project control system.

1. The special-project control system is a particular type of the job-order control system. Most of the principles and characteristics which apply to the job-order control system would apply to the special-project control system also.
2. Process and product planning must be done for each order.
3. Scheduling generally must be based on past experience.
   3.1 The principle problem is lack of experience on which to make good estimates for production and costs.
   3.2 Detailed schedules are practically impossible and are usually established in the form of target dates.
4. In some cases, it is practically impossible to do much of the process planning until after the work starts.
5. The follow-up and corrective action phases of the Production Control system are of primary importance.
   5.1 Good communication systems are extremely important.
   5.2 Complete and up-to-date information must be kept at all times.
6. Unless good Production Control systems are developed for the special project type of operation, a great amount of cost is incurred in expediting.

Other principles and characteristics which will be described throughout the text relative to job-order control, will also apply to the special project control. It must be emphasized, that this is one of the most complex types of operations to control. Where the activity is of a sizeable nature involving special projects a great amount of confusion and cost can be eliminated by developing effective Production Control systems.

Summary   The would-be designer of a Production Control system is now familiar in a general way with some of the basic tools with which to work. He has a set of rules or principles associated with the type of Production Control system which applies to a given type of activity or "production" process. He also has been exposed to all of the functions within Production Control.

The next step is to acquire a more complete understanding of each of the functions of Production Control and the techniques used to perform them. The following chapters will be devoted to this task.

6

Forecasting Techniques

The five preceding chapters have been devoted to a broad description of the functions of Production Control and the development of fundamental principles of Production Control as applied to any type of activity. In this and the following chapters attention will be directed to detailed discussions of each of the specific functions.

Referring to Table 3-1, page 18, the reader will note that the planning phase is divided into two parts—prior-planning and action planning. The first function in the prior-planning part is forecasting.

Forecasting is defined as the estimation of future activities. Obviously, the better management is able to estimate the future, the better it can prepare for the future. In this chapter the important aspects of forecasting are considered and the techniques which are commonly used in developing a forecast are described. These are as follows:

1. Need for forecasting, and its uses.
2. Types of forecasting.

3. Techniques of forecasting:
  3.1 Historical estimate—presuming that what has happened in the past is a good indicator of what will continue to happen in the future.
  3.2 Sales force estimate—presuming that the people closest to the market can best determine its future behavior.
  3.3 Trend lines—projecting into the future the increases or decreases which have occurred in the past.
  3.4 Market surveys—sampling to determine the potential sales for a product for which there is no history on which to base a forecast.
  3.5 Correlation—deriving the forecast for a particular activity by relating the activity to a given industry or industries, or to the economy as a whole.
  3.6 Prior knowledge—the substitution of information provided by higher management levels in order to forecast at the local level.
4. The use of judgment in forecasting.
5. When the forecast should be made.
6. Analyzing the forecasting function.

NEED FOR FORECASTING, AND ITS USES

As will be explained in later chapters, the effectiveness with which each function can be performed depends to a great extent upon the manner in which the preceding functions were carried out. All activities, formal and informal, base their plans on a forecast. Frequently, serious difficulties are encountered because forecasts are developed with little thought and few facts or because a tacit assumption is made that "things will continue about the way they are."

If there is to be consistency in planning and control by the top management of an organization, it is essential that all the elements within the organization develop their plans from the same forecast. Figure 6-1 illustrates the many uses of the forecast. Some of the most important planning data developed from the forecast are:

1. Inventory schedules
2. Production schedules
3. Material requirements
4. Purchasing schedules
5. Subcontracting schedules
6. Manpower requirements
7. Equipment requirements
8. Cash requirements

Figure 6-1 also indicates the basic steps involved in translating a typical forecast into a plan of activity:

1. The general sales forecast is developed on a dollar basis for either the entire line of products or for the different product groups.
2. After the sales dollar figure is determined, the forecast is converted into the number of units of each product that is anticipated to be sold.

2.1 This conversion from dollars into units is extremely important for planning purposes, especially in the area of material planning, which will be discussed in Chapter 11.

2.2 The product breakdown is especially important where varying marginal income exists between the products.

3. The product sales forecast is compared with the forecast inventory balances (based on the forecast for the preceding period) and the production requirements are determined.

   3.1 Material, subcontracting, manpower, and equipment requirements are all developed from the production schedule.

4. These requirements determine the cash that will be needed during the future period.

TYPES OF FORECASTING

A differentiation should be made between long- and short-range forecasting. In broad general terms, long-range forecasting is usually based upon analysis of the national or even the international economic situa-

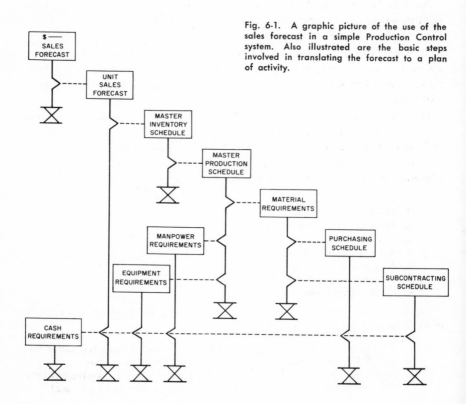

Fig. 6-1. A graphic picture of the use of the sales forecast in a simple Production Control system. Also illustrated are the basic steps involved in translating the forecast to a plan of activity.

tion, and is concerned with periods of two, three, five, or even ten years in the future. Short-range forecasting is concerned with the potential sales of a single product or a product group during a given period in the near future. This period may be a year, a half year, or even as short as a week or a day. The Production Control functions are much more dependent upon the accuracy of the short-range forecast than upon that of the long-range forecast. For this reason, we will limit the discussion to the short-range forecasting techniques.

Before starting our discussion of the more important forecasting techniques, it should be re-emphasized that all forecasts are only estimates. It is just as important to know when the estimate needs to be revised as it is to develop a valid forecast originally. We are going to discuss methods of determining the need for forecast revision in the following chapter.

FORECASTING TECHNIQUES

The Historical Estimate   The historical estimate is probably the most common forecasting technique. This technique is based upon the assumption that what has happened in the past is the best indication of what will occur in the future. For example:

1. Last year we sold 10,000 units, therefore we will sell 10,000 units this year.
2. In previous years, the sales for Christmas began in the middle of November, therefore this year the sales for Christmas will begin in the middle of November.
3. Last year we used 100 tons of steel, therefore this year we will use 100 tons of steel.
4. In previous years 10 per cent of the models sold were blue, so this year 10 per cent of the models sold will be blue.

The historical estimate is particularly useful for situations where the activity is effected by patterns of seasonality. For example, the per cent of total annual sales received for a given product in each month can be calculated for a preceding period of three to five years. An average per cent for each month can then be calculated and the anticipated sales for each month of the coming year can be calculated as a per cent of the forecast annual sales. It is also very useful for determining model, size, or color distribution when the original forecast has been in more general terms such as total sales dollars or sales of product groups. The historical estimate will provide a basis for the total sales forecast, but it should be modified by the other techniques discussed later. To assume that the pattern of events that has occurred in the past will continue unchanged in the future involves assuming that the economy is static. This is rarely true except for short periods of time.

The Sales Force Estimate   Frequently it is felt that the people within the organization who are closest to the market (the salesmen) know best how the market will behave in the future. The sales force estimate technique is based on this assumption. Figure 6-2 contains a schematic diagram which explains how the sales force estimate is developed:

1. The individual salesmen within each territory estimate the number of units they expect to sell during the forecast period.
2. The territory or district sales manager consolidates these estimates and adjusts them on the basis of past experience. Sometimes each salesman's estimate is multiplied by a percentage factor based upon the error between his estimates in previous years and his actual sales.
3. The district estimates are then forwarded to the general sales manager, who combines all the district data, making modifications where he feels it is necessary, to develop the sales force estimate.

Fig. 6-2. A graphic picture of the development of the sales force estimate.

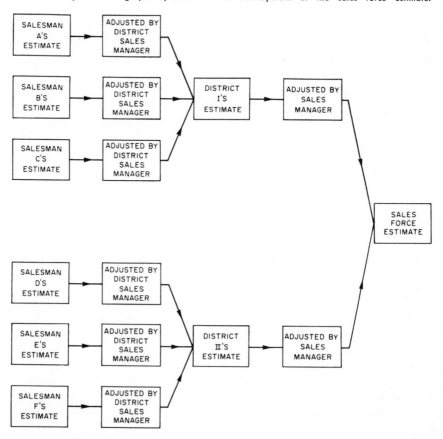

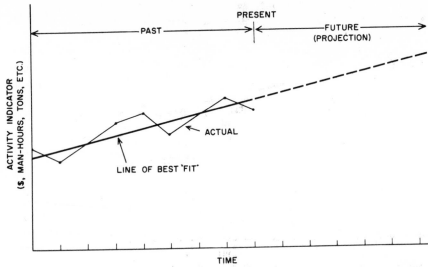

Fig. 6-3. The basic trend line diagram. Time is plotted on one axis, and activity on the other. A line is fitted to the past points and projected into the future.

A slight modification of this technique is used by the maintenance department in a large plant. When it is time to prepare the forecast, the maintenance supervisor asks the manager of each cost center to describe any anticipated maintenance requirements. As in the case of the salesmen, the information is adjusted on the basis of experience. This approach is not perfect, because emergency work will never repeat itself exactly, but it has been found to be much more satisfactory than just presuming that the past will be repeated. This adaptation has also been applied in other service activities.

The sales force estimate is best used when there are a limited number of products and only a few large customers. In such situations, salesmen handle only one or two accounts and are familiar with the customers' future plans. An example of such an activity is the manufacture of commercial power generating equipment.

Trend Lines   An important forecasting tool that can be used when there is an appreciable amount of historical data is the trend line. Figure 6-3 illustrates a simple trend line diagram. The activity indicator (dollars, man-hours, tons, and so on) is plotted on the vertical axis and time is plotted on the horizontal. The actual level of activity of the preceding years is established by using the historical data; then a single line which best represents the pattern of past activity is drawn. The trend line is then projected into the future and the forecast established from it.

When the activity indicator is in terms of units of production or consumption, this technique is relatively simple. This is illustrated in Figure 6-4 which shows the trend in electrical energy consumption in the United States over a 20-year period.

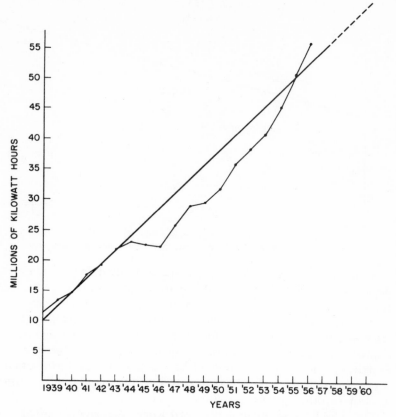

Fig. 6-4. The electric power industry has grown very rapidly, and continued expansion is indicated by this trend line analysis. When units are plotted rather than dollars, adjustments need not be made for price increases.

Frequently in a business situation the only readily available statistic is dollars of sales. Under such conditions, an adjustment must be made for selling price changes if a realistic picture is to be obtained. The necessity for this adjustment is illustrated in Figure 6-5. If only dollar revenue from passenger fares were plotted for this railroad, a slight declining trend would be indicated. However, when these dollar figures are adjusted for the rate increases, a serious reduction in passenger volume becomes very apparent.

When sales figures must be adjusted for price changes, a time period, frequently the current year, is selected as the base or norm and the following computations are made:

1. The actual price at which the product was sold during the various periods is divided by the base price in order to establish an adjustment factor for each period.

2. The actual sales, expressed in dollar volume, for the periods are divided by this adjustment factor to indicate the "normalized" volume of
sales in terms of the base year's selling price. For example:

    2.1 The selling price in a particular year was $1.25 per unit and the
total sales revenue was $500,000.

    2.2 The selling price in the established base year was $1.00.

    2.3 The adjustment factor would therefore be 1.25/1.00 or 1.25.

    2.4 The adjusted sales in terms of the base year's selling price would
be $500,000/1.25 or $400,000.

Once the points for the adjusted sales dollars have been plotted on
the graph, the single line of "best fit," the trend line, must be drawn
through them and then extended into the future time periods. If it is to
be assumed that this line is linear in nature, it can simply be drawn by
"eye", e.g., a straight edge can be lined up until it appears to be centered
as best as possible with all the points. A line can then be drawn. The
result is a picture of the general trend based upon past sales activity.
The disadvantage of this technique is that if another person attempts to
line the points up with the straight edge by "eye" it will probably result
in a new trend line with perhaps a significantly different slope. A more
precise method of establishing the trend line is to use one of the sta-

Fig. 6-5. The projection for this railroad company indicates a continuing decrease
in passenger fare revenue, but the adjustments for rate increases give the most
accurate picture of the decline in volume.

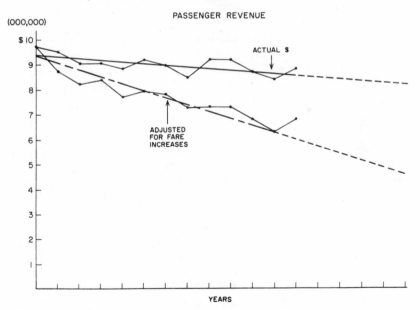

tistical techniques for analyzing the data and establish a regression line based on the "least squares" technique. The formula and general procedure for this technique is in Appendix I. Although the statistical approach will result in a mathematically correct trend line, there are two major disadvantages in using it. The first is the obvious factor of the amount of time necessary to do the rather long mathematical calculation. The second is the fact that many persons studying the trend line during the development of a forecast will put undue faith in the trend because it has been established "scientifically." It must be remembered at all times that the extension of the trend line into the future periods is still only an estimate. Obviously, trends can, and on many occasions have, changed and even reversed. Because of this, care must be exercised when predicting the future. This point cannot be over-emphasized.

One important limitation on the trend line, as well as some of the other mathematical techniques of forecasting is that it assumes an "infinite population" of relatively small customers. This means that no one individual customer decision can have an appreciable effect on the total demand for the product. Where the market for an organization's product is made up of a relatively few large customers, this "infinite population" assumption is not valid and hence the mathematical techniques have negligible value.

Market Survey   When an organization decides to introduce a new product group into the market, usually there will not be the necessary historical information available upon which an estimate can be made or trends developed. In such situations, market surveys may be effectively used to determine the potential sales for the item. The market research may be very informal, utilizing the sales force to "feel out" the potential customers in order to establish the extent of the market, or it may be a systematically conducted survey utilizing special mathematical techniques.

The most common method of market research is, however, to introduce the product in a relatively small trial area and project the total sales from the results obtained. An example of the results of such a survey may be as follows:

> $500,000 sales are realized in the trial area during the time that an estimated total of $10,000,000 was spent by consumers in the area on this type of product. Thus, it appears that the initial acceptance for the product will be equal to approximately 5% of the total market. If the total annual market for this type of product averages $100,000,000, the forecast estimate of $5,000,000 sales per year would be made for the new product.

Of course, the preceding example has been greatly simplified. Some of the most important problems confronted when the survey approach is used are:

1. Determining total sales of the product group in the trial area **during** the period that the new product is being introduced.
2. Selecting an area for the trial which can be considered "typical" **of** the total market so that valid inferences can be made.
3. Knowing the sales of the entire product group in the total **market.**

Correlation   Correlation is another common forecasting technique.
   When correlation is used, the forecast for a particular organization is derived by relating the organization's activity to the activity of a given industry or industries, or to the economy as a whole.

There are two types of correlators which can be effectively used in forecasting:

1. Leading correlators
2. Current correlators

Leading correlators are the most desirable if they can be established. A classical example of leading correlation is in the automobile replacement battery market. In Figure 6-6, the number of replacement batteries sold in a given month is plotted along the ordinate ("*Y*" axis), while the number of new cars sold eighteen months prior to the same given month is plotted along the abscissa ("*X*" axis). The line of best fit is drawn

Fig. 6-6. Correlation technique used by battery manufacturers to forecast replacement sales. For example: 800,000 new cars sold 18 months prior to month under consideration would correspond to 500,000 estimated replacement sales.

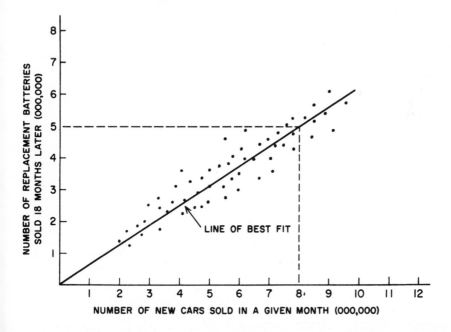

through these points in the same manner as described in the discussion on trend lines. The type of formulae and an example calculation are given in Appendix I. The battery replacement sales for the current month can then be estimated. This is done by drawing a perpendicular line up from the abscissa value equal to the number of new cars sold eighteen months ago. This perpendicular line is extended until it crosses the line of best fit. From this point a horizontal line is drawn to the left until it reaches the ordinate. The value on the ordinate is the forecasted battery sales.

Current correlators can be used whenever leading correlators are not available. With current correlators, the total sales for an industry are estimated from trends and other indicators and then a forecast for the individual organization is established based on a predicted share of the industry market. When current correlators are used, the forecast for the particular organization is based upon another forecast which has been made for the industry or the economy as a whole. One of the most common current correlators used by industrial concerns is gross national product. The ' G.N.P." is estimated on a continuing basis by the economists working for the Federal government and is readily available in newspapers, business magazines and government publications. The main advantage in using a current correlator is that it is the result of considerable research and scientific analysis by the government, educational institutions, and industrial associations. An organization can improve its forecasting ability by utilizing this information as much as possible.

When an organization analyzes its previous sales in terms of available correlators, it frequently finds that the activity seems to be related to several different correlators. Mathematical techniques of multiple correlation can be used to determine the weightings for each of these correlators. In general, the composite relationship will provide a better forecast than will a single correlator. For example, one organization found that excellent results were achieved with a combined correlator made up of 32 different correlators. This is probably an extreme example and most companies could secure a good forecast with a lot fewer factors in the combined correlators. Generally speaking, the more variables fit into the correlation the shorter the run will be in which one may be confident of a given percentage of deviation.

Some of the most important sources of correlation data are:

   1. *Economic Data*
      1.1 *Survey of Current Business,* Office of Business Economics, U.S. Department of Commerce, Washington 25, D.C.[1] A monthly survey comprising 2,500 different statistical services, including

---

[1] All federal government publications may be purchased from the Superintendent of Documents, U.S. Government Printing Office, Washington 25, D.C.

national income, personal and farm income, retail and wholesale sales indices, manufactures orders and inventory, production prices, and shipments.

1.2 *Monthly Labor Review,* Bureau of Labor Statistics, U.S. Department of Labor, Washington, D.C. Statistics by month and by
industry, for employment, payroll, hours; weekly and hourly earn-
ings by months, cities, and commodities; work stoppages; building and construction.

1.3 Business magazines such as *Business Week* and the *Kiplinger
Letter.*

2. *Industry Data*

2.1 *Annual Survey of Manufacturers,* Bureau of the Census, U.S.
Department of Commerce, Washington 25, D.C. Annual production and estimated dollar sales for most major industries.

2.2 Industrial trade associations.

2.3 Industrial trade journals.

Correlation can be used successfully in estimating the requirements of
an organization for service activities such as shipping, materials-handling,
billing, process engineering, and equipment repair. The correlation technique, when used in these situations, is often described by the term "staffing pattern" or "semi-variable budgeting." However, the work in the
support activity is actually correlated with the basic production work
load and therefore can be predicted once the basic work load has been
estimated.

The Army Ordnance Corps uses correlation to determine the repair
parts requirements for its many different kinds of vehicles. The original
forecast is made in terms of troop strength and disposition. The total
number of each type of vehicle that will be in use is determined from
Tables of Organization and Equipment which list the type and number
of vehicles authorized for each type of unit. The repair parts requirements are determined by multiplying the number of vehicles assigned
to Army units by a use factor that has been developed from past experience. Of course, the procedure for analysis is complicated by the adjustments which must be made for geographic location, type of unit (National
Guard, regular Army, reserve, and so on) and the projected type of activity of the unit—war, maneuvers, or stand-by. All of the calculations
are carried out on high speed computors and the accuracy of the forecast is remarkable.

Prior Knowledge as a   When an activity is part of a much larger
Substitute For Forecasting   organization such as a large multi-plant
                                   corporation, or the various governmental
agencies, little, if any, forecasting has to be done at the local or plant
level. The activity is supplied with established forecast information or
even firm orders, and all that remains to be done is to complete the

remaining planning functions. Even when the activity has been supplied with prior knowledge for the major production elements, some degree of forecasting must be carried out at the local level to establish the requirements for the auxiliary service activities. In such situations the forecasting function has not been eliminated, but the function has been accomplished by a higher level in the organizational structure.

USE OF JUDGEMENT IN FORECASTING

The forecasting techniques which have been discussed cannot be used as a substitute for sound judgment. The forecaster must always keep in mind the possible actions of competitors within his industry and within similar industries. This factor is very difficult to reduce to numerical data, yet cannot be ignored. There are naturally many other intangible factors which must be considered and in many cases only the judgment of the forecaster can effectively cope with them.

In general, the "best" forecasting technique is one that uses as many of the basic techniques as possible, compares the estimates indicated by each technique and then applies judgment to produce the final forecast. This is a judgment based on all of the available facts—not wild guesses. Furthermore, if at all possible, this judgment should be a group effort of the higher management of the organization. The management should be presented with the results of the different techniques and the final forecast developed from this base.

WHEN THE FORECAST SHOULD BE MADE

An important consideration in the forecasting function is the determination of *when* the forecast is to be made. It can be prepared basically in one of two ways:

1. Periodic forecasting
2. Continuous forecasting

These basic time phasings are for establishing the forecast and not simply for revising an existing forecast. Revising the forecast is discussed in detail in the next chapter. In this chapter we are only concerned with the original forecast.

The periodic forecast, as its name implies, is established for a given length of time, usually a year and at the end of this period, a new forecast is prepared for the next period. Periodic forecasting is best applied in situations where:

1. There are annual or semi-annual style or model changes.
2. The activity is part of a much larger organization which plans and budgets on a periodic basis.

In continuous forecasting, the forecast time period (normally a year) is divided into shorter time intervals (normally months). When one time interval has elapsed, it is dropped from the forecast and another interval is added. In this way there is always a current 12 month forecast available. For example, on the first of March an organization has a forecast for each of the following 12 months, up to the end of February. At the end of March, the forecast is made for March of the following year, so that the organization still has a twelve-month forecast. If alterations in the previously anticipated activity level are indicated, the forecast for the other eleven months is revised.

The advantages of the continuous forecast are:

1. Forecasting is carried on as a *regular* part of the planning procedure and is not just an annual "extra" job.
2. Seasonality factors can be more easily considered because of the information secured during the time interval just completed.

ANALYZING THE FORECASTING FUNCTION

When designing a Production Control system, there is a tendency to overlook the forecasting function. This is usually due to the fact that the forecast is prepared in the sales or market research department which is organizationally separated from Production Control. It is important that the designer be fully cognizant of the importance of forecasting and that he carefully consider the function's place in the system. As was previously emphasized, forecasting is the cornerstone upon which all the planning functions are built. Without a sound forecast, the effectiveness of the entire Production Control program is seriously impaired.

When the forecasting function is analyzed, there are certain aspects that must be given careful attention:

1. If the forecast is being made in a department other than Production Control, it must be prepared so as to include data useful for production planning purposes. Forecasts prepared in the sales department are frequently nothing more than quotas expressed in terms of sales dollars. Factors such as product mix and seasonality are of particular concern to production planning, and therefore must be included in the forecast. In many companies this means that the current sales forecast technique must be completely revised in order to make it meaningful to Production Control.
2. Frequently it is found that a number of different forecasts are made within the same organization. The financial activities are established on one forecast, sales on another, and Production Control on a third. Not only does this situation result in needless duplication of work but it also creates inconsistency in plans. There must be one forecast used by all elements of the organization in order to insure uniformity in planning.

3. The responsibility for forecasting should be assigned to a person or persons who have the capabilities, the desire, and the time to do a thorough and competent job.
4. Situations in which periodic forecasts are made should be analyzed to determine whether continuing forecasts would not be more beneficial.
5. The accuracy of the current forecasts must be determined. When there are compensating errors, actual total dollar sales may come out very close to the original forecast even though the forecasts for individual products may be off by as much as 50 or 60 per cent. Such a situation might not be evident until a complete analysis has been made. Needless to say, inaccuracies of this type would create gross errors in the plans of Production Control and must be eliminated.

Only after careful consideration of all of these factors can the most desirable techniques be selected for application to the organization.

It should also be pointed out, that forecasts for a given activity may turn out to be over-optimistic when prepared by those responsible for the activity. This is to be expected. After all, every good manager wants to show what he believes he is capable of doing. However, the most useful forecast is the one which realistically indicates the future activity. This forecast will provide a warning signal to management to be prepared to reduce costs, expand budgets, diversify product lines, embark upon a sales campaign to secure a larger portion of the market, or other indicated management action.

Summary   Forecasting, which is defined as *the estimation of future activities,* is the first step in the Production Control functions. The forecast is the cornerstone from which the requirements for facilities, manpower, and material are derived.

Long-range forecasting is usually concerned with broad, general terms while short-range forecasting is concerned with the potential sales of a single product or group of products during a given period of time in the near future. The following techniques of forecasting have been discussed:

1. Historical estimate
2. Sales force estimate
3. Trend lines
4. Market survey
5. Correlation
6. Prior knowledge

No matter what technique is used, judgment must still be employed, based upon sufficient objective data to provide valid results. The end result—the forecast—is in itself an estimate. For this reason considerable "art" is required in forecasting. In the next chapter, techniques will be discussed for controlling the forecast and determining when it should be revised.

7

Forecasting Control

In forecasting, as in all other areas of management, we must compare actual progress with a plan, and take any corrective action indicated, if we are to have control. Forecast control is exercised when actual orders received are compared with those forecast for a period and corrective action is taken by revision of the forecast or modification of the sales effort.

If forecasting control is to be effective, the corrective action should be based upon comparisons made *during* the forecast period. The frequency of the comparisons depends upon the duration of the forecast period and the nature of the business. Where there is a dominant seasonality pattern, there will be a need for comparisons at shorter intervals in the peak period than in the low period of sales. Following are some examples of this condition:

1. In a major steel company, detailed comparisons are made on a semiannual basis and general comparisons are made on a monthly basis.

2. In a basic chemical company, comparisons are made on a quarterly basis.
3. In a mail-order house, comparisons are made on a biweekly basis during the six-week peak Christmas season, on a weekly basis during the six weeks prior to the peak season, and monthly during the remainder of the year.
4. In a pharmaceutical firm, comparisons are made every four hours during the peak two week period for flu vaccine, because once the vaccine has been rebottled in vials it cannot be stored for more than a few days.

The first forecast comparison usually made is between total sales forecast and actual sales for the period under analysis. This is important primarily because much top management planning is based on the total sales figure. However, even if actual sales compare favorably to forecast sales further analysis is usually required. If the forecast is based on a specified share of the market, the total market sales must be compared to company sales to determine if the organization's competitive position is improving or declining. Furthermore, the actual sales must be analyzed to determine if each product or product group sales are in the planned proportion to total sales. The analysis of product mix is very important. Varying amounts of marginal income and manpower and material requirements are more directly related to individual products than to total sales dollars.

When the comparison between the forecast and the actual sales indicates a need for corrective action, the most common approach is to revise the forecast. However, before the forecast is revised, an attempt should be made to determine the cause of the deviation. Deviations may be caused by:

1. Errors caused by the basic forecasting techniques used.
2. Errors in judgment or modification of available facts by management.
3. Poor pricing procedures.
4. Ineffective sales or advertising program.
5. Declining or increasing economic conditions.
6. The nature or design of the product.
7. Actions of competitors.

After a thorough analysis of the causes of deviations from forecast, corrective action may be to make positive efforts to correct the actual sales situation rather than to change the forecast. For example, it may be decided to begin an extensive sales and advertising campaign when total sales have been unsatisfactory or to modify the pricing formula when the product mix has been unfavorable from a marginal income point of view.

WHEN THE FORECAST SHOULD BE REVISED

Since a "perfect" forecasting method has not yet been devised, there will invariably be discrepancies when the comparison is made between the actual orders received and the forecast. Traditionally, the task of evaluating current order activity against the forecast activity has been based on "judicious interpretation of data by seasoned managers." However, for forecasting control to be most effective, it is very important for an organization to have an objective means of determining if the activity merits continuance of the present forecast or if the forecast should be revised.

The need for this objective basis for forecast revision can be emphasized by considering the costs associated with improper action in such revision. These costs may be divided into two areas:

1. *Costs associated with excessive adjustment.* These costs are created by losses in production time, increases in set-up time, inefficiencies of small lot production, and premiums involved in overtime, unemployment taxes, and fringe benefit costs.
2. *Costs associated with insufficient adjustment.* These costs are created through loss of profit opportunities, cutomer ill-will due to poor delivery and excessive inventory buildup when production exceeds demand.

To provide this objective basis of forecast revision, statistical control chart techniques similar to those commonly used in other areas of management may be applied. The statistical techniques relieve the manager of the time consuming and laborious job of analysis and place the task with clerical personnel. The manager is advised when a significant shift in sales activity has occurred so that he may take corrective action. This approach has the advantages of "management by exception."

In the remainder of this chapter we will describe the application of a statistical control-limit technique to the problem of determining an objective basis for forecast revision.[1]

CONTROL-LIMIT APPLICATION

As was mentioned in the preceding chapter, forecasts may be prepared on unit or dollar bases, for total company sales, for different product groups, or for individual products. In the following example the forecast and subsequent analysis will be considered on the basis of the units in a particular product group. The reader will be able to determine the slight modifications that would have to be made in the technique in order to apply it to other data.

In order to reduce the effect of exceptionally large orders received (or lost) during any month, it is necessary to make all comparisons of actual activity vs. anticipated or forecast activity on a cumulative basis.

[1] Based upon W. L. Westerman, *Investigation into the Applicability of Control Limit Techniques as an Objective Basis for Forecast Revision,* Master's Thesis, State University of Iowa, 1958.

Most organizations experience seasonal variations in the receipt of orders. For example, 5 per cent of the annual orders may be received in January, 10 per cent in February and 8 per cent in March. When this data is converted to a cumulative basis, it means that 5 per cent of the annual orders have been received by the end of January, 15 per cent by the end of February and 23 per cent by the end of March. Mathematically these cumulative orders represent a cumulative probability distribution. Kolmogoroff has derived a formula to determine the confidence limits for such a cumulative distribution.[2] Table 7-1 lists various confidence coefficients and factors which have been derived from the Kolmogoroff formula.[3] These factors divided by the square root of the sample size establish a confidence limit.

TABLE 7-1

FACTORS ASSOCIATED WITH COMMON CONFIDENCE COEFFICIENTS FOR A CUMULATIVE PROBABILITY DISTRIBUTION

Confidence Coefficients	Factor
.50	.835
.60	.896
.70	.971
.80	1.07
.90	1.30
.95	1.35
.98	1.52
.99	1.63
.995	1.73
.998	1.86
.999	1.95

Confidence limit expressed as a per-cent of total sample $= \dfrac{\text{Factor} \times 100}{\sqrt{\text{Sample size}}}$

EXAMPLE

To illustrate the application of the Kolmogoroff formula and the derived factors to an actual situation, the following example is presented.

First, is it necessary to determine the pattern of seasonality for the product group of interest. This is done by analyzing the pattern of order receipts experienced by the group during previous years. For illustrative

---

[2] A. Kolmogoroff, "Confidence Limits for an Unknown Distribution Function," *The Annals of Mathematical Statistics*, Volume 12 (1941), pages 461-463.

[3] W. L. Westerman, *op. cit.*

purposes, assume the product group has been in existence for four years and the order receipts have had the same seasonality pattern for each year (see Table 7-2). The same seasonal activity pattern is expected to continue into the future. In Table 7-2 the cumulative orders, in units, are converted to cumulative orders in percentages for each of the four years. The expected monthly cumulative probability distribution is determined by averaging on a monthly basis.

According to Kolmogoroff's theorem, the confidence limit is a function of the factor based on the confidence coefficient and the square root of the sample size. In this application, the "sample" is the total number of orders which are forecast to be received during the year. Thus, the sample size is actually a hypothetical quantity because the forecast in itself is an estimate. It is very important to note that the sample is represented by the forecast of *total number of orders,* not total units ordered. This is because the variability occurs in the receipt or loss of an order which may consist of one or more units.

The next step is to determine the average order size which is done by

LE 7-2

RMINATION OF CUMULATIVE PROBABILITY DISTRIBUTION

	\multicolumn{8}{c}{Cumulative Orders by End of Month}		
	\multicolumn{8}{c}{Year}		
nth	1		2		3		4		Total	Av.
	Units (100)	%	Units (100)	%	Units (100)	%	Units (100)	%	%	%
.	42	3.6	50	4.1	34	2.7	30	2.2	12.6	3.2
.	94	8.1	109	9.0	99	7.8	118	8.7	33.6	8.4
.	142	12.2	168	13.9	162	12.7	193	14.2	53.0	13.2
.	264	22.7	260	21.5	274	21.5	350	25.8	91.5	22.9
y	419	36.1	442	36.6	438	34.4	551	40.6	147.7	36.9
.	525	45.2	556	46.0	550	43.2	654	48.2	182.6	45.6
.	627	54.0	699	57.8	761	59.8	750	55.3	226.9	56.7
.	864	74.4	876	72.5	983	77.2	993	73.2	297.3	74.3
.	956	82.3	997	82.5	1063	83.5	1099	81.0	329.3	82.3
.	1034	89.1	1082	89.5	1179	92.6	1269	93.5	364.7	91.2
.	1120	96.5	1138	94.1	1238	97.3	1331	98.1	386.0	96.5
.	1161	100.0	1209	100.0	1273	100.0	1357	100.0	400.0	100.0

dividing the total units ordered in the previous year by the total orders received. Assume that the sample in our example has indicated that the average order is one gross—144 units.

Next, it is necessary to determine the confidence coefficients upon which to base the control limits. These coefficients do not necessarily have to be the same for every time period (month in this example) or the same for the upper and lower control limits of the forecast. The actual selection of the confidence coefficient can be an arbitrary management decision or can be determined in a breakeven type of analysis. We will assume that management has selected a confidence coefficient of .95 for the lower control limit and .90 for the upper control limit during the first six months of the year. A confidence coefficient of .90 for the lower limit and .80 for the upper limit during the last six months of the year is selected.

A total yearly sales of 144,000 units has been forecast for the product group. This completes the information required to establish the forecast control limits.

The first step in establishing the forecast control limits is to divide the total activity forecast, in units, by the average order size and thus convert the unit forecast into an order number forecast:

144,000 units forecasted ÷ 144 units per order = 1000 orders forecasted

The second step is to take the square root of the orders forecast:

$$\sqrt{1000} = 31.5$$

The next step is to determine the confidence limit for each month. Selecting February as an illustration the confidence coefficients of .95 and .90 would be used. Table 7-1 gives the factors 1.35 and 1.30 for the .95 and .90 confidence coefficients respectively. Dividing these factors by the square root of the sample size and multiplying by 100 establishes the confidence limits as a per cent of the total forecast.

$(1.35 \div 31.5) \times 100 = 4.3\%$ (lower confidence limit for February)
$(1.30 \div 31.5) \times 100 = 4.1\%$ (upper confidence limit for February)

Since units and not percentages are used in the actual analysis of the validity of the forecast, the percentages multipled times the total annual forecast will establish the confidence limits on a unit basis:

$4.3\% \times 144,000 = 6192$ units (lower confidence limit for February)
$4.1\% \times 144,000 = 5904$ units (upper confidence limit for February)

The average cumulative percentages shown in Table 7-2 have been reproduced in Table 7-3 in the column titled *"Historical Cumulative Per Cent Ordered."* Multiplying the historical cumulative per cent for the month (for February this value is 8.4%), times the annual forecast in units, the forecast cumulative units ordered is obtained.

8.4% × 144,000 = 12,096 units ordered by the end of February

If the total annual order was perfect and the seasonality pattern was identical with that determined by averaging the four preceding years, by the end of February orders for 12,096 units should have been received. Since the possibility of such a situation is very remote, the confidence limits are placed around the forecast cumulative units ordered to establish the upper and lower control limits:

Lower control limit = 12,096 − 6192 =   5,904 units
Upper control limit = 12,096 + 5904 = 18,000 units

For each individual month these calculations are repeated to establish the appropriate upper and lower control limits. As long as the actual cumulative orders received fall between the upper and lower control limits, the forecast is considered to be "in control" and no corrective action is taken. Table 7-3 shows the calculated control limits for each month.

In the right hand column of Table 7-3 the actual cumulative units ordered during the forecast year are given. The forecast for the product group was "in control" until the end of July. At that time, orders for 75,110 units had been received. Since the lower control limit was set at 75,774 units the forecast had to be revised. The important aid in preparing the new forecast is the fact that 56.7% of the total orders are normally received by the end of July. Thus the actual orders of 75,110 units should approximately equal 56.7% of the total units to be ordered, or 132,469 units should be the total ordered during the forecast year.

This figure was rounded off to 132,000 units. In Table 7-3, the calculations for the control limits had to be revised in August because of the change in "sample size" due to the change in the forecast. The revised forecast kept the actual monthly orders "in control" for the remainder of the year.

Other than the obvious conclusion that the forecast is incorrect when an "out of control" situation occurs, two other possibilities exist that should be investigated:

1. The seasonality pattern may have changed. The total forecast for the year might still be satisfactory in this case.
2. The receipt (or loss) of one exceptionally large order may have caused the "out of control" situation.

Control charts, similar to statistical quality control charts, may be drawn so that the analysis can be performed graphically. If the data is charted, quick presentations can be made to top management and trends will become apparent even though no "out of control" situation may exist. Figure 7-1 illustrates the control chart for the original forecast in our example and Figure 7-2 shows the chart for the revised forecast.

DETERMINATION OF CONTROL LIMITS AND ANALYSIS OF ACTUAL ORDERS

Original and Revised Annual Forecast (units)	Months	Current Forecast (orders)	√ Current Forecast (orders)	Confidence Coefficient .95 / .90 / .80 — Factor 1.35 — Confidence Limit %	Units	1.30 %	Units	1.07 %	Units	Historical Cumulative Per Cent Ordered	Forecast Cumulative Units Ordered	Control Limits (Units) Lower	Upper	Actual Cumulative Units Ordered
144,000	Jan.	1000	31.5	4.3	6192	4.1	5904	X	X	3.2	4,608	—	10,512	4,738
"	Feb.	"	"	"	"	"	"	X	X	8.4	12,096	5,904	18,000	11,782
"	Mar.	"	"	"	"	"	"	X	X	13.2	19,008	12,816	24,912	19,856
"	Apr.	"	"	"	"	"	"	X	X	22.9	32,976	26,784	38,880	28,311
"	May	"	"	"	"	"	"	X	X	36.9	53,136	46,944	59,040	49,236
"	Jun.	"	"	"	"	"	"	X	X	45.6	65,664	59,472	71,568	62,547
"	Jul.	"	"	X	"	3.4	"	3.4	4896	56.7	81,648	75,744	86,544	75,110
132,000	Aug.	916	30.3	X	X	4.3	5676	3.5	4620	74.3	98,076	92,400	102,696	100,034
"	Sep.	"	"	X	X	"	"	"	"	82.3	108,636	102,960	113,256	107,485
"	Oct.	"	"	X	X	"	"	"	"	91.2	120,384	114,708	125,004	118,793
"	Nov.	"	"	X	X	"	"	"	"	96.5	127,380	121,704	132,000	129,374
"	Dec.	"	"	X	X	X	X	X	X	100.0	132,000	X	X	133,643

TABLE 7-3

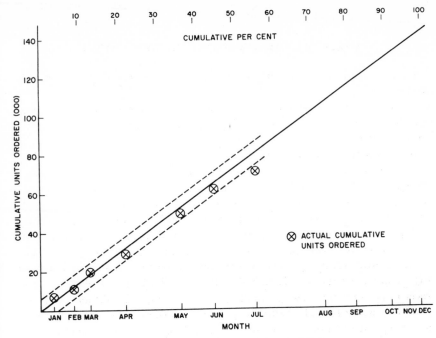

Fig. 7-1. Original forecast and control limits.

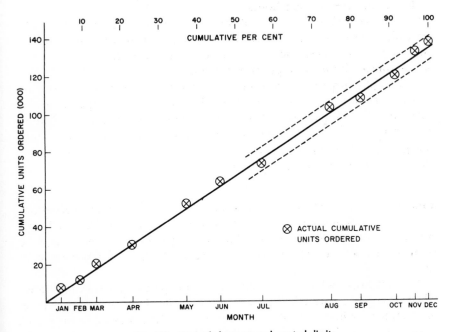

Fig. 7-2. Revised forecast and control limits.

LIMITATIONS OF CONTROL-LIMIT APPLICATION

Certain limitations of the application of this technique should be understood:

1. The control-limit technique is not a substitute for a good original forecast. Since some time must elapse in the forecast period before the control-limit analysis can be made, a portion of all planning must be based upon the original forecast.

2. The control-limit technique should be applied only to products which have been sold by the organization for several years so that a definite seasonal pattern can be established. The more data available, the better the estimates of the monthly cumulative percentage of order receipts for each product or group. The technique can be applied, with due caution, to a new product if there is reason to believe that the new product order receipt pattern will follow the pattern of some other existing product, or if there is sufficient data available to develop an order receipt pattern for new products in each group. In either case, new products will probably evidence excessive variability in order receipts, and the control-limit analysis may lead to frequent forecast revision.

3. In custom production and some job order production, the technique can be applied by using some "common denominator" basis such as dollars of sales or man hours.

4. The technique is most applicable in situations in which the average order size is relatively consistent and small in relation to total annual sales. The larger the average order size, the wider the placement of the control limits (sample size decreases).

5. In applying the control limit technique to analysis of order receipts versus a forecast, it is assumed that the order receipt pattern is fairly stable. Thus, an "out of control" situation can be attributed to an inaccurate forecast. The user of the technique must be cognizant of changes in marketing activities in order to recognize situations which may affect the order receipt pattern.

Summary   After the forecast has been prepared based on the various techniques in Chapter 6, comparisons must be made between the actual activity and the forecast. The process of making these comparisons and taking the necessary corrective action is called forecasting control.

The comparisons should be made during the period of the forecast and, when necessary, corrective action—revising the forecast or modifying management policies—should be taken as quickly as possible to improve undesirable conditions and maintain planning data on a current basis.

The forecast should be revised when there appears to be a significant deviation between the actual and the forecasted activity. In this chapter, a procedure based upon control-limit applications has been proposed. The basic steps in this procedure are as follows:

1. Determine the seasonality pattern.
2. Determine the average order size.
3. Select the confidence coefficients for the control limits.
4. Convert the unit forecast to an order forecast.
5. Determine the confidence limits.
6. For each month:
   6.1 Determine the control limits.
   6.2 Compare the cumulative activity with the control limits.
   6.3 Take corrective action when indicated.

Forecasting control should be considered an adjunct to forecasting, but not a substitute for an accurate original forecast.

8

Work Authorization

Work authorization is defined as: *the authority given to an individual or a group of individuals to undertake a specific objective.* The reader at this point should again refer to Table 3-1, page 18, and note that work authorization is the second basic function in the prior-planning phase and follows forecasting.

Although the forecast is the cornerstone from which all planning develops and hence is the "first" function in any Production Control system, we normally think of "Work Authorization" as the function which actually initiates action in the system. In a formal system, this is obvious because the flow of "paper" begins with the written work order. Even in the informal situations, the remaining functions are dependent upon a person or a group being given authority to undertake an objective.

The *authority* which is mentioned in the definition may occur in several forms such as:

1. The written work order authorizing an individual or a crew to perform a particular job.

2. A verbal request which later may or may not be formalized in writing.
3. In informal situations, as standard operating procedures, departmental or company policy statements, job descriptions, or simply tradition.

The other key word in the definition is *objective*. The objective which the individual or the group is to undertake may be a specific task, such as unloading a truck. It may be the creation of a certain end-product such as a plan for a component in a rocket engine. The objective may even be to provide some form of service such as answering inquiries.

There is a fine point of distinction between work authorization and the function of dispatching which is discussed in Chapter 18. Dispatching is the *actual assignment* of a specific task to the individual who is to perform it, whereas work authorization simply provides the overall "permission" to perform the work.

BASIC REQUIREMENT

The basic requirement of the work authorization function is that the *individuals who are responsible for coordinating the work within an organization must have control of the authorization of work.* The person charged with the responsibility of coordination must have control over this function in order to do an effective job of planning and scheduling work. He must be able to authorize overtime and initiate reductions in the work force. He must be able to insure that all elements of the work load are placed in the proper perspective. Priorities must be determined and, even more important, proper control must be maintained to insure that individuals do not take tasks or assume responsibilities for functions which they are not authorized to perform.

In the informal situations the coordinator will usually be the foreman while in formal systems the coordination will be done by the Production Control staff.

RELATIONSHIP BETWEEN WORK AUTHORIZATION AND OTHER MANAGEMENT CONTROLS

There is a close relationship between work authorization, performance evaluation, and accounting. Performance evaluation can be described as the comparison of the actual time taken to accomplish a task to a standard amount of time set for the task. To be of value, the standard should be determined in advance and *credit should be given only for work which has been properly authorized.* Realistic standards for authorized work and insistence that these standards be met is one of the best ways to insure that unauthorized work will not be performed.

An accounting system based upon strict adherence to budgets for all "overhead" accounts will also aid in insuring that only authorized work is performed. This is done by controlling the labor and material dollars which can be charged to the nonproduction accounts. Also, if accounting data are to be meaningful for future budgeting and management control, they must be generated in a situation where there has been a positive form of work authorization supported by realistic performance standards.

EXTERNALLY-GENERATED WORK

An externally-generated work situation is one in which the authorization stems from an organization which is separate from the individual or individuals performing the work. It is external to the person performing the work, not necessarily external to the plant. In organizations producing to the customer's order, the work authorization is generated "externally" by a sales or production order department. In organizations producing for stock, the work is authorized "externally" by inventory or production control.

From the point of view of the designer of the Production Control system, an externally-generated work situation has the following advantages:

1. Writing of the order provides a firm point of departure for the Production Control system.
2. Some degree of formalization of the Production Control system is required to insure that the authorized work is performed—and performed on time.

INTERNALLY-GENERATED WORK

An internally-generated work situation exists when the authorization stems from an individual within the organization which is performing the work. Internally-generated work is very common in all activities other than manufacturing and usually develops from three sources:

1. The supervisor
2. A person to whom the "production" elements provide a staff service
3. The worker himself

Following are some examples to illustrate further what we mean by internally-generated work:

1. A supervisor may generate work internally by assigning a machine operator to repair a machine or by assigning a typist to clean out and check the files.
2. Work can be generated by persons to whom the "production" elements are staffed in such forms as:
    2.1 A request for emergency repair service directly from the area foreman to a maintenance man who happens to be in the area at the time.

    2.2 A request for clerical service by a teaching staff member to a clerk in the typing pool of a school's administrative unit.

    2.3 A request for a change in a pilot model by a project engineer directly to a machinist in the pilot model shop.

3. An individual worker generates work when:

    3.1 An engineer decides to pursue a new approach to a problem.

    3.2 A janitor determines that he shall wash windows (when normally the windows are washed by a special crew with equipment for this operation).

    3.3 A secretary decides to scrub and clean her desk top.

    3.4 A fiscal clerk keeps a "private" record of all the vouchers processed.

Needless to say, many problems arise in situations where such internally-generated work authorization systems prevail. Some of these problems are as follows:

1. The capacity of the work group or department cannot be determined. This makes it difficult to load accurately or schedule realistically. (See Chapter 15.)

2. There is no way of assuring that effort is directed most effectively because unnecessary work may be performed such as:

    2.1 The machine which is being repaired by the machine operator may already be scheduled by the maintenance department for a complete overhaul.

    2.2 The emergency maintenance job may not be an emergency at all because the piece of equipment which has broken down may have a substitute.

    2.3 The course that the engineer decides to pursue may already have been investigated by another member of the department.

3. Necessary work may *not* be performed and priorities may be misdirected such as:

    3.1 The pilot model change requested by the engineer may cause another more important product to get behind schedule.

    3.2 The secretary may be cleaning her desk to avoid a job that she dislikes, such as collating.

4. It is very difficult to budget and perform the other planning functions and almost impossible to achieve any objective measure of performance.

In spite of the problems just mentioned, it would not be practicable to force all work through a formal Production Control function with an authorization for every externally-controlled activity. The Production Control system would quickly be bogged down with many small and insignificant jobs. The result would be control costs perhaps equal to or greater than the value of the end product being controlled.

There are instances in almost every work situation where the designer may want to ignore or bypass the formal system. The important point to keep in mind, is that *the decision on what jobs should bypass the system must come from the coordinating medium* if an adequate degree of control is to be maintained. There are several effective methods in which this problem can be handled:

1. Establish a small separate organizational element to handle miscellaneous or nonstandard work.
2. Set aside a certain portion of the day for the miscellaneous work.
3. Establish a service element on a staffing pattern basis (such as a private secretary) with no formal control over this staff.
4. Set up a budget for this type of work. As long as a supervisor does not exceed his budget, no other attempt is made to control.

No matter which method is used, it is exceedingly important to determine in advance how much time and money should be represented by this form of work. Current reports on the time and dollars spent must be maintained so that tight control can be exercised.

ANALYZING THE WORK AUTHORIZATION FUNCTION

When a Production Control system is being designed or analyzed, one of the first steps is to determine the organizational responsibility for coordination of work authorizations. It must be determined if it is possible to retain this function in the line organization or if it should be assigned to a staff. See Chapter 27 for a discussion of organization planning.

The Production Control system must be designed so that all work authorizations stem from the coordinating medium and the coordinating medium has control over *what* work may bypass the system. In situations where miscellaneous work can be kept out of the formal system, the magnitude of this work must be determined in advance, and measures taken to insure that the planned amount will not be exceeded. This is usually achieved through budgetary control. Furthermore, the method of performance evaluation and the standards upon which the performance evaluation system is based must provide quick and simple indication when unauthorized work is being performed.

In the manufacturing activity, the work authorization is usually generated in an order department. This function is not a source of many problems unless salesmen or other personnel consistently bypass the control system. However, in other types of activities—particularly those of clerical and service activities—the absence of proper control over the authorization and generation of work may be one of the most troublesome weaknesses of the system. "Overhead" activities and costs increase greatly when there is insufficient control of the work authorization function because of the unnecessary work that is generated. One of the most striking examples of this is the preparation of reports that simply duplicate information which is already available. Without proper control of work authorization, many of the remaining planning phases cannot be properly performed. Inefficient methods of operation carry on throughout the activity.

Summary   The paperwork which initiates action in any formal Production Control system is generated in the work authorization function. Thus, the analysis of the day-by-day Production Control system will usually begin with investigation of the work authorization function.

When work is generated externally, a formal Production Control system usually exists and the analysis of the work authorization function is relatively simple. However, when work is generated internally, the problem of analyzing and controlling the work authorization becomes difficult and the lack of adequate control of this function may be the fundamental weakness of the entire system.

The basic requirement for successful performance of the work authorization function is that the individuals who are responsible for coordinating work within an organization must have control of the authorization of work. If it appears desirable to bypass formal authorization in conjunction with miscellaneous jobs, the magnitude of this work must be determined and adequate controls over time devoted to such work must be established.

The work authorization information plus data generated in the product planning function which is to be discussed in the next chapter, are required in order to plan properly the process and perform the remaining Production Control functions.

9

Product Planning

The third and last function in the prior-planning phase is product planning. *Product planning is defined as the extension of the original product design to information which can be used to plan the production process and procure the materials which are required to make the end product.* Before considering this function, it must be assumed that the forecasting function has been completed as described in Chapters 6 and 7, and that the forecast has indicated that there is a need for the product and that production has been authorized.

The original product design may reach the product planning function in many different forms. It may be a detailed blueprint from a customer or from research engineering; it may be in the form of a sketch giving a general indication of what is required; or it may even be received as a verbal or written assignment such as: "Give me a report," or "Fix the motor."

EXTENDING THE ORIGINAL PRODUCT INFORMATION

1. *In all activities, the product must be broken down into sub-assemblies and component parts.* With

manufacturing items such as motors, this is very obvious. For example, the sub-assemblies are the frame, rotor, and so forth, and the component parts are the washers, bolts, coils, and so on. However, this analysis is equally true of the nonmanufacturing type of activities. For example, if the end product is a report, the sub-assemblies would be the sections of the report and the component parts the tables, supporting sketches, or appendices. When determining the component parts and sub-assemblies, the product planner should utilize standard parts and assemblies to the greatest extent possible in order to reduce the total cost of the product.

2. *The tolerances and specifications for each of the component parts and sub-assemblies must be established.* To do this, the planner should be familiar with the production methods which will be used and the costs associated with various degrees of tolerance. There is a common tendency on the part of the designer, as well as the product planners, to establish specifications which are much more demanding than is really required, considering the function of the end product. This is true in the nonmanufacturing areas as well as in manufacturing. For example, a management control information report will frequently supply the information required if all figures are rounded off to the nearest $10 or even $100, yet we sometimes find them being prepared with all figures accurate to the last penny.

3. *The "bill of materials" is prepared.* The bill of materials is the end product of product planning. The bill of materials lists each part and the quantity required for one assembly. Figure 9-1 gives an example of a bill of materials arranged in tabular form. In some cases the bill of materials is prepared so as to indicate the order in which the parts are used to produce the sub-assemblies and the end product. The tabular bill of materials in Figure 9-1 has been arranged in assembly order in Figure 9-2. The bill of materials is the basic document used in material planning. In Chapter 11, we will discuss how this document is used to determine the material requirements.

VALUE ANALYSIS

One of the most important considerations in the product planning function is to determine whether the component part or sub-assembly is to be made or purchased. Many factors go into this decision, such as:

1. The capability of the organization to produce the item to the required specifications.
2. The available capacity to produce it within the required time.
3. The reliability of vendors in meeting specifications and delivery dates.
4. The investment required to obtain equipment to produce the item.
5. The utilization of excess manpower and equipment.

BILL OF MATERIAL

DESCRIPTION: <u>Power Unit Assembly -- Heavy Duty</u>   B. M. NO. <u>313022</u>

<u>-- 115 Volts -- 60 Cycles</u>   DWG. NO. <u>C-313022</u>

PART NO. <u>313022</u>

ITEM	PART NO.	DESCRIPTION	QUAN, PER UNIT
1.	17058	Filter Condenser	3
2.	34600	Name Plate	1
3.	40024	Socket -- 8-prong tube	1
4.	40025	Socket -- switch, 2-prong	1
5.	40026	Socket -- phonograph, 3 prong	1
6.	40027	Socket -- amplifier, 4 prong	1
7.	40028	Socket -- tuner, 5 prong	1
8.	40685	Receptacle -- line cord plug	1
9.	41729	Fuse Holder	1
10.	63650	Power Transformer	1
11.	64452	Filter Choke	2
12.	75880	Voltage Divider	1
13.	86501	Chassis	1
14.	86502	Cover	1
		Hardware	
	44100	Rivet	12
	44141	Drive Screw	4
	44701	Screw	8
	44801	Screw	4
	44910	Nut	9
	44930	Lockwasher	9
	44940	Grounding Lug	1

WRITTEN BY: <u>W.T.</u>   DATE ISSUED: <u>9/14</u>   LOT SIZE: <u>100</u>   PAGE NO.: <u>1/1</u>

ENG. CHANGE NO.:				
DATE ISSUED:				
APPROVED BY:				

Fig. 9-1. A typical bill of material arranged in tabular form.

However, in the final analysis, the question must be resolved in terms of total cost to the organization. In order to consider the many variables involved, techniques of mathematical programming have been applied to assist in developing the best solution.[1] The organized approach to the solution of this problem is called *value analysis.*

---

[1] Niles Reinfeld and William Vogel, *Mathematical Programming* (Englewood Cliffs, N. J.: Prentice-Hall, Inc., 1958).

BILL OF MATERIAL

DESCRIPTION: Power Unit Assembly -- Heavy Duty   B.M. NO. 313022

-- 115 Volts -- 60 Cycles   DWG. NO. C-313022

PART NO. 313022

OPER. NO.	DEPT. NO.	GROUP NO.	PART NO.	DESCRIPTION	QUAN. PER UNIT
1.	180	112	40024	Socket -- 8-prong tube	1
			40025	Socket -- switch, 2-prong	1
			40026	Socket -- phonograph, 3-prong	1
			40027	Socket -- amplifier, 4-prong	1
			40028	Socket -- tuner, 5-prong	1
			40685	Receptacle -- line cord plug	1
			44100	Rivet	12
			86501	Chassis	1
2.	180	120	34600	Name Plate	1
			41729	Fuse Holder	1
			44141	Drive Screw	4
			44701	Screw	2
			44910	Nut	3
			44930	Lockwasher	3
			44940	Grounding Lug	1
			75880	Voltage Divider	1
3.	180	122	44701	Screw	6
			44910	Nut	6
			44930	Lockwasher	6
			63650	Power Transformer	1
			64452	Filter Choke	2
4.	180	123	17058	Filter Condenser	3
5.	180	123	44801	Screw	4
			86502	Cover	1

WRITTEN BY: W.T.   DATE ISSUED: 9/14   LOT SIZE: 100   PAGE NO.: 1/1

ENG. CHANGE NO.:					
DATE ISSUED:					
APPROVED BY:					

Fig. 9-2. The bill of materials arranged in assembly order facilitates stock planning and handling.

Value analysis is the term used to designate the process of applying creative thinking and the scientific method of decision-making to the product planning function. When the value-analysis approach is used, five basic questions are asked in regard to each part and sub-assembly:

1. What does it do?
2. What does it cost to make (or buy) it?
3. What else will do the same thing?
4. What does it cost to make (or buy) the alternative?
5. Which will result in the least cost to the organization?

In studying these questions it can be seen that the most important factor is that the part is being investigated from the point of view of what it *does* rather than what it *is*. Many organizations have achieved significant savings by using this approach in the product planning functions.

Actually, the questions asked in the value-analysis "technique" should always be asked and investigated as part of the product planning function. Thus it appears that the responsibility for such analysis should rest with whoever performs this function. However, this individual or group of individuals is usually so busy with the detail work of transferring sketches and research drawings to production drawings, and preparing bills of materials and other plans, that there is no time left to perform such an analysis.

Some organizations have found it desirable to establish a separate group to carry out the value analysis. Usually this group is a part of the engineering organization and is composed of individuals who have a good engineering background and considerable imagination or creative thinking ability. If such a group is set up, it is very important to avoid simply processing through it all the plans which go through engineering, and thus making it another bottleneck in the planning phase. Procedures must be developed so that the "exception principle" is applied and the value analysis group must devote its time to the parts which are the most costly to the organization.

Many companies have established committees composed of representatives from production, engineering, research, purchasing, and sometimes sales, to apply the value analysis approach to the most important parts which the organization makes or buys. The advantage in using the committee approach is that a variety of backgrounds is represented with the result that suggested changes usually will be more readily accepted throughout the organization. The disadvantage in this approach is that the committees usually concern themselves only with products after they are in production. By that time purchase orders may have been let or sizeable expenditures in equipment may have been made.

PROBLEMS CREATED BY LACK OF PRODUCT PLANNING

In the nonmanufacturing activities it is frequently found that insufficient time is devoted to product planning. Sometimes this lack of product planning can be explained by the nature of the end product which, in itself, is not fully defined initially. This is true in certain research activities where the end product is a prototype which will do "something." In maintenance work also, the end product may be fairly obvious —a repaired piece of equipment. However, unless the cause of the equipment breakdown is also known, the planner still would not have much information upon which to base the remaining Production Control functions for the activity of repair. In office work, it is frequently found that the individual clerk is left to determine the actual format of the end product as it is being prepared. As was mentioned in the preceding chapter, management has failed to recognize the importance of the planning functions in the clerical activities. Gross inefficiency is the obvious result. Actually, much of the product planning function which is performed in clerical operations takes place in the design of forms since the completed form in many cases is the end product of the office activity.

Whenever product planning is incomplete, problems will be encountered. Some of the main problems are:

1. Material requirements can't be determined and hence material is not on hand to insure that production can meet its goals with a minimum of delay.
2. It is very difficult to plan the process by which the end product is to be created.
3. The planner cannot determine what facilities will be required.
4. A reliable estimate cannot be made of the time that will be required to perform operations. This makes scheduling difficult. For this reason, scheduling in custom-built types of activities simply amounts to establishing target dates for certain phases, recognizing that these dates will probably be revised many times before the project is completed.
5. Performance evaluation based on time standards becomes almost impossible. Unless the product has been standardized, it is unrealistic to assume a standard procedure for creating the product and without a standard procedure, standard times cannot be established.

Summary   Product planning has been discussed as a separate function in the sequence of functions necessary to sound Production Control. However, product planning cannot be performed independently of the function of process planning. In some cases, both product and process planning are performed by the same individual within the organization. This individual is simply called a "planner." The value-analysis technique is an attempt to draw the two functions of product and process planning closer together. The lack of sufficient realization of the importance of this relationship is one of the reasons why value analysis has attracted so much attention in recent years.

The Production Control system designer will often find that it is impossible for complete product specifications to be developed in the product planning function. The final design can only be determined after production has begun. This is especially true in the areas of custom-built activities such as research and development, and maintenance. In such situations the designer must develop a system based on check points and target dates and the system must be extremely flexible to accommodate the many changes which will be likely to occur.

In many other nonmanufacturing areas, the product planning function has not been given sufficient attention, nor has its importance been recognized. Thus the designer of the Production Control system will often find that some action must be taken to formalize this function in order to develop an effective Production Control system.

10

Process Planning and Routing

Process planning is defined as the determination of how the work is to be done. As mentioned in the preceding chapter, there must of necessity be considerable overlap between the process and product planning functions. This is due to the fact that process planning is dependent upon *prior* determination of *what* is to be made, along with the detailed product information. The function of process planning is performed in the action planning phase of Production Control. (See Table 3-1, page 18.)

Routing is defined as the determination of where the work is to be done. When there are no alternative "routes," process planning automatically includes the function of routing. For example, if the process plan calls for an operation which can only be performed by one individual in the organization, the work is also automatically routed to that individual; if the process plan calls for an X-ray inspection and the organization has one X-ray machine, the work is automatically routed to the department in which the machine is located and routed within the department to the machine itself.

If there are alternative ways in which an operation

can be performed, these alternatives, as well as their costs or times, should be shown on the process plan. The determination of which of the alternatives shall be used is usually performed in the scheduling function. Certain mathematical programming techniques have been developed to aid the scheduler in this situation. These will be discussed in Chapter 16.

If there are alternative people or machines that can perform the operation and there is little or no difference in the desirability of the alternatives, the routing function is performed as a part of dispatching. For example, a straight typing job which may be assigned to any one of several girls in a typing pool will be routed to whoever is free when it is time to do the job. If a department contains several similar milling machines, all of which can do a certain operation in the same amount of time, the job will be routed to the first available machine.

The objective of process planning is to plan the most economical methods of performing an activity, all factors considered. In many situations an item will only be produced once, making it necessary to produce a profit the first time because there is no second chance.

PREREQUISITE INFORMATION NEEDED FOR PROCESS PLANNING

In order to do a competent job, the process planner must have available certain basic information. Some of this information is developed in the work authorization function, some in the product planning function and some is basic to the organization and its capabilities. The process planner must know the following:

1. *The volume that must be produced.* This includes the volume of existing orders as well as a forecast of possible repeat orders. This information is usually included as a part of work authorization and is developed in the forecasting function.

2. *All the information which is developed in the product planning function.* The quality requirements are particularly important because of the fact that these determine to a large extent the type of equipment that will be required. The process planner can then compare these requirements with the machine capabilities and anticipate the scrap which will occur. If it appears that there will be a high scrap rate, he normally would refer this fact back to product planning to determine if the quality requirements may be reduced. This comparison of machine capabilities with quality requirements is a vital and often overlooked part of the process planner's job. If he can anticipate such problems, serious and costly production delays can be avoided. The quality requirements also determine the need and degree of inspection necessary by operators, inspectors, and the quality control department. Every piece or dimension inspected by anyone costs money.

3. *Equipment and personnel available within the organization.* The process planner must know the types and capabilities of machinery in

terms of size and production accuracy. He must know in which departments the different types of equipment are located and he must also
know if special purpose tooling is available for the operations which
are to be performed. He should be capable of determining alternative
ways in which an operation may be performed—i.e., on a lathe, a hand
screw, or an automatic screw machine. Knowledge of the types and
degrees of skills available is also very important to the process planner
because this will determine the amount of detail which he must develop.

4. *The time available to perform the work or "delivery" date.* Process
   planning can develop only as much information as time permits. The
   process planner may wish to specify special tooling for certain operations, but this can be done only if there is sufficient time available for
   the tools to be designed and produced.

STEPS IN PROCESS PLANNING

There are three basic steps that must be performed in process planning
regardless of the type of activity and the end product:

1. The process planner must list all of the work that needs to be performed on a given product in terms of work elements or suboperations.
2. He must group the work into operations based upon the capabilities of
   the available equipment and personnel and the volume of end products
   to be processed.
3. The process planner must arrange the operations in the most economical sequence, taking into consideration what work must be performed
   first in order to control quality and maintain minimum operational
   costs.

The process planning function is frequently referred to as "standard
operating procedure preparation" in manufacturing activities and "procedure writing" in nonmanufacturing activities.

INFORMATION GENERATED AS A RESULT OF PROCESS PLANNING AND ITS RELATED ACTIVITIES

The amount of information generated and formally recorded as a
result of the process planning function will vary considerably. It depends
upon the nature of the organization, the type of work, the quantity to
be produced, the skill of the process planner and the skill of the workers
who are to perform the operations. Obviously, the more information
available to Production Control, the better the planning and control.
There are practical limitations, however. The following list might well
be entitled "A Production Control Manager's Dream," because of its
completeness. It contains all of the information which could result from
the process planning function. The designer of the Production Control
must determine the practical and economic limits in providing the
information.

1. *The operations to be performed and the sequence of these operations.* For each operation the following information may be provided:

   1.1 *Routing (where the operation is to be performed).* The department name or number is listed for each operation and if there are alternative departments in which the operation can be performed, these should also be indicated. If the operation requires a machine, the type of machine is stated as well as the alternatives.

   1.2 *Instructions for performing the operation.* The amount of detail required depends upon the skill of the persons performing the work. Where there is a considerable volume of work and the workers are semiskilled, a separate sheet is frequently prepared for each operation. The sheet contains a methods description for the manual operations, the workplace layout, machine speeds, feeds, welding currents, and so on, and the operator inspection frequency.

   1.3 *Tooling required.* The type of tool and tool number if applicable. Usually a separate specialized function in the organization has the responsibility for designing new tools.

   1.4 *Personnel required.* The type of skill—for example, machinist, machine operator, toolmaker—and the labor grade within the skill where such labor classification systems exist. Detailed personnel requirements are essential because labor is frequently the largest single factor contributing to the cost of the product. Accurate cost estimates and control of costs are necessary in order to insure that the product is produced at a minimum cost. The actual hourly rate of the lowest skilled individual who can perform the job must be known to make such estimates. It is not uncommon to find that bids are made on jobs in which estimates are made for a reasonable profit margin, only to find that the expected profit does not exist because many operations were performed by highly skilled (and highly paid) individuals when the estimator assumed they would be performed by semi-skilled workers at lower wages.

   1.5 *Time requirements.* The setup time and the operating time to produce the quantity in the order or the lot. These time values may be based on estimates, standard time data, or historical data for the same or similar jobs. (See Chapter 14 for a complete discussion of work measurement.) It should be pointed out, that the more detailed and complete the process plan, the more accurate the time estimates tend to be. This information is exceedingly important because it provides the basis for the scheduling and loading functions. As with the tool design information, the work measurement time data is frequently prepared by a separate organizational element such as the industrial engineering or time study departments.

2. *Materials-handling procedures.*
   2.1 The type of equipment to be used to transport the material between operations.
   2.2 The lot size—that is, the quantity to be processed and moved between operations as an integral unit.
   2.3 The type of containers to be used to store the material at the workplace and in which the material is to be placed when it is transported between operations.

Various organizations have different names for the document upon which this process planning information is recorded. Some of the most common are: "Route Sheet," "Process Sheet" (see Figure 10-1). Some

Fig. 10-1. A typical route sheet for a manufacturing operation.

.O. NO.	QTY. TO RUN	DATE TO FIN.	NAME	DATE STARTED	MIN. MFG. QTY.	PATT. NO.	STOCK NO.
58229	50	11 - 7	RS25A OPER ARM		50		
ATE OF ORDER	**QTY. RQD.**	**SCHED. BY**		**DATE FINISHED**	**PROCESS TIME**	**UNIT CODE**	**DWG. NO.**
10 - 26	50	S. V.					A 358229

QUANT.	WEIGHT	ISSUED	SHORT	MAT'L. SPECIF. — STOCK NO. — UNIT CODE	QUANT. — PER	% SCRAP	WT. — PER
50				358243	1	-	
50				358246	1	-	

TOCK NO.	PREPARED BY	APPROVED BY	DATE	REVISED BY	APPROVED BY	DATE
RS 25A	J. K.	W. W.	9-27			

ATE OVED	GOOD PIECES	DEPT.	OPR. NO.	OPERATION	MACH. OR GROUP & NUMBER	GUAGES - FIX. - TOOLS	SET-UP HOURS	P. H. PER		TOTAL HRS. FOR JOB	%	START DATE
			5	Pick up at 236 Stores								
		23	10	To Dept 756								
		75	15	Braze	75-8	358229 A	1	40		2 .25		
		99	20	Grind excess braze	99-3			100		.5		
		76	25	Drill	76-4	358229 B	2	75		2 .66		
		76	30	Burr	76-3		1	200		1 .25		
		76	32	Ream 503 inch holes	76-5		1	250		1 .2		
		97	35	Clean				200		.25		
		97	40	FS 53 BLK Enamel				200		.25		
		15	95	Inspect								
		15	97	To Dept 94								

ATE	INSPECTOR	OPR. NO.	PCS. RCD.	PCS. OK.	SCRAP	SCRAP TAG NO.	REWORK	REWORK TAG NO.	FACTORY ORDER
									SHEET NO. 1 OF 1

organizations also record the product information on the same sheet with the process planning data. The raw material required at each operation is listed and the quantity and specifications set forth. Almost always, the dimensions and tolerances to be produced at the operation are indicated either directly on the process planning document or by attaching a blueprint of the part to the process plan.

BREAKEVEN ANALYSIS

An analytical tool which is of considerable value to the process planner is the technique of breakeven analysis. This tool can be used to determine when it is desirable to go from a process type to a product type layout, when it is desirable to invest in special purpose tooling or when to invest in special purpose equipment. The basic concept involved is that total cost at any volume of units produced consists of two components:

1. The fixed portion, which is a constant through a rather wide range of output possibilities.
2. The variable portion which is based on the cost to produce each unit exclusive of the fixed cost. Figure 10-2 illustrates this concept graphically.

This technique can be illustrated by the following actual case: Assume that an order has been received to produce 50,000 items. These items can be produced in the present "job shop" type of layout or by setting up a production line. Furthermore, it has been established that there is no likelihood of repeat orders. The costs have been estimated as follows:

Producing with present "job shop" conditions:

Set-up costs—$30,000

Variable unit cost, including material handling—$5.00 per item

Producing on "production line":

Set-up cost—$70,000

Variable unit cost, including material handling—$3.60 per item.

The process planner must chose between the two alternatives. He can use the breakeven graph as is illustrated in Figure 10-2 or he can solve the problem algebraically as follows:

Let:

N = the number of units at which there is no difference in total cost between methods one and two—the breakeven point.
F_1 = the fixed cost method one ("job shop") = $30,000
V_1 = the variable cost method one ("job shop") = $5.00/unit
F_2 = the fixed cost method two (production line) = $70,000
V_2 = the variable cost method two (production line) = $3.60/unit

Then the breakeven point can be determined with the following equation:

$$F_1 + V_1N = F_2 + V_2N$$
$$\$30,000 + \$5.00N = \$70,000 + \$3.60N$$
$$1.4N = 40,000$$
$$N = 28,571$$

So, in this situation, the process planner should select the production line process since the breakeven point is well below the order quantity of 50,000 items.

Fig. 10-2. The breakeven analysis indicates that the production line process should be used for this order of 50,000 units.

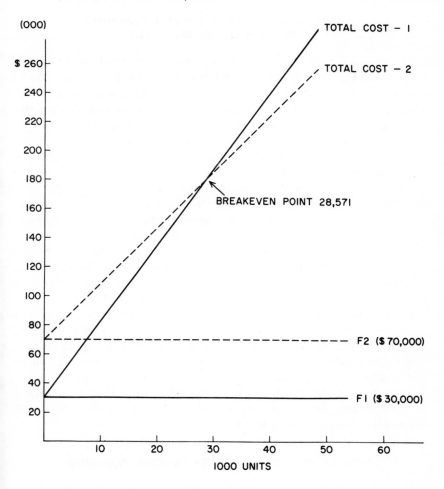

There may have been many other alternatives than the two just mentioned. A large number of them could have been eliminated on the basis of a cursory examination of the costs and the others can be analyzed on the basis of total cost for the order quantity using the fixed and variable concept.

The breakeven technique, like many other management techniques, is only as good as its data. Thus to make effective objective decisions between alternatives, accurate data must be kept and a sound work measurement system must be used. A small error in the estimated unit time for a large order could cause a significant difference in the "answer" obtained with this technique.

RELATIONSHIP BETWEEN PROCESS PLANNING
AND METHODS IMPROVEMENT

The goal for every process planner should be to plan the process by which an end product is to be created so well that methods improvements will never be required. However, this is an ideal which is seldom achieved. After production is initiated, there will be problems and changes that were not anticipated in the planning phase. Thus, some follow-up analysis by the methods analyst will almost always be required.

There is little doubt, that the "Production Control Manager's Dream" of the information which he could have available as the result of process planning would certainly increase the time and detail required in the process planning function. Hence, the cost of performing this function would also increase. Yet it is reasonable to assume that increased expenditures in process planning should reduce the amount of effort which must be expended on methods improvement. Furthermore, the activity will benefit from the "improved methods" from the start of the operation rather than after the job has been in progress for some time and the methods analyst has had an opportunity to study it.

Every process planner should have a thorough knowledge of the principles and techniques of methods improvement. When he has developed his plan, he should analyze it as if it were an existing operation and look for any flaws that the methods analyst might find after the operation is in process. Even after the job has actually been started the process planner can attack bottlenecks in the planned operation by analyzing the methods and determining ways in which the operation can be simplified and the time and cost reduced.

ADVANTAGES OF THOROUGH PROCESS PLANNING

As is the case with the other functions of Production Control, the process planning function will always exist in any activity. If the function is not carried out on a staff basis in an Engineering or Planning department, it will be performed by the line supervisor or even the worker himself. These are certain very important advantages that occur when a thorough job of process planning is done by a staff group prior to the initiation of the job.

1. Supervisors are freed from these detail tasks so that they can concentrate on the more important duties of managing their personnel. When the majority of the process planning has been performed by the foreman in the past and a staff group is established to handle this function, complaints are frequently heard that the supervisors' "prerogatives" are being violated. There can be no denial of the fact that a certain portion of the supervisors' workload is being redistributed to the staff function. If the change is introduced too quickly there frequently will be considerable resentment and opposition by the supervisors because of this supposed intervention. However, if the change is initiated slowly and a cooperative attitude is developed between supervisors and the persons performing the process planning function, the question of prerogatives is soon forgotten. Everyone soon discovers that the net result is a more effective and efficient operation.
2. The information generated in process planning facilitates coordination of all the activities in the organization by improving the information flowing to the other Production Control functions. Thus, the designer finds that he has a system which is more readily controllable.
3. A competent process planner familiar with the techniques of methods improvement will more likely design the operations the first time in "the one best way" than one who does not have such knowledge. The end result, of course, is lower total cost of operation.

ANALYSING THE PROCESS PLANNING FUNCTION

The production control system designer must determine where and how the process planning function should be performed. The function may be established on a formalized basis as a staff activity, it may be located in the line organization, or it may be divided between the two.

If the activity is performed by a staff, adequate personnel of sufficient skill must be assigned to the job. It is often difficult to persuade management to staff a process planning function adequately because it is an "overhead" function. In the constant battle to cut costs, management fails to realize that this additional "overhead cost" can be saved many times over in the increased efficiency of the productive operations. Also, in many organizations the process planning functions have unfortunately become a place to which older workers are "retired" when they no longer perform satisfactorly in the production operations.

The plans prepared by the process planning staff should be complete and not just sketchy lists of operations with all the real process planning left up to the supervision in the line organization.

When there is no staff assigned to the process planning function, it must be performed by the line organization—the supervisors, set-up men, and even the individual workers. If the job is assigned to the supervisors, the systems designer should survey the work situation to determine if the supervisors are devoting sufficient time to this function or if most methods are left up to the individual worker. If the supervisors are spending considerable time on this function, it may prevent them from performing some of their other important duties. In cases where the supervisors and set-up men are doing the major portion of the process planning, they should be thoroughly trained in methods analysis so that they can recognize a poor method when they see it and take corrective action immediately.

Under any circumstances, but particularly when the process planning is left up to the individual worker, the designer should familiarize himself with the quality of the methods being used. An experienced methods analyst should study a sample of the jobs to determine the effectiveness of the methods.

The amount of process planning done as a staff function and the amount done within the line depends upon the relative abilities of the personnel and the nature of the work. Where highly skilled workers—e.g., toolmakers—or technicians or scientists are involved, the worker is being paid for his ability to plan his own operation. For example, in many research organizations, when a project is undertaken, the persons who are going to make a study or experiment get together in planning meetings to outline their method of attack, the facilities which they will need and a tentative time schedule. In the case of a toolroom where one special purpose tool is to be made, the toolmaker who is assigned to the job will perform the process planning function. The fact that only one item is to be made would not merit the cost of planning the process on a formalized basis.

Although process planning and product planning have been discussed as separate functions, the designer of the Production Control system may have difficulty in distinguishing them when he is analyzing an organization. This is particularly true in the nonmanufacturing activities where both may be performed simultaneously by the same individual. This individual may also be the one who actually does the job. The designer may find that the differentiation may have to be a synthetic one for analysis purposes. However, it is important that each function be analyzed thoroughly. If both do not generate sufficient information, the performance of the remaining Production Control functions may be severely handicapped. On the other hand, there may be a very sharp distinction between the two in some manufacturing organizations. This may result in insufficient change of information, ideas, and problems. The most important single question to be answered is whether or not the product planning function is producing sufficient information to enable the process planning function to be performed effectively.

Summary   Both process planning and routing have been discussed in this chapter, because normally they are performed at the same time. Occasionally routing is performed in conjunction with the scheduling or dispatching functions.

To perform his job adequately, the process planner must have available certain basic information:

1. Volume to be produced
2. Product data
3. Equipment and personnel availability
4. Time availability

Much of this information is developed in the forecasting, work authorization, and product planning functions. Hence, the manner in which process planning is performed will depend upon the degree of refinement with which the preceding functions are carried out. As the reader studies subsequent chapters, he will realize that this is true of all the functions—each link in the Production Control system is dependent upon all the preceding ones.

The amount of information generated in the process planning function will vary considerably from organization to organization. Expenditures of time and effort in process planning will usually reduce the over-all costs of an organization and particularly the need for methods analysis and improvement after operations have been initiated.

In this chapter, the reader has been introduced to the breakeven analysis technique. The basic concept employed in this technique is that the total cost at any volume of units produced consists of a fixed and a variable component. In later chapters the application of this technique to other problem areas of Production Control will be discussed.

11

Material Control

All activities are engaged in the creation of an end product. Sometimes this "product" may be a service with little or no material required to provide the service. However, most activities are engaged in *modifying material* to create an end product. Actually, the modification of material is the basic reason for the existence of all manufacturing organizations. Since material is one of the fundamental components of most activities, it is extremely important that the designer of the Production Control system provide the necessary elements of planning and control of material to insure that the right material will be available in the right quantities when needed.

Since material control is a greater problem in manufacturing activities than in other types of activities, terms and examples in this and the following chapter will be taken from manufacturing operations.

In order to set the basis for the discussion of material control, the following definitions and relationships must be fully understood: The function of *material control* consists of two main sections:

1. The physical control of the material
2. The maintenance of records concerning the material

Material management is concerned with two basic problems:

1. The amount of material to be ordered—the order quantity
2. When the material is to be ordered—the reorder point

An adequate material control system must exist before there can be any effective application of material management. Therefore, material control will be presented in this chapter and material management in the next.

The first step in understanding the function of material control is to understand the classifications of inventory. There are three major classes of inventory:

1. Raw materials and purchased parts
2. In-process material
3. Finished goods

Material-in-process control is essentially a problem of the dispatching and progress reporting functions and will be discussed in Chapters 18 and 19. The finished goods of one activity simply become the raw material of another and, therefore, the approach taken for analyzing and controlling raw materials can be applied to finished goods as well. We will concentrate our attention on the material control problems associated with raw materials and purchased parts.

It is very important for the designer of the material control system to recognize ways in which the problems of control can be minimized through improvements in related activities. There are two main areas to which attention should be given:

1. Material nomenclature and identification procedures
2. Standardization of materials

The first will reduce the possibiltiy of duplication of materials and errors in record keeping. Standardization will reduce the number of items to be controlled, eliminate obsolete items and provide a systematic method of parts substitutions to avoid production delays. The material control function has a responsibility either directly or indirectly in assuring that proper effort and attention are given to the above mentioned areas.

USE OF THE BILL OF MATERIALS

The bill of materials, which is the end product of product planning, is the basic document in material planning. To use the bill of materials, the material planner (in some organizations he may be the same individual as the product planner or even the process planner) must "explode" the information. "Exploding" consists of determining the requirements for a given quantity of finished assemblies by breaking the finished assembly into sub-assemblies, sub-assemblies into parts, and parts into the raw material and purchased parts requirements. This provides the material requirements for some future period of time. The requirements are first determined for each product that is to be produced and then these requirements are combined to establish the procurement plan.

"Exploding" is primarily a data processing activity. If there are not too many parts in the system and requirements do not have to be determined at frequent intervals, the process may be performed manually. As the volume of parts and the frequency for calculating requirements increase, needle-sort equipment may be utilized. A still larger volume of parts and greater frequency of requirements change may justify the use of tabulating equipment (electrical accounting machines). Figure 11-1 illustrates a procedure for performing the "explosion" for determining material requirements.

This procedure can be summarized in the following steps:

1. A requirement card is prepared for each end product.
2. Sub-assembly cards are selected from the master file for each end product that is to be produced during the planning period. (This master file contains a card for each sub-assembly used in the end product.)
3. The sub-assembly cards are combined with the requirements cards and a new card is produced containing all the information on the sub-assembly card and the quantity on the requirements card.
4. Following this, the original sub-assembly cards are returned to the master file.
5. Then the new sub-assembly cards are sorted so that all the cards for the same sub-assembly are together and the "net" requirement for each sub-assembly is automatically punched into a new card or printed on a report.
6. Steps 2 through 5 are then repeated to determine the requirements for purchased parts and raw material.

Some very large organizations find that the number of items contained in this material control system is so great and the frequency with which the anticipated demand changes is so rapid that electrical accounting machines are incapable of satisfactorily performing the task of "exploding." In these cases it is necessary to utilize electronic data processing equipment.

An important part of the information required on the bill of materials is the scrap allowance. Specifications and machine capabilities are prepared by the product planner or process planner at which time the potential scrap loss is determined. Normally the scrap allowance should not be a fixed per cent of the total quantity to be produced, but should be a given number of pieces or per cent for an initial run of the lot, a smaller number of pieces or per cent for the next quantity, and then an

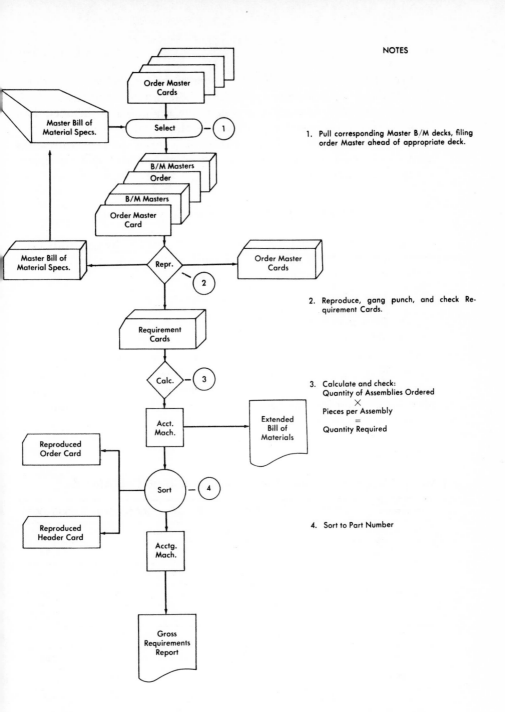

1. Pull corresponding Master B/M decks, filing order Master ahead of appropriate deck.

2. Reproduce, gang punch, and check Requirement Cards.

3. Calculate and check:
   Quantity of Assemblies Ordered
   ×
   Pieces per Assembly
   =
   Quantity Required

4. Sort to Part Number

Fig. 11-1. Schematic diagram of punched card procedure for "exploding" bill of materials. (Reprinted by permission of International Business Machines Corp., New York.)

even smaller percentage for the remainder of the lot. When this is done, the planning data closely approximate the actual conditions. It is usually found that most scrap occurs at the beginning of an operation. Table 11-1 illustrates a scrap allowance chart which incorporates this approach. In addition to the allowance for scrap or spoilage, allowances must also be made for such things as cut off on bar stock and for clamping in certain machine operations.

TYPES OF MATERIAL CONTROL SITUATIONS

The types of material control situations may be divided into categories similar to those used to differentiate the work activities as described in Chapter 4. These are:

1. Flow control
2. Similar process control
3. Custom-order control
4. Job-order control

Flow control is found in those organizations which are mass producing according to a predetermined master schedule—for example, the automobile and chemical industries. In such situations, there is a minimum need for communications between the material control and production control organizations. Usually two departments are established and the

TABLE 11-1

SCRAP ALLOWANCE TABLE FOR MILLING MACHINE OPERATIONS

Tolorance	.050	.01-.049	.005-.01	.001-.0049	.0005-.001
Lot Size	Quantity to Start				
10	10	10	11	12	14
25	26	28	29	30	34
50	52	54	56	59	66
100	103	105	108	115	127
200	205	208	215	225	245
500	510	516	530	550	600
1,000	1,015	1,025	1,045	1,070	1,180
5,000	5,025	5,075	5,200	5,400	5,750
10,000	10,040	10,100	10,300	10,500	11,000

Quantity to start depends upon tolorance, and allowance percentage varies with lot size.

coordination is obtained through the use of a master production schedule. Material control frequently includes the function of procurement. The normal procedure is for material control to place open purchases orders in order to obtain quantity discounts and to issue shipping schedules weekly, bi-weekly or monthly to the vendors. It is then necessary to follow-up and expedite as required in order to insure that material arrives within the allowed time. Production control must be informed immediately when any delay in the receipt of material is anticipated. A material shortage may result in considerable lost production time and excessive costs. The idea which is sought in mass production is to unload the material from the freight car or truck that has delivered it and move it directly onto the production line. Due to careful planning and painstaking attention to details most automobile manufacturers have been able to accomplish this goal, and thus hold their inventories to a minimum.

In the *similar process* industries, such as the textile and shoe industries, the material control situations are mixtures of the flow control and the other types which will be discussed. However, the material control function is very similar to that in the flow or mass production activity. The slight product changes which characterize similar process activities may make little or no difference in the material requirements.

In the *custom type* of production such as the construction industry, there may be little or no raw material stock maintained. Materials are purchased for a specific order after the order has been received. Where such situations exist, it is extremely important to maintain very close cooperation and coordination between the persons performing the material control functions and those handling production control. Frequently these two functions as well as purchasing are combined in the same department. Where the functions are separated organizationally there must be a positive means of follow-up so that the planners in production control can be sure that material will be on hand when production is scheduled to begin.

This follow-up may be achieved by means of a "tickle file" procedure such as the following one described:

1. The person making the requisition keeps one copy of the requisition filed in a "tickle file" on the date which will allow for the normal procurement process elapsed time. The desired delivery date is also written on the requisition.
2. After the order has been placed by the purchasing department, one copy of the requisition is returned to the originator marked with the vendor's name, order number, and the expected receipt date if different from date originally requested.
3. In the event the person initiating the requisition has not received the return copy from purchasing by the "tickle" date, he contacts the purchasing department to determine the cause for the delay.

Procurement may take up the major portion of the production cycle when the items being purchased are also custom built. It is important to estimate this procurement lead time as accurately as possible. Past experience with similar types of items can be very valuable guides when such conditions prevail. A method of formalizing the knowledge gained through past experience is to construct a table. This gives the person who is setting up the over-all master plan or schedule some guide-lines by which to establish the procurement "lead time." An example is Table 11-2.

TABLE 11-2

TABLE FOR ESTIMATING PROCUREMENT LEAD TIMES FOR CUSTOM-BUILT TYPE OF ACTIVITY

Metal Form	Procurement Time—Weeks					
	Steel Normal	Steel Special Alloys	Bronze	Aluminum	Magnesium	Copper
Sheet Stock	2	20	5	4	7	1
Bar Stock	2	22	6	3	7	2
Castings	10	26	15	22	25	12
Forgings	12	26	20	—	—	—
Machined	20	35	23	27	31	18

Even in organizations where the end product is custom built, it is very unusual for all items to be purchased only after an order has been received. All, or a large portion of the standard materials used in production are stocked. This situation exists widely in the *job-order* type of activity and constitutes the fourth general classification of material control. Most references to material control refer to this fourth general type.

In job-order control it is very important to have close cooperation and possibly even integration between the material control and production control functions. Orders are usually accepted on the basis of a requested completion date and thus the production control department must know the status of material availability. There are two important ingredients for a successful material control system in such situations:

1. Adequate physical control of the material
2. Accurate maintenance of the material records

These factors are discussed in detail in the following pages. Their importance cannot be overemphasized.

PHYSICAL CONTROL

The first requirement of a sound material control program is adequate physical control. This consists of several important factors:

1. Adequate physical facilities for proper storage
2. Adequate stockroom security
3. Effective requisitioning procedure

It is virtually impossible in any activity to maintain reliable material control records unless these factors are effectively applied.

All sound and effective material control starts with the provision of adequate storage facilities. Even with minimum record keeping it may be possible to maintain reasonably effective material control under certain conditions, with good physical facilities. This also reduces the possibility of "lost" material, minimizes material locating costs and maintenance, and permits taking physical counts more easily.

The physical facilities should also incorporate adequate security within the storage area, particularly where high pilferage items are stored. Losses due to pilferage are substantial in many industries. However, security is important for other reasons. It reduces discrepancies in record keeping in cases where unauthorized personnel help themselves to material instead of following established procedures. It also gives better assurance that good housekeeping can be maintained and places responsibility on the authorized personnel entirely.

Effective security depends greatly upon a sound requisitioning system for withdrawing materials from storage. Delays in obtaining materials are costly to the company and aggravating to the person drawing material. Poor requisitioning procedure eventually will break down the security, and thus the physical control. The requisition is an important document in the design of the material control system. The form and procedure for use of the form must be carefully designed.

In addition to there being adequate physical control of the material, procedures must exist to insure that *all* necessary information is transmitted from the stock room to the material records. All withdrawals must be recorded on a requisition. When material is received, quantities must be checked against orders and vendor's invoices, and a receiving report filled out. In addition, all returns from production to the stock room must be recorded.

A troublesome problem which the designer of a material control system must encounter is the organizational relationship of the persons operating the stockroom with those maintaining the material records. According to the internal control procedures usually recommended by accounting, these two activities must be separated to prevent collusion in the theft of material. Another important consideration is that in many cases the stockroom area is not conducive to accurate record-keeping due to excessive heat, dirt, or noise. It is often desirable to locate the material records close to the Production Control office, but it may not be convenient to have the Production Control department located near the stockroom. The disadvantage in separating the material records from the stockroom is that such an arrangement usually leads to some duplication of records. This is due to the fact that the stockroom must keep a separate location card for each item in stock and frequently duplicate sets of the entire records will be maintained.

RECORD KEEPING

Maintaining the material records is a vital phase of the material control function. This activity contributes significantly to the cost of material control. To be of any value the material records must be posted accurately and must be maintained on a *current* basis.

In order to keep postings on a current basis, it is often found desirable to establish a simple Production Control system for the material control function—the "end product" being the posted card. Accuracy in the record keeping can be improved by assigning each clerk to a definite group of records. The postings can be sampled periodically to determine the quality level, and even quality control charts may be maintained. If a clerk is not maintaining an acceptable quality level, then corrective action such as further training or replacement should be taken. The quality rating may be the basis of advancement or bonus payment. When this is done, the quality should be combined with a measurement of production.

Careful attention must be given to the design of the material record. It must provide all the necessary information and be arranged so as to keep posting time to a minimum. One portion of the card is usually allotted for the constant information. This constant information segment of the card is illustrated in Figure 11-2. Not all organizations choose to

Fig. 11-2. Constant information portion of the material records.

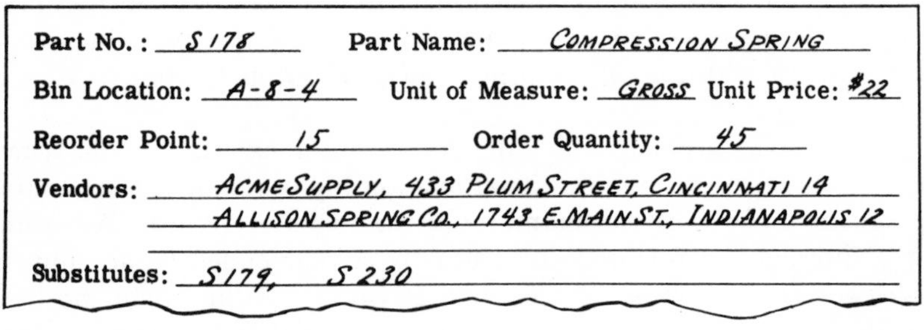

Part No. : *S 178*   Part Name: *COMPRESSION SPRING*

Bin Location: *A-8-4*   Unit of Measure: *GROSS* Unit Price: *$22*

Reorder Point: *15*   Order Quantity: *45*

Vendors: *ACME SUPPLY, 433 PLUM STREET, CINCINNATI 14*
ALLISON SPRING CO., 1743 E.MAIN ST., INDIANAPOLIS 12

Substitutes: *S 179,   S 230*

maintain all this information on the material control card, but it must then be recorded somewhere else, usually in Production Control or purchasing.

A second portion of the card is allocated for the variable information. The amount of variable information will differ in various activities depending upon the material control requirements and past practice. Almost always there should be columns for "Receipts," "Issues," and "Balance on Hand." A card containing only this minimum information is illustrated in Figure 11-3. Usually it is desirable to determine the frequency of usage of an item and to identify the authorization of the withdrawals in case a posting must be checked. As is illustrated in Figure 11-4, columns can be added for "date" and "purchase" or "factory order number" to show the activity of the item and maintain a record of the transaction authorization.

Frequently, in planning the production of an item, more information is required than can be obtained from the cards previously illustrated. There must be assurance that there will be sufficient material available when the job actually goes into production. To accomplish this, two more columns are added to the card. When the job is planned, the quantity of material required is determined and posted to the "allocated"

Fig. 11-3. Basic variable information appearing on the material record.

Part No. : _S 178_   Part Name: ___COMPRESSION SPRING___

Bin Location: _A-8-4_   Unit of Measure: _GROSS_ Unit Price: _$22_

Reorder Point: _15_   Order Quantity: _45_

Vendors: ___ACME SUPPLY, 433 PLUM STREET, CINCINNATI 14___
___ALLISON SPRING Co., 1743 E. MAIN ST., INDIANAPOLIS 12___

Substitutes: _S 179, S 230_

RECEIPTS	ISSUES	BALANCE
	10	30
	5	25
	10	15
25		40
20		60

Part No. : _S 178_   Part Name: _COMPRESSION SPRING_

Bin Location: _A-8-4_   Unit of Measure: _GROSS_ Unit Price: $22

Reorder Point: _15_   Order Quantity: _45_

Vendors: _ACME SUPPLY, 433 PLUM STREET, CINCINNATI 14_
ALLISON SPRING Co., 1743 E. MAIN ST, INDIANAPOLIS 12

Substitutes: _S 179, S 230_

DATE	PURCHASE/SHOP ORDER NO:	RECEIPTS	ISSUES	BALANCE
4-23	S.O. 1612		10	30
5-2	S.O. 1783		5	25
6-2	S.O. 2017		10	15
6-3	P.O. 1213	25		40
6-10	P.O. 1213	20		60

Fig. 11-4. Material record with columns added to identify transaction and indicate usage.

column. Even though the material is still in the stockroom, the required amount is deducted from the "available" column. By referring to the "available" column, the planner can immediately determine if there is sufficient material on hand to initiate production. Figure 11-5 illustrates a card with these two columns added. With this procedure it is possible that excessive stocks may be built up because material is being reserved for orders which will not be produced until some future time. To avoid such a situation, production orders should not be released to material control for processing any further in advance of the start date than the time required for the procurement cycle.

In some situations, it is important to maintain a record of the quantity of material that has been received against a given purchase order. This knowledge is especially desirable when orders are filled by partial shipments. This information can be recorded by adding two more columns to the card, one for "amount purchased" and the other for "total amount due." When a purchase order is sent to a vendor, the quantity on the order is posted to the "amount purchased" column and also added to any balance which may exist in the "total amount due" column. When material is received, the amount received is posted to the "receipts" column and deducted from the "total amount due" column.

The question frequently arises as to whether or not quantities on order should be considered "available." If the procurement cycle is short in comparison to the planning cycle, this practice may be acceptable. Figure 11-6 illustrates a card with columns for "amount purchased" and

Part No. : _S 178_   Part Name: _COMPRESSION SPRING_

Bin Location: _A-8-4_   Unit of Measure: _GROSS_ Unit Price: _$22_

Reorder Point: _15_   Order Quantity: _45_

Vendors: _ACME SUPPLY, 433 PLUM STREET, CINCINNATI 14_
_____ALLISON SPRING CO., 1743 E. MAIN ST., INDIANAPOLIS 12_

Substitutes: _S179, S230_

| DATE | PURCHASE/SHOP ORDER NO.: | RECEIPTS | ISSUES | BALANCE | ALLOCATIONS | |
					ALLOCATED	AVAILABLE
4-13	S.O. 1783				5	25
4-23	S.O. 1612		10	30		25
5-2	S.O. 1783		5	25		25
5-19	S.O. 1891				10	15
6-2	S.O. 2017		10	15		15
6-3	P.O. 1213	25		40		40
6-10	P.O. 1213	30		60		60

Fig. 11-5. Material record with columns added to reserve material for planned production.

Part No. : _S 178_   Part Name: _COMPRESSION SPRING_

Bin Location: _A-8-4_   Unit of Measure: _GROSS_ Unit Price: _$22_

Reorder Point: _15_   Order Quantity: _45_

Vendors: _ACME SUPPLY, 433 PLUM STREET, CINCINNATI 14_
_____ALLISON SPRING CO., 1743 E. MAIN ST., INDIANAPOLIS 12_

Substitutes: _S179, S230_

| DATE | PURCHASE/SHOP ORDER NO.: | PURCHASES | | RECEIPTS | ISSUES | BALANCE | ALLOCATIONS | |
		AMOUNT PURCHASED	TOTAL DUE IN				ALLOCATED	AVAILABLE
4-13	S.O. 1783						5	25
4-23	S.O. 1612				10	30		25
5-2	S.O. 1783				5	25		25
5-19	S.O. 1891						10	15
5-21	P.O. 1213	45	45					15 (60)
6-2	S.O. 2017		45		10	15		15 (50)
6-3	P.O. 1213		20	25		40		40 (50)
6-10	P.O. 1213			20		60		60 (50)

Fig. 11-6. Material record with columns added to record material on order.

"total amount due." The figures in parentheses indicate the balance that would appear in the "available" column when the amount on order is considered available. The numbers outside the parentheses in the "available" column of Figure 11-6 indicate the balance that would appear if the amount on order is not considered available. If this approach is used, it is a simple matter for the planner to determine if material is on order when he is referring to the card to make an allocation. If additional material is required for production of the order, the planner can then expedite the delivery of the material so that the desired start date of the job can be met.

PART NO: _S-178_				
Month	19__	19__	19__	19__	19__
Jan.	5	6	8	7	
Feb.	9	8	11	12	
Mar.	6	7	17	15	
Apr.	8	12	13	10	
May	7	9	8	5	
June	13	14	17	20	
July	10	11	16		
Aug.	9	12	12		
Sept.	9	8	10		
Oct.	12	14	15		
Nov.	14	16	19		
Dec.	13	16	17		

Fig. 11-7. Usage recap sheet to provide historical demand data.

Figure 11-7 illustrates the third important portion of the material record—the "Usage Recap Sheet." The recap sheet shows the quantity of the item issued each month. It is the basis upon which usage and variability of demand are determined. Knowledge of these figures is a prerequisite for material management, which will be discussed in the next chapter.

Many office equipment supply firms market visible index inventory card equipment. This equipment provides a means for placing the cards in a tray in such a way that about a quarter-inch of each card is visible. It is illustrated in Figure 11-8. Colored slides and tabs on the visible edge indicate at a glance such information as stock level, material on order, and stock level below the reorder point. Some of the most common trade names for these cards are: Kardex, Visi-Record, Sig-Na-Lok, and Acme Visible Record.

Punched card equipment may also be used to maintain the material records. Such equipment is best applicable when there are many items in the material control system or frequent issues of the items from vari-

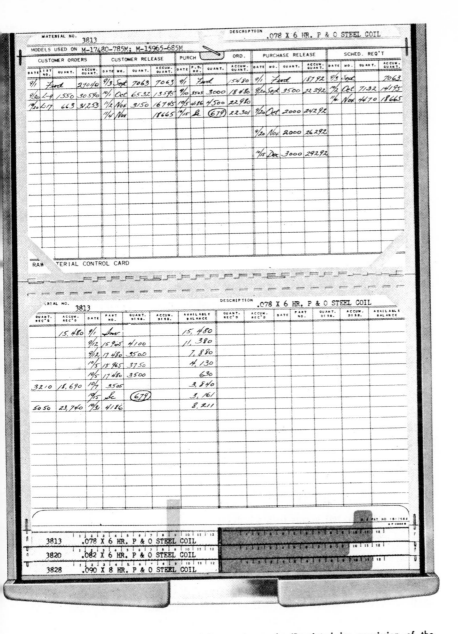

Fig. 11-8. Visible index material control record. (Reprinted by permission of the Remington Rand Division of Sperry Rand Corporation, New York.)

ous locations. When the frequency of transactions becomes large and an accurate material record must be maintained at all times, high-speed digital computors may be justified for this operation.

A punched card material control procedure is described in Chapter 21.

TWO-BIN MATERIAL CONTROL SYSTEM

There is another approach to the material control problem which is relatively new as a formal concept, but which has been used informally for years. This approach is usually described as the "two-bin" or "traveling requisition" system. It is based on the idea that the material itself will serve as the record of the quantity on hand.

The stock location for each item is divided into two sections, hence the title "two-bin" system. Until the first section is emptied, all withdrawals are made from it. The second section contains an amount of material equal to the reorder-point quantity. A "traveling requisition" is attached to this second section and when the first piece is withdrawn from the second bin, the requisition is detached and used to initiate purchasing action. The requisition form may be the regular purchase requisition or a special form may be designed just for this purpose. Normally, a new requisition is prepared by material control or purchasing as soon as an order is placed. The requisition is sent to the stockroom and when the material is received, the reorder-point quantity is counted, placed in the second bin and the traveling requisition attached.

An actual second bin is not always required. The reorder-point quantity can be separated from the normal issue stock by wrapping it, keeping it in the original container, inserting a divider in the bin, placing a separator between packages or even by wiring together the required quantity. This approach has been applied successfully in many different types of situations. Some of these are:

1. Paper supply in a job printing shop
2. All hardware in a manufacturing plant
3. Expendable supplies such as rags, and cleaning agents
4. Office supplies
5. All finished goods in a wholesale house filling many small orders

Although the material itself serves as the record of the quantity on hand, a material card must still be maintained to record the constant information which is needed for identification, purchasing, and substitutions. Also, the quantity purchased and the date are usually posted to a card in order that the usage may be determined. The operations that are eliminated are the posting of each individual withdrawal and the calculations of balances. This recording of each individual issue usually is the largest single operation contributing to the cost of the material control function. By eliminating it, significant savings can be realized. Furthermore, since the material serves as the "record," the problems caused by posting errors are eliminated.

The two-bin system, however, cannot function properly unless there is tight physical control of the material. If the "second bin" is used to fill past experience. The result was that some departments ended up with excessive amounts of certain items and ran out of others. Under the an order and the "traveling requisition" is not detached, there will almost certainly be a "stock-out" (out-of-stock situation). Stockrooms must be closed to all but authorized personnel. Stockkeepers must understand the importance of immediate transmission of the "traveling requisition."

There are certain situations which limit the applicability of the two-bin approach to material control. At times the bulk of the material is so great that it is impracticable to maintain a "second bin" for the re-order-point quantity. In some organizations rigid accountability policies require that each issue must be charged to a particular account; so the transaction must be recorded. However, even under such conditions, it is desirable to consider elimination of posting to stock record cards and to maintain control on a dollar basis.

THE SUPERMARKET CONCEPT

A recent innovation in the area of material control is the "supermarket system," which can best be explained by describing some applications.

One company applied this system to the material control function for office and expendable supplies. Prior to applying the supermarket concept, each department was allowed a given amount of expendable supplies such as paper, pencils, and so on, based on "yardsticks" developed from supermarket system, each department is given so much money to spend on expendable supplies each month. Periodically, a representative of the department makes up a "shopping list" by checking with others in the department and inspecting the supply cabinet. He then goes to the "market," which is a converted storage area, to select the items which he has on his list. The items are laid out on counters with the prices marked similar to a five and ten cent store. He collects the items and takes his order to a check-out counter where the total bill is recorded and charged to the department's account for supplies. In this particular organization each department has an addressograph "charge plate" which is used to facilitate the charging and billing.

The advantage obtained by this system is that no time is consumed in filling out requisitions, listing each item by name and number and doing the subsequent accounting and paper processing. Furthermore, the tendency to "hoard" items is reduced because each department head knows that he can send a representative to the market whenever something is needed.

Another example is that of an organization which does repair and rebuild work. A parts cost charge is attached to each piece of equipment which is being repaired. As in the office application, the charge card is in the form of an addressograph plate. When a mechanic determines that a repair part or hardware is needed, he detaches the plate from the equipment and goes to the stockroom. He selects the part which is needed and checks out with a clerk who uses the addressograph plate to charge the material to the job order. (If the mechanic cannot find the material when he goes to the stock area, he may determine a substitute and select it immediately.) No time is wasted while the stockroom attendant brings out possible alternative parts for approval.

Frequently the "supermarket" system operates in conjunction with a "two-bin" system for maintenance of the supply in the "market."

Summary   Since the basic activity of most organizations is the modification of material, the material control function is of extreme importance. The right material must be available at the right time, and costs must be minimized.

There are three major classes of inventory:

1. Raw materials and purchased parts
2. In-process material
3. Finished goods

The bill of materials is the basic document used in material planning. The material planner "explodes" the bill of materials to establish subassembly, purchased parts, and raw material requirements. This operation can be performed manually with needle-sort equipment, punched card equipment, and, when volume warrants, computors. An important part of the bill of materials is the scrap allowance.

There are four principal types of material control situations:

1. Flow control
2. Similar process control
3. Custom-order control
4. Job-order control

When most people discuss material control, they are referring to the last of these situations, and there are two important ingredients of a successful material control system under such circumstances:

1. Adequate physical control of the material which involves:
   1.1 Adequate physical facilities for proper storage
   1.2 Adequate stockroom security
   1.3 Effective requisitioning procedure
2. An adequate record keeping system which requires:
   2.1 Maintaining records on a current basis
   2.2 Accurate posting of information to the records

A relatively new approach to material control is the two-bin system which is based on the principle that the material itself will serve as the record of the quantity on hand. The supermarket system which was described in this chapter is also a recent innovation in the material control field.

Before any of the profit optimizing techniques of material management can be applied, an effective material control system must be in existence.

12

Material Management

In the preceding chapter it was stated that the most common material control situations were those in which material was kept in stock and replenished periodically. There are two basic quantities that must be determined for stocked materials:

1. The quantity of new stock to order at one time
2. The quantity of old stock at which to order new

Each of these questions will be discussed in detail. Figure 12-1 illustrates these two considerations graphically.

Material management is defined as the determination of reorder points and order quantities. Before any of the scientific techniques of material management can be applied, there must be an effective material control system in operation which will provide the necessary data. The data must be both accurate and timely if it is to be of value. Just as in the material control function, material management is vital to effective production in manufacturing type activities where material is a signifi-

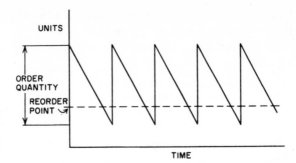

UNITS

ORDER
QUANTITY

REORDER
POINT

TIME

Fig. 12-1. Material management is concerned with the determination of two basic quantities: the order quantity, the recorder point quantity.

cant portion of the total cost of the product. For this reason, our examples will again be drawn from manufacturing activities. The reader should relate the principles to other types of activities as desired.

The "scientific" techniques of material management have generally received a great amount of attention in industry because management has recognized the existence of a fertile field for substantial reductions in costs associated with maintaining material inventories. Some of the most significant costs are:

1. Investment costs
2. Storage costs
3. Clerical costs
4. Set-up or procurement costs
5. Out-of-stock costs

Material management is a popular field of study for those interested in operations research. Many advances have been made in the application of mathematical techniques for solving material management problems. However, in this chapter, mathematics will be kept to a minimum and nothing more advanced than algebra will be used. The reader who is interested in pursuing the mathematical aspects of the subject should consult other texts in the field.[1]

ECONOMIC ORDER QUANTITY

The mathematical determination of the quantity of an item to be ordered at any one time was one of the first subjects of investigation by the pioneers of scientific management. By the early 1900's many formulas

---

[1] Thomson M. Whitten, *The Theory of Inventory Management* (Princeton, N.J.: Princeton University Press, 1953); W. Evert Welch, *Tested Scientific Inventory Control* (Greenwich, Conn.: Management Publishing Company, 1956); General Services Administration, Federal Supply Service, *Scientific Inventory Management Simplified*, Revised Edition, Federal Stock Number 7610-543-6765 (Washington, D.C.: United States Government Printing Office, 1957).

had been developed, but until after World War II the application of these formulas was very limited.

In attempting to establish an economic order quantity, two extreme views are encountered:

1. The production-oriented solution: "Produce in very large lots in order to minimize set-up and procurement costs."
2. The treasurer-, comptroller-, or accountant-oriented solution: "Produce in very small lots in order to minimize the investment in stock."

Which approach is right? The answer is not to be found in either extreme but somewhere between the two approaches. The economic order quantity should be established so as to balance all the variable costs of inventory. The variable costs of inventory are defined as: *Those costs which vary with the size of the order quantity.* They are made up of two principal components:

1. The costs that vary with the number of times the order is produced (set-up or procurement costs).
2. The costs that vary with the average inventory (storage or carrying costs).

The same factors will apply whether the items are purchased or manufactured.

Procurement or Set-up Costs   Procurement or set-up costs are those which vary with the number of times the item is produced. Procurement costs apply when the item is to be bought, and comprise the following factors:

1. Cost of processing the requisition:
    1.1 Preparing the requisition
    1.2 Determining the order quantity
    1.3 Requesting suppliers' quotations
    1.4 Maintaining information on material record cards
2. Cost of following-up and expediting the order
3. Cost of receiving the material:
    3.1 Unloading and storing the material
    3.2 Inspecting the material
    3.3 Preparing the receiving reports
    3.4 Entering information on material record cards
4. Cost of processing the vendor's invoice:
    4.1 Auditing and comparing with receiving reports and original orders
    4.2 Preparing checks for payment
    4.3 Disbursing the checks and later auditing

Set-up costs apply when the item is being produced within the organization, and comprise the following factors:

1. Cost of setting-up the process:
    1.1 Setting-up the necessary equipment and making tool changes

1.2 Changing operations (punching in and out on job time cards and other necessary paper work)

1.3 Instructing the operator and following-up to assure that the job is being performed correctly

1.4 Building up skill on the operation (Even on jobs which have been performed in the past, it takes time for the operator to get into the "rhythm" of the job.)

2. Cost of scrap which occurs primarily at the beginning of a new operation

3. Cost of planning production and controlling costs:

3.1 Preparing the necessary production control paper work

3.2 Scheduling the work

3.3 Following-up and expediting the job

3.4 Accounting for job costs

Procurement or set-up costs increase as the order quantity decreases. In determining these costs, the salaries of supervisory personnel generally should not be included. It is usually better to estimate such costs than to try to apply accounting data.

Another factor which should be taken into consideration when estimating procurement costs, even though it is very difficult to reduce it to a dollars-and-cents figure, is "stock-out." As the order quantity decreases or the order rate increases, the risk of a stock-out increases. For this reason, procurement costs should usually be estimated on the high side.

Carrying Cost   Carrying costs are those which vary with the size of the inventory. The carrying costs are based on average inventory. This is one-half the order quantity, as graphically illustrated in Figure 12-2. Carrying costs are expressed as a per cent of the average inventory valuation in dollars.

The factors which should be included in the determination of the carrying cost are as follows:

1. The cost of storage space. Items must be classified for significant differences in the space occupancy per dollar of inventory and for specific storage requirements (such as temperature and humidity).

2. The cost of upkeep of material in stock and allowances for possible deterioration.

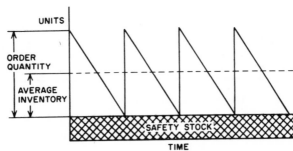

Fig. 12-2. Not considering safety stock, average inventory equals one half the order quantity.

3. The cost of keeping the stock records:
   3.1 Posting the records
   3.2 Taking physical counts
4. The cost of insurance. This cost can usually be determined directly from current insurance policies.
5. The cost of obsolescence. This cost depends upon the type of items being stored—for example, it would be insignificant for hardware items, but very high for style goods. Items must be classified if there is a significant difference in the obsolescence factor.
6. The cost of the money invested in inventory (interest).
7. The cost of taxes on the items being stored. In many states there is a property tax on inventories.

Carrying costs decrease as the order quantity decreases. The industrial average for this charge is between 15 and 25 per cent of the average yearly inventory.

Objective of Determining the   The objective of economic order quan-
Economic Order Quantity   tity calculations is to determine an
                                     order quantity so that the total vari-
able cost of inventory (procurement or set-up costs and carrying costs) is minimized. To illustrate this objective, an actual case will be used to determine the economic order quantity by trial-and-error. The following facts have been established about the item under consideration:

1. The procurement cost is $10. (The item is purchased rather than made.)
2. About $100 worth of the item is used each month.
3. The carrying charge for this item is equal to 20 per cent of the average yearly inventory.

TABLE 12-1

"TRIAL-AND-ERROR" APPROACH TO
ECONOMIC ORDER QUANTITY DETERMINATION

Number of Months Supply Ordered	½	1	2	3	3½	4	5	6	12
Average Inventory ($)	25	50	100	150	175	200	250	300	600
Annual Storage Cost ($)	5	10	20	30	35	40	50	60	120
Annual Procurement Cost ($)	240	120	60	40	34	30	24	20	10
Total Annual Cost ($)	245	130	80	70	69	70	74	80	130

In Table 12-1 the total annual variable cost is calculated for different order quantities in terms of monthly supply. It appears that the economic order quantity is approximately 3½ months' supply.

Formula Determination   It would not be feasible to go through a
series of calculations such as those in Table
12-1 every time an economic order quantity had to be established. If the
figures determined in the trial-and-error solution are plotted, it becomes
apparent that the minimum cost occurs when the annual procurement
cost equals the annual carrying cost. This is illustrated graphically in
Figures 12-3, 12-4, and 12-5.

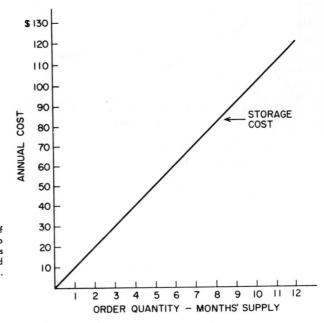

Fig. 12-3. Relationship of total annual storage cost to various order quantities as determined in the "trial and error" solution in Table 12-1.

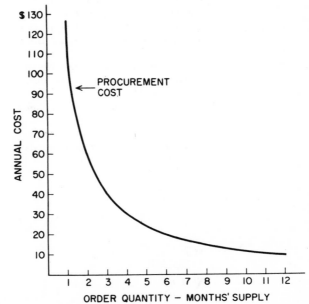

Fig. 12-4. Relationship of total annual procurement cost to various order quantities as determined in the "trial and error" solution in Table 12-1.

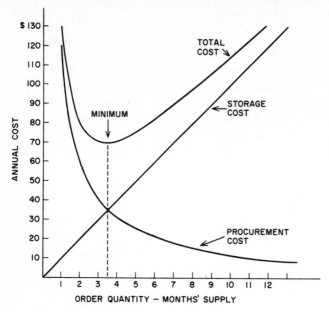

Fig. 12-5. When the annual procurement and storage costs associated with various order quantities are combined graphically, the total cost curve is at a minimum at the point where storage and procurement costs are equal.

The reader should recognize that the problem is resolved into a break-even point determination such as that discussed in Chapter 10 on Process Planning. The following notation is used in developing the economic order quantity formula:

N_m = the number of months' supply to be ordered at one time (economic order quantity)
M_d = the monthly usage of the item in *dollars*
P = the procurement cost (or set-up cost) in dollars
I = the inventory carrying charge per year expressed as a *decimal*

The formula is derived as follows:

1. Procurement cost:
   1.1 If three months' supply is ordered at one time, orders would be placed four times per year. Therefore, $12/N_m$ equals the number of orders that would be placed per year.
   1.2 Thus, $12/N_m \times P$ will equal the total yearly procurement cost.
2. Carrying cost:
   2.1 The order quantity equals the number of months' supply ordered at one time, times the monthly usage, or:
   $$N_m \times M_d$$
   2.2 The average inventory equals one-half the order quantity:
   $$\frac{N_m \times M_d}{2}$$
   2.3 The annual carrying cost equals the average inventory times the carrying charge:
   $$\frac{N_m \times M_d}{2} \times I$$
3. Solving for the breakeven point:
   $$\frac{N_m \times M_d}{2} \times I = 12/N_m \times P$$

$$N^2{}_m = \frac{24P}{M_d I}$$

$$N_m = \sqrt{\frac{24P}{M_d I}}$$

If this formula were applied to the previous example, the solution would appear as follows:

$$N_m = \sqrt{\frac{24P}{M_dI}}$$

$$N_m = \sqrt{\frac{24 \times 10}{100 \times .20}}$$

$$N_m = 3.46 \text{ months' supply}$$

By referring to the formula, the reader can note the effect of any errors made in the calculation of the procurement and carrying costs. If the estimates of the procurement cost and carrying costs are both high, they will tend to cancel each other. Also because the formula involves a square root, an absolute error of 50 per cent in the determination of the variable costs of inventory will result in an error of approximately seven per cent in the final solution.

Simplification of Calculations   If clerical personnel are to determine order quantities, the calculations must be further simplified.
One method of providing this simplification is illustrated in Table 12-2. Tables are developed for each of the appropriate carrying charges used in the organization. The two remaining variables (procurement cost and monthly usage) are listed in logarithmic incre-

TABLE 12-2

ECONOMIC ORDER QUANTITIES
FOR A CARRYING CHARGE OF 24 PER CENT

P\M	3	5	8	11	15	20	25	35	50	75	100	150	200	300	500	750	1000
3	10	8	6	5	4.5	3.9	3.5	3.0	2.5	2.0	1.8	1.4	1.2	.3	.25	.2	.17
5	12	10	8	6.7	5.8	5	4.5	3.7	3.1	2.6	2.2	1.8	1.6	1.3	1.	.8	.7
8		12	10	8.5	7.3	6.4	5.7	4.8	4.	3.4	2.8	2.3	2.	1.6	1.3	1.1	.9
11			12	10	8.5	7.4	6.7	5.6	4.7	3.8	3.3	2.7	2.4	1.9	1.5	1.2	1.0
15				12	10	8.7	7.8	6.6	5.5	4.5	3.9	3.2	2.7	2.2	1.8	1.4	1.2
20					12	10	9.	7.5	6.3	5.2	4.5	3.7	3.2	2.6	2.	1.7	1.4
25						11	10	8.5	7.1	5.8	5.	4.8	3.8	2.9	2.2	1.8	1.6
35						12	12	10	8.5	6.9	5.9	4.8	4.2	3.4	2.8	2.2	1.8
50							12	10	8.2	7.1	5.8	5.	4.1	3.2	2.6	2.2
75								12	10	8.7	7.1	6.1	5.	3.9	3.2	2.7
100									12	10	8.2	7.1	5.8	4.5	3.7	3.2
150			ONE YEAR'S SUPPLY								12	10	8.7	7.1	5.5	4.5	3.9
200												12	10	8.2	6.3	5.2	4.5
300													12	10	7.7	6.3	5.5
500														12	10	8.3	7.1
750															12	10	8.6
1000																12	10

ments (due to the fact that the formula involves a second-degree equation). When it is time to place an order the following simple steps are performed:

1. The usage is checked.
2. The procurement cost (which should be listed on the control card) is determined.
3. The table is referred to and the order quantity determined.

One of the rules of thumb often applied in material management is illustrated in Table 12-2. Because obsolescence increases significantly for periods in excess of one year, a twelve-month supply is established as the maximum which will be ordered regardless of the results obtained with the formula.

Nomographs can also be used to reduce individual calculations. An example of a nomograph is given in Figure 12-6. The advantage of nomographs is that they provide continuous values. However, they are slightly more difficult to use than the tables, and a higher likelihood of error exists.

Modifications of the Basic Formula   There are many modifications of the basic formula. All of them developed in the same manner as the basic formula. The only difference is the manner in which the variables are expressed.

Some of the most frequently used modifications are as follows:

1. The order quantity and the monthly usage are expressed in units:

$$N_u = \sqrt{\frac{24\ P\ M_u}{I\ C_u}}$$

Where:

N_u = the economic order quantity in units
M_u = the monthly usage in units
C_u = the cost per unit
P = procurement cost
I = carrying charge as a decimal

2. The order quantity and the *annual* usage are expressed in units:

$$N_u = \sqrt{\frac{2\ P\ A_u}{I\ C_u}}$$

Where:

N_u, C_u, P, and I are same as in the preceding formula
A_u = the annual usage in units

CLASSIFICATION OF INVENTORY BY DOLLAR USAGE

The first step in any scientific inventory analysis is to classify the inventory according to dollar usage. The term *dollar usage* refers to the number of units of a raw material item required annually or the number of units of a finished goods item sold annually multiplied by the unit cost. The discussion of this problem has been deferred until now in order that the reader can better appreciate the need for being selective when determining the refinement of the variables involved in the economic order quantity formula. The cause of failure of scientific material management is often the lack of recognition of the need for treating various segments of the inventory in different ways.

Fig. 12-6. Nomograph for determining economic order quantity in terms of month's supply. (Reprinted by permission of The National Institute of Management, Cleveland, Ohio.)

ECONOMIC ORDER QUANTITY CHART

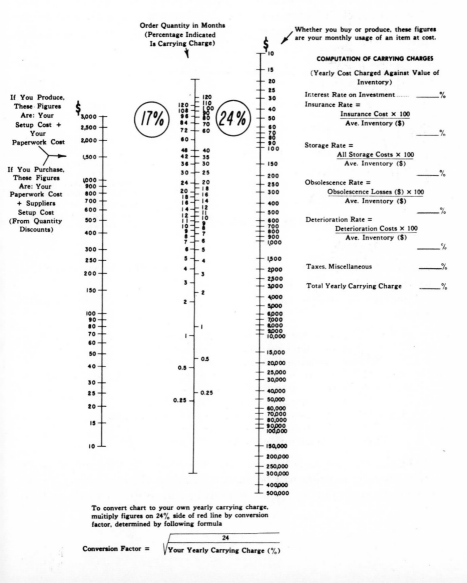

The items carried in inventory should be divided into classes based upon the annual dollar usage. Attention should then be focused upon the high dollar usage items, and only a minimum amount of effort should be devoted to determining order quantities for low dollar usage items. It is common to find that about 15 per cent of the items carried in stock account for approximately 85 per cent of the total dollar usage.

Common practice is to divide the inventory items into three classes as follows:

"A" items—those with high dollar usage
"B" items—those with medium dollar usage
"C" items—those with low dollar usage

To determine the boundaries of each classification, a sample of all the items carried in the inventory system is taken, the items arranged according to dollar usage, and a graph drawn such as shown in Figure 12-7.

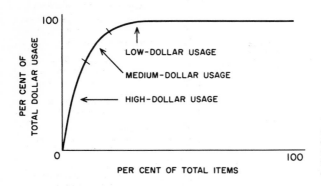

Fig. 12-7. Many organizations find relatively few items constitute a large proportion of their material expenditures, and when the distribution is analyzed, the result is a graph such as this.

The points of inflection on the curve (as determined by visual inspection) are then used to establish the class boundaries.

This type of analysis made in one company produced the following results:

Class "A" items—those with more than $2000 annual usage:
6 % of the items carried in inventory
62% of the annual inventory dollar volume
Class "B" items—those with between $500 and $2000 annual usage:
15% of the items carried in inventory
23% of the annual inventory dollar volume
Class "C" items—those with less than $500 annual usage:
79% of the items carried in inventory
15% of annual inventory dollar volume

The classification indicates the frequency with which the factors of the order quantity formula will be analyzed and by which the order quantities will be recomputed. "A" items must be controlled on a current basis, but less control and some errors can be afforded in the "C" items. The procedure for the application of this concept is as follows:

"A" items:

1. Procurement cost and monthly usage for the order period is checked each time the item is ordered and the order quantity determined.
2. Purchase orders are approved by the Chief of Inventory Control.

"B" items:

1. The monthly usage for the following six months is determined for the first order placed after January 1st and for the first order placed after July 1st (or any other six-month interval).
2. The order quantities determined in (1) are checked and approved by the Chief of Inventory Control.
3. The order quantities determined in (1) are used for all subsequent orders placed in the six-month period. Purchase orders do not require the approval of the Chief of Inventory Control.

"C" items:

1. The usage and procurement costs are checked every two years and order quantities determined.
2. Order quantities determined in (1) are checked and approved by the Chief of Inventory Control.
3. Purchase orders do not require the approval of the Chief of Inventory Control.

MODIFICATIONS FOR LIMITED QUANTITIES

Another common reason for failure with the scientific approach to material management is that the designers of the system fail to recognize that certain restrictions may exist regardless of the "answers" obtained with any economic order quantity formula. Application of the formulas will frequently result in a large increase in average inventory and thus require an increase in storage space and capital requirements. Application of the formulas may also result in a large increase in the order frequency; and purchasing, receiving, warehousing, and inventory control sections may not be prepared to absorb the load.

The proper procedure is to determine the potential changes in the storage and capital requirements or in the order frequency by applying the formulas to a sample. If the application to the sample indicates a significant increase in storage and capital requirements or order processing frequency, an investigation must be made to determine if such changes are feasible. When this procedure is not followed, the resulting chaos eliminates future consideration of any attempt at scientific material management.

In the event the organization is not in a position to absorb the changes indicated in the sample, there are alternative formulas which may be applied. These alternative formulas will not result in an absolute minimum variable cost of inventory. However, they will usually produce significant savings by apportioning storage space and capital or the number of orders which can be processed in the most economical manner.

The formula used when working with restricting conditions is as follows:[2]

$$N_d = CA$$

Where:

N_d = the order quantity in dollars
A = the total annual dollar usage of the item
C = a constant depending upon the restrictions and determined from a sample of the items in the inventory as illustrated in Table 12-3.

TABLE 12-3

DETERMINATION OF CONSTANTS WHEN USING FORMULA FOR RESTRICTING CONDITIONS

Item	Annual Usage	$\sqrt{\text{Annual Usage}}$	Orders Per Year	Size of Order	Average Inventory
1	$5200	$ 72	5	$1040	$520
2	485	22	3	162	81
3	1720	11	4	430	215
	$8405	$\Sigma\sqrt{A} = \$135$	$\Sigma N = 12$	$\Sigma Q = \$1632$	$816

For Constant Number of Orders: $C_1 = \dfrac{\Sigma\sqrt{A}}{N} = \dfrac{135}{12} = 11.2$

For Constant Inventory: $C_2 = \dfrac{\Sigma Q}{\Sigma\sqrt{A}} = \dfrac{1632}{135} = 12.1$

1. If the number of orders (order rate) cannot be increased:

$$C_1 = \frac{\Sigma\sqrt{A},}{\Sigma N}$$

where we get $\Sigma\sqrt{A}$ and ΣN from the sample and they are defined as follows:

$\Sigma\sqrt{A}$ is the sum of the square roots of the annual dollar usage of the items in the sample, and
ΣN is the sum of the orders placed annually.

Applying this constant in the restriction formula will result in a minimum inventory without increasing the order rate.

2. If the storage space or capital is limited—i.e. inventory cannot be increased:

$$C_2 = \frac{\Sigma Q}{\Sigma\sqrt{A}}$$

---

[2] The proof of this formula is given in Welch, *Tested Scientific Inventory Control,* page 32.

where we get ΣQ and $\Sigma\sqrt{A}$ from the sample and they are defined as follows:

$\Sigma\sqrt{A}$ is the sum of the square roots of the annual dollar usage of the items in the sample, and

ΣQ is the sum of the order quantities of the items in the sample in dollars.

Applying this constant in the restriction formula will result in a minimum number of orders without increasing the size of the inventory.

The application of the restriction formula under each limiting condition is illustrated in Tables 12-4 and 12-5 and the resulting savings indicated.

TABLE 12-4

APPLICATION OF ORDER CONSTANT

Item	Annual Usage	$\sqrt{\text{Annual Usage}}$	C_1	Size of Order	Orders Per Year	Average Inventory
1	$5200	72	11.2	805	6.4	402
2	485	22	11.2	247	1.9	124
3	1720	41	11.2	460	3.7	230
					12.0	756

Using the formula $N_d = C_1\sqrt{A}$, we will place 12 orders per year as at present, but the distribution of the orders will be such that the average inventory will be reduced by $60 (See Table 12.3). If the carrying charge were 20 per cent, this would result in a net saving of $12.

TABLE 12-5

APPLICATION OF INVENTORY CONSTANT

Item	Annual Usage	$\sqrt{\text{Annual Usage}}$	C_2	Size of Order	Order Per Year	Average Inventory
1	$5200	72	12.1	875	5.9	438
2	485	22	12.1	257	1.9	129
3	1720	41	12.1	497	3.4	249
					11.2	$816

Using the formula $N_d = C_2\sqrt{A}$, we will obtain an average inventory of $816 as at present, but the distribution of orders will be such that the total number of orders will be reduced by .8 (See Table 12.3). If the procurement cost were $30 this would result in a net saving of $24.

MODIFICATIONS FOR QUANTITY DISCOUNT AND VARIABLE USAGE

Quantity Discount   When vendors offer a discount for buying in quantity, the application of the economic order quantity using the highest unit price may produce a solution between price "breaks." In such cases a determination must be made on increasing the order quantity to receive the price discount. This problem and the method of solution are illustrated in the following example:

1. The vendor quotes the following prices:
    100–999 units: $ .90 each
    1000 units and over: $ .85 each
2. Application of the economic order quantity formula indicates that 700 units should be ordered.
3. A decision must be made on whether the order quantity should be increased to 1000 units in order to take advantage of the quantity discount. This depends upon the comparison between the savings in cost resulting from the reduction in the unit price and the additional inventory cost created by an increase in the average inventory.

A precise answer would have to consider the following factors:

1. The cost of carrying inventory.
2. The cost of placing orders.
3. The unit price for quantities indicated by application of the basic formula.
4. The price for quantity discounts.
5. The monthly usage of the item.

If the cost of placing the order is considered to be negligible in comparison with the other factors, the solution to the problem is greatly simplified.

First, the effect of the increased order quantity on the storage cost must be considered:

> If one month's usage of an item is added to a given order quantity, the average dollar inventory value of the item is increased by one-half of a month's, or 1/24th of a year's usage. If "M" months' usage is added to the order quantity, the increase in the average inventory for the item will be $\dfrac{M A}{24}$, where "A" is the annual usage of the item in dollars.
> Then the increased carrying cost will be:
> $\dfrac{M A I}{24}$, where I is the carrying charge expressed as a decimal.

This increased carrying cost must be compared with the saving gained from the quantity discount which is expressed as follows:

> If there is an $X\%$ reduction in the unit cost by purchasing the increased quantity, the annual saving will be:
> XA, where X is expressed as a decimal.

Comparing these two factors algebraically, the decision is determined as follows:

> If $XA > \dfrac{M A I}{24}$ or $X > \dfrac{M I}{24}$ (since A is common to both sides) purchase the increased quantity to take advantage of the price reduction.

The case problem is then solved as follows:

1. $X = \dfrac{.05}{.90} = .055$

2. Assume that the monthly usage is 150 units. Then M equals

$$\frac{1000\text{-}700}{150} = \frac{300}{150} = 2$$

(1000 units must be ordered to receive the price reduction, whereas the original economic order quantity determination indicated that 700 units should be purchased.)

2.1 Assume that the Inventory Carrying Charge $(I) = 16\%$

2.2 Then $\dfrac{M\,I}{24} = \dfrac{2\,(.16)}{24} = .013$

The saving resulting from the price reduction is greater than the increased carrying cost. Therefore, the order quantity should be revised to 1000 units. (.055 is greater than .013).

Variable Usage   Another factor that complicates the application of the economical order quantity formula is that frequently the monthly usage of an item is effected by a seasonal demand. Thus, the monthly usage factor in the formula, is not a constant. The following procedure is recommended in such situations:

1. For "C" (low dollar volume) items:
   Select a value for the monthly usage equal to the average monthly usage during the year regardless of any seasonal influence since the money involved does not merit more precise considerations.
2. For "B" (medium dollar volume) items:
   Determine the average monthly usage during the "fast" and "slow" seasons, and use whichever one is applicable when determining the individual order quantity.
3. For "A" (high dollar volume) items:
   Use the procedure illustrated in Table 12-6.
   3.1 Project the cumulative monthly usage month by month, starting from the expected date of receipt of the order.
   3.2 Determine the average monthly usage month by month, based on the cumulative figure.
   3.3 Determine the economical order quantity month by month, using the average usage determined in the preceding step.
   3.4 When the order quantity determined in step 3.3 is equal (or approximately equal) to the cumulative monthly usage, select this as the economical order quantity.

INTRODUCTION TO REORDER POINT DETERMINATION

So far we have discussed the first basic quantity involved in material management—determination of the basic economical order quantity formula. The second basic quantity determined is the reorder point. This problem is illustrated in Figure 12-8.

TABLE 12-6

DETERMINATION OF ORDER QUANTITY FOR ITEM AFFECTED BY
VARIABLE USAGE

Month	Monthly Usage	Cumulative Monthly Usage	Average Monthly Usage	Order Quantity
1	400	400	400	1700
2	350	750	375	1650
3	300	1050	350	1580
4	200	1250	312	1490
5	150	1400	286	1420
6	150	1550	258	1360
7	200			
8	250			
9	250			
10	350			
11	400			
12	450			

$$I = .20 \quad P = 30 \quad C_u = .50$$

$$N_u = \sqrt{\frac{24\ P\ M_u}{C_u\ I}} = \sqrt{\frac{24 \times 30\ M_u}{.50 \times .2}} = 85\ \sqrt{M_u}$$

Orders 1400 units

When seasonality must be considered, the economic order quantity is that point at which
the order quantity (determined by using the average monthly usage in the formula)
equals the cumulative monthly usage.

Referring to Figure 12-8, Page 128, the reader will note that the reorder point is established by consideration of the following factors:

    1. The usage during the procurement lead time.
    2. The safety stock or "cushion."

Following are some of the most common procedures used in determining the reorder point:

    1. Determine the usage during lead time and increase by a fixed per-
centage, such as 125% or 133%, to determine the reorder point. For
example, in a case where a 33% "safety factor" is used with a lead
time of 4 weeks and usage of 30 units per week, the reorder point is:
1.33 × 4 × 30 = 160 units.

    2. Determine the usage during the lead time and add to this quantity
the usage during a fixed period of time such as 2 or 3 weeks, to determine the reorder point:
For example:
    Where a 2-week safety factor is used with the same usage and lead
time as in the preceding example, the reorder point is:
4 × 30   +   2 × 30   =   180 units
(usage during lead time) + (2 weeks' usage) = (reorder point)

    3. Reorder when the stock gets to a certain level regardless of the usage
during lead time for the particular item. This level is usually the

maximum lead time for any item in the inventory system:
For example:
Where a 5-week level is the basis of the reorder point in a situation where the usage is 30 units per week as in the other examples, the reorder point is:
5 weeks × 30 units per week = 150 units (reorder point)

There are disadvantages to all these common procedures. One weakness is that all items in the system are considered equally important. There is as much chance of stock-out on critical items as on the relatively unimportant items. Because of the different degrees of importance of items as well as fluctuations in lead times and usage, such procedures usually result in too much protection for some items and insufficient protection for others with an excessive investment in safety stock. In spite of this excessive investment, there is inadequate protection of "critical" items.

The safety stock that will be required to protect a given item will depend upon:

1. The accuracy of the estimate of lead time.
2. The probability of variations in lead time.
3. The accuracy of the estimates of usage.
4. The probability of variations in usage.
5. The cost of carrying safety stock.
6. The cost of stock-outs.[3]

"SCIENTIFIC" REORDER POINT FORMULA

Assuming that these factors are the main basis for safety-stock requirements, a formula can be developed for determining the reorder point that is superior to the commonly used procedures described in the preceding section.

Procurement Lead Time   The procurement lead time includes the time for performing all necessary functions of procurement from the time that the reorder point has been reached until the material has been received and is placed in the storeroom. Thus, lead time takes into consideration the following points:

1. Time to process the requisition and place the order.
2. Time to deliver the order to the vendor.
3. Time for the vendor to fill the order.
4. Time for the order to be shipped from the vendor.

It is important to determine accurately the lead time for "A" items. The "B" and "C" items do not require such accuracy. Lead time should

---

[3] Welch, *Tested Scientific Inventory Control*, page 124.

be based on the *average* time taken in the past. The common tendency is to base lead time on the maximum time historically taken to receive an item. This tendency results in providing an extra safety factor and should be avoided because of the distortion caused in the proration of the safety stock dollar.

Experience has shown that lead time usually varies from its normal length by no more than the square root of that length. For example, it may take as long as 6 weeks to receive an item which normally has a lead time of 4 weeks:

$$4 + \sqrt{4} = 6$$

An item which normally has a lead time of 9 weeks may require 12 weeks for delivery:

$$9 + \sqrt{9} = 12$$

Usage During Lead Time   The usage during the procurement lead time depends upon two factors:

1. The procurement lead time (discussed in the preceding section)
2. The average monthly or weekly usage

The usage should be determined as accurately as possible for "A" items and should be analyzed to determine the necessity for revision each time an order is placed.

The usage need not be determined as accurately for the "B" and "C" items. For "C" items, the usage figure is usually based on the maximum usage during the "fast" season. For "B" items, the usage figure should be based upon the maximum usage during the season when the next order must be placed.

Here, as with procurement lead time, experience has shown that usage varies with the square root of its own average.

Cost Considerations   Theoretically, the safety-stock level should be so established that the cost of maintaining it equals the cost of stock-outs. However, this criterion is not normally used because of the difficulty of putting an accurate dollar figure on the cost of stock-outs. Even when the stock-out cost can be determined, the common tendency is to demand a greater safety factor than that which is justified in the cost analysis because of the many intangible factors which cannot be reduced to a definite dollar figure.

The alternate approach to a straight cost analysis is to base the safety stock on the "acceptable frequency of stock-outs." However, when this approach is suggested, the immediate reaction is that the "acceptable frequency" is *never*. It must be understood that only with an infinite investment in inventory can 100 per cent protection against stock-outs be assured. The acceptable frequency is normally expressed as once every so many years; for example, once in three years, once in five years, and so on.

When the frequency approach is used, another variable must be considered. This variable is the order frequency—the number of times each year an order is placed for the item. The following example illustrates this point:

Assume that once in five years is used as the acceptable frequency, and the order quantity is such that 10 orders are placed each year. This means that once every fifty times an order is placed the stock will run out before the reordered material is received. With five years as still acceptable frequency, but with an order quantity such that only five orders are placed each year, stock will run out before the reordered material is received once out of every 25 times an order is placed.

Reorder Point Formula   Based upon the considerations discussed in the preceding sections, the reorder point formula is developed as follows:

$$R = U_L + P\sqrt{U_L},$$

Where:

R = the reorder point in units

U_L = the usage during lead time

P = a factor based upon the acceptable frequency of stock-outs in occurrences every given number of years with a given order frequency

The factor P is based upon the characteristics of the Poisson Distribution. The Poisson Distribution is a completely random distribution. Such a distribution can be described as being similar to the dispersion of holes around a bull's-eye resulting from firing a large number of bullets from a carefully aimed rifle held firmly in a vise. If the stock level at the time purchases are received is considered to be similar to the "bull's-eye," as illustrated in Figure 12-8, the distribution of stock levels around this theoretical level (based upon average usage and average lead time) will have the properties of a Poisson Distribution.

Table 12-7 lists values for the factor P depending upon the order quantity for the item and the acceptable frequency of stock-outs for that item. The application of this table is illustrated in the following two examples:

1. Item "X":
    1.1 The usage during lead time is 100 units.
    1.2 The economic order quantity is 2 months' supply—thus the order frequency is 6 times per year.
    1.3 The acceptable frequency of stock-outs is once in 5 years. Thus, referring to Table 12-7, $P = 1.8$.

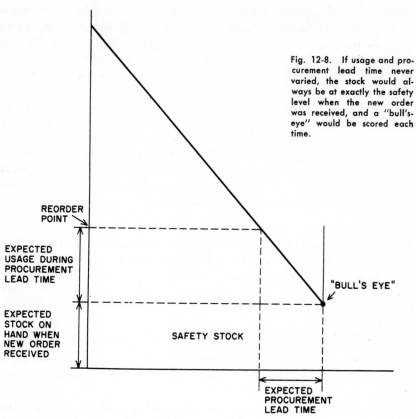

Fig. 12-8. If usage and procurement lead time never varied, the stock would always be at exactly the safety level when the new order was received, and a "bull's-eye" would be scored each time.

TABLE 12-7

VALUES OF *P* FOR REORDER POINT FORMULA

E.O.Q.		Acceptable Frequency of Stockouts							
N Months	Orders Per Year	Once Per Year	Once Per 2 Years	Once Per 3 Years	Once Per 5 Years	Once Per 7 Years	Once Per 10 Years	Once Per 20 Years	N
12	1	0	0	.4	.9	1.1	1.2	1.6	
6	2	0	.7	1.0	1.3	1.4	1.6	1.9	
4	3	.4	1.0	1.2	1.5	1.6	1.8	2.1	
3	4	.7	1.1	1.3	1.7	1.8	1.9	2.2	
2	6	1.0	1.3	1.6	1.8	2.0	2.1	2.4	
1.5	8	1.1	1.5	1.7	2.0	2.1	2.2	2.5	
1	12	1.3	1.7	1.9	2.1	2.2	2.4	2.6	

Safety Stock = P $\sqrt{\text{Usage during lead time}}$
Reorder Point = Usage during lead time + Safety Stock
Usage during lead time = Lead time (Months) × Monthly Usage

1.4 The reorder point is equal to:
$$R = U_L + P \sqrt{U_L}$$
$$= 100 + 1.8 \sqrt{100} = 118 \text{ units}$$

2. Item "Y":
   2.1 The lead time equals 2 months.
   2.2 The monthly usage equals 200 units. Therefore, the usage during lead time is 400 units.
   2.3 The economic order quantity is 4 months' supply. Therefore, the order frequency is three times per year.
   2.4 The acceptable frequency of stock-outs is once in seven years. Referring to Table 12-7, $P = 1.6$.
   2.5 The reorder point is equal to:
   $$R = U_L + P \sqrt{U_L}$$
   $$= 400 + 1.6 \sqrt{400} = 432 \text{ units.}$$

Classification of Inventory For Reorder Point Determination   The ideal situation would be to operate so that no stock-out would ever occur. However, as was stated previously, unless an infinite amount of capital and storage space is available, this goal is virtually impossible to achieve. There are usually some items whose stock-outs have greater adverse effects than those of others. Therefore, it is desirable to classify all the items carried in stock in terms of the effects of stock-outs. These classifications are as follows:

1. "*Critical*"—Stock-outs of these items will cause major production delays and excessive costs. Emergency quantities of these items or substitutes cannot be obtained locally.
2. "*Major*"—Stock-outs of these items will cause minor production delays and excessive costs. Emergency quantities of these items or substitutes can be obtained locally, but at additional cost.
3. "*Minor*"—Stock-outs of these items will cause little inconvenience and no extra costs.

Once the items carried in stock have been classified, the next step is to determine the acceptable frequency of stock-out for each classification. Various criteria can be applied to a sample of the total inventory and the total investment in safety stock can be calculated. Following are examples of such criteria:

	CRITICAL	*MAJOR*	*MINOR*
1.	Once every 5 years	Once every 3 years	Once each year
2.	Once every 7 years	Once every 5 years	Once every 2 years
3.	Once every 10 years	Once every 7 years	Once every 3 years
4.	Never	Once every 10 years	Once every 3 years

The results of applying the various criteria to the sample can be compared with the total present investment in safety stock or a desired investment as defined by management.

In the event that management feels that there is no difference between items as far as the effect of a stock-out is concerned, the problem is greatly simplified. The maximum dollars to be invested in safety stock is determined and then various acceptable frequencies applied to the sample until the calculated amount approximates the desired amount.

Summary   Material management is defined as: *The determination of reorder points and order quantities.*

In establishing order quantities, it has been illustrated that the minimum variable cost associated with inventory will be realized when the procurement or set-up costs equal the carrying costs, and the breakeven concept has been used to develop suitable formulas. It is extremely important that all the items in an inventory system be separated into classes based on the annual dollar volume, so that the effort spent in establishing the variables which go into the order quantity formula may be properly prorated.

Frequently, application of the basic reorder point formula to all the items in an organization's inventory system would result in an impractical situation even though theoretically costs would be minimized. Alternative formulas have been indicated which can be used when the total inventory investment or the total number of orders processed may not be changed. Modified formulas must also be used when there are quantity discounts or extreme seasonal usage conditions.

The reorder point is based upon usage during the procurement lead time plus a "cushion" or safety factor. Procurement lead time includes the time for performing all necessary functions of procurement from the time that the reorder point has been reached until the material has been received and is placed in the storeroom. There are many different rules of thumb employed for determining the safety factor or "cushion." The one recommended in this chapter employs the properties of the Poisson Distribution and is based upon the concept of an acceptable probability of stock-out expressed as once every so many years. When this approach is used, all items in the inventory system are normally classified according to the consequences of an out-of-stock situation.

This and the preceding chapter have been devoted to material, one of the basic resources brought together in the coordination of the Production Control system to create an end product. In the next chapter, tools, another one of the basic resources, will be discussed.

13

Tool Control

The function of determining tool requirements, procuring necessary tooling, and controlling tools once they have been procured is called tool control. In Chapter 10, it was pointed out that tool requirements are usually determined as a part of the process planning function. It is very important to assign this function to one specific person or to a group within the process planning activity specializing in tool design. In any organization, one or more individuals should specialize in the analysis of operations to determine the tooling requirements. There must be assurance that adequate attention is devoted to this analysis prior to actually performing the operations.

Someone, at some point in the processing of a job, must determine the tool requirements. This person may be the tool or process planner, the set-up man, the foreman, or the individual operator. It is more desirable that this determination be made prior to the time of production to insure that the proper tools will be used and to avoid lost time in procuring tools. Lost time resulting from incomplete tool planning during the action-planning phase can be expensive as well as causing schedules to be missed.

Tooling is a more important function in manufacturing activities than it is in most other types of activities, and has been carried to its most effective application there. However, there are several nonmanufacturing areas which have considerable cost reduction possibilities through better tooling. Investigation and analysis of tooling in the maintenance, janitorial, and sanitation areas can result in significant improvements.

In the clerical and engineering areas, tooling has minor or no significance since tools are only of negligible importance. Therefore, in this chapter as in other chapters, discussions will center on the manufacturing type of activity.

In order to facilitate planning and to limit the investment in tool inventory, it is important to standardize wherever possible all the tools within an organization. Standardization is especially important in the methods of grinding cutting tools. One of F. W. Taylor's first "scientific management investigations" was in the field of metal cutting. He recognized the fact that there is one best way to do the job in terms of feed, speed, and tool form. Furthermore, the standard operating procedures which are used to determine operation times for scheduling purposes are based upon specified machine speeds and feeds, which in turn are dependent upon the type of tool and the way in which it has been ground.

PROCUREMENT OF TOOLS

In activities where special-purpose tools are required, the tools should be procured according to Production Control plans. This situation normally exists in custom-built types of activities, but planning and control must be exercised whether the tools are produced within the company requiring the tools or by a special tool supplier. The production of special purpose tools usually creates severe process planning problems, which are, in turn, primarily the result of poor product planning because the *product* is not always fully defined. The process planning function is frequently delegated to the toolmaker himself. This is another skill which causes toolmaking to be such a highly valued craft.

Accurate scheduling is a difficult task in activities concerned primarily with making special tools. This difficulty is caused chiefly by an inability to estimate the time required to produce the tools. In spite of this difficulty, it is extremely important to maintain work load data that are as realistic as possible. These data are usually based on estimates by the foremen. It is also worthwhile to establish check points or audits for following up on the production schedule of any special-purpose tools.

The problems relating to planning and scheduling for special-purpose tools will not be discussed in detail in this chapter since the procurement of tools constitutes a complete Production Control problem in itself. It should be analyzed from the design approach to each of the Production Control functions discussed in other chapters of the book.

Standard tools (such as blanks for single-point tools or milling cutters) are procured and stocked in conformance to the standard procedures similar to those discussed in Chapter 11 on material control.

CONTROL OF TOOLS

After the procurement of tools, whether special-purpose or standard, adequate controls must be established to insure against loss through theft or negligence and production delays through nonavailability of tools. Controls are also necessary to insure that the investment in tool inventories is minimized consistent with proper tool availability.

A common problem in tool control is that of determining the centralization or decentralization of tool cribs. The costs of the decentralized tool cribs must be analyzed by comparison with the potential savings in production time which may be realized by greater accessibility of tools. Decentralization of tool cribs usually results in increased tool control problems and increased tool inventory. The service problems of centralization can often be overcome by the use of a delivery system—which in effect becomes a decentralized tool crib system.

There are two common procedures used for tool control. These are explained in the following pages:

1. *The Brass Ring System.* In the brass ring system, rings are made up with small plates containing each worker's time clock number. These rings are issued to the individual workers at the time of employment. When the worker draws a tool from the crib, he gives one of his rings to the attendant, and the ring is hung on a peg at the tool bin. When the worker returns the tool, the ring is returned to him. The advantage of this system is that it is very simple. However, it invites dishonesty because of the ease by which counterfeit rings can be made. It also does not provide any means of determining tool usage.

2. *The McCaskey System.* The McCaskey system is based upon a three-part carbon backed form (Figure 13-1). The worker fills it out and presents it to the tool-crib attendant when he wishes to withdraw a tool. As illustrated in Figure 13-2, one copy of the form is maintained under a clip with the worker's name or clock number and a second copy under a clip for the tool number. The third copy is given to the worker for identificaton of the tool and is returned when the tool is turned in.

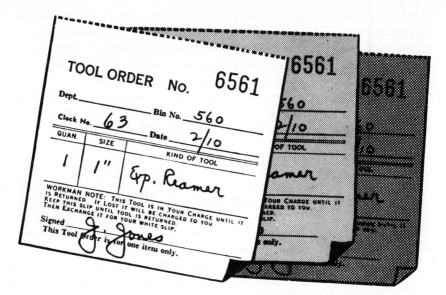

Fig. 13-1. Typical set of tool order slips. (Reprinted by permission of Victor Adding Machine Co., McCaskey Industrial Division.)

The copies filed under the tool number provide ready reference in case the tools are not available when a later request is made for them. Periodic checks of the slips under the worker's clip will indicate if tools are being "hoarded" or held an excessively long time.

When the tool is returned with the third copy of the form which the worker has kept to provide identification, the copy under the worker's clip is removed and returned to him; the copy under the tool clip is removed and placed behind the tool inventory card in the back of each clip. Monthly (or at other time intervals if desired), the slips behind the card are counted, the amount posted to the card to indicate usage and the individual slips thrown away. Figure 13-3 illustrates the inventory card which is used to record the inventory of the number of tools assigned to the crib; the back of the inventory card is designed to provide information on the activity of the particular tool for a period of seven years.

The McCaskey system is widely used in manufacturing establishments because of its simplicity of application and excellent control features.

It is often practical to utilize embossed metal or plastic plates with the McCaskey system. These plates insure proper nomenclature in the processing of tool orders between a centralized tool procurement and control activity and the individual tool cribs. The plates are also used to reduce the amount of worker delay at the tool cribs. With this modification, a plate for each tool is prepared which bears all the pertinent tool information. The plate is filed by tool number or under the tool clip. In addition, each worker is also given a "charge-a-plate" bearing his name and the clock number. When he withdraws a tool from the crib, the two plates are placed together to produce the three-part form.

Fig. 13-2. Wall boards and white slips provide essential visibility of employees' records and permit speedy issuing and receiving of tools. (Reprinted by permission of the Victor Adding Machine Co., McCaskey Industrial Division.)

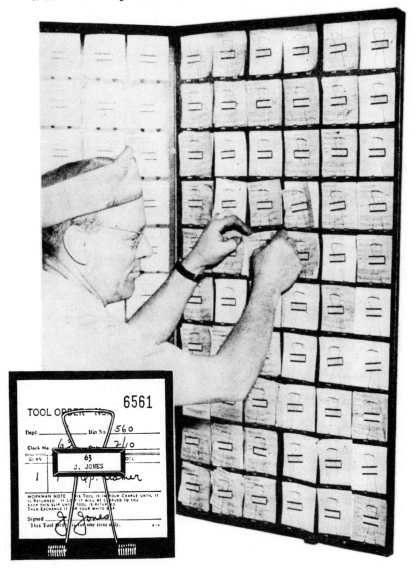

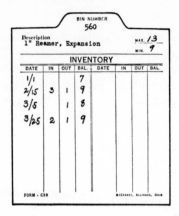

Fig. 13-3. Inventory card records on one side the number of tools assigned to crib; on the other, the activity of each tool during 7-year period. (Reprinted by permission of Victor Adding Machine Co., McCaskey Industrial Division.)

ANALYZING THE TOOL CONTROL FUNCTION

The designer of the Production Control system must determine the extent to which tools are used in the activity which he is analyzing. Some types of activities have few, if any, tool requirements. Hence the tool control function can be disregarded in his analysis. In manufacturing, maintenance, construction and other activities, however, tooling and tool control will constitute an important management problem.

The designer must first of all determine whether sufficient attention is devoted to tooling in the process planning function. He must ask such questions as the following:

1. Are the tool requirements preplanned in conjunction with the process planning function, or are they determined at the time of performing the activity?
2. Has the organization developed standard tooling nomenclature?
3. Is there a tool standardization program?
4. Is there sufficient control over the way in which tools are maintained and ground?
5. Are there standard production procedures to follow?
6. Are the production times associated with this procedure maintained relatively constant?

Whether procurement of tools is made from within the organization's own tool room or from an outside vendor is another source of problems. The designer must determine if a Production Control system is in existence to insure that tools are obtained on time. If the analysis of the Production Control system is a group effort, one individual can be given the responsibility of designing the subsystem for tool procurement.

Finally, the designer must determine whether or not there is proper physical control of the tools. If there are tool cribs, he must determine whether or not they are properly maintained and protected for issuance only to authorized personnel. The greatest costs do not occur as a result of theft, but because tools are mislaid or lost. It is also very important to determine whether or not there are sufficient cribs and personnel to insure the minimum of lost production time. If tools are issued to individuals, the designer must establish an adequate system for effecting their proper return.

Summary   The tool control function, which includes the determining of tool requirements, procuring necessary tooling and controlling tools once they have been procured, is most important in manufacturing activities.

Someone must determine the tool requirements. It is most desirable that this determination be made prior to the time of production. In order to facilitate planning, it is important that tools, tool nomenclature and grinding methods be standardized.

When standard tools such as single point and mill blanks are involved, the tools are procured and stored through procedures similar to those discussed in Chapter 11 for material. When special-purpose tools are to be procured or produced within the organization, a custom-built activity exists and an adequate production control system should be established to insure timely receipt of these tools, *in accordance with minimum cost of production.*

When tools are issued to employees, procedures must be established to insure that the tools will be returned at the proper time. The widespread acceptance of the McCaskey system indicates the desirability of systems which are adequate but also relatively simple.

14

Work Measurement

Work measurement is concerned with economical use of human resources. It is defined as *"The procedure for determining the amount of time required, under certain standard conditions of measurement, for tasks involving some human activity."*[1]

If the functions of scheduling and loading, which will be discussed in the next chapter, are to be performed effectively, realistic time values must be available. Aside from the specific requirement of accurate time values to be used as the basis of establishing schedules and loading facilities, work measurement in general is a requirement for successful management control of an activity.

Some of the most important uses of work measurement are as follows:

1. Planning Activities:
    1.1 Scheduling and loading: Before facilities can be re-

---

[1] Marvin E. Mundel, *Motion and Time Study* (Englewood Cliffs, N. J.: Prentice-Hall, Inc., 1955), p. 1.

served for the performance of an operation, or a completion date for the operation can be determined, there must be some determination of the time required for the operation.

1.2 Determining manpower requirements: As was illustrated in Figure 6-1 (page 000), the product forecast is analyzed to determine the manpower requirements. In order to do this, the number of manhours required to complete each product must be known or estimated.

1.3 Budgeting: Once the work measurement data have been used to establish the manpower requirements, the labor dollar requirements are determined by extending the manhours needed by the hourly wage rate.

2. Establishing management objectives:

2.1 Establishing a goal: Management control depends upon the establishment of goals. In many cases the goal is to complete a task or tasks in a given amount of time.

2.2 Measuring effectiveness: Once a goal has been established, it is important to measure the effectiveness of the organization in achieving this goal. If the goal is in terms of the time requirement, the effectiveness determination can be made by comparing the time actually taken with the time specified in the goal or the standard time.

2.3 Measuring utilization: It is exceedingly important to the manager to know how much of the time of each element within his organization is devoted to the accomplishment of the specific task or goal which is the prime function of the element. This is known as the utilization of the personnel or equipment. Unless there is a sound work measurement system which provides the times for the goals (the standards); it is almost impossible to determine how effectively the facilities are being utilized.

3. Comparing alternate methods:
Methods are usually compared on the basis of labor cost. If the time is known for each alternate method, this comparison can be made very quickly.

4. Paying wage incentives:
In many organizations, wage incentives are paid in order to stimulate the workers to increase productivity. These incentive payments are usually based on the amount by which the worker exceeds a standard, and the standard in turn is based upon the time which has been established for the operation or task.

Since a great deal of management effort is expended in methods improvement, the relationship between methods improvement and work measurement should be noted. Work measurement indicates the areas in which most time is spent; these areas or operations should be the subject of the methods analyst.

Once an improved method has been established, it is important that a time standard based upon the improved method be developed. Then if the actual time required for the operation should exceed the standard, an investigation can be made to determine if the improved method is still being used. Thus, work measurement provides a practical control technique.

Further, a time standard has meaning only in terms of a given method. Time standards should be associated with a specific sequence of movements by the operator rather than simply with the creation of a certain end product. If this concept is thoroughly understood by all parties involved, many disputes over rates or standards which are changed when a method is changed will be avoided. It stands to reason that the motions upon which the standard is based should be the "best" or most effective method for the given conditions.

STOP-WATCH TIME STANDARDS

We will discuss stop-watch standards and the manner in which they are developed in some detail because these steps must be performed in the derivation of any standard which is to be considered truly "engineered." The other techniques of work measurement which will be discussed will be compared with the stop-watch procedure. The steps are as follows:

1. *Define what is meant by standard time or standard performance.*
   This definition need be established only once for an organization. The standards developed for all operations will be based on this concept of standard. Standard may be defined in several ways:
   1.1 Verbal definitions:
       Simply a statement such as: "Standards will be such that a fair day's work will be expected for a fair day's pay," or "the standard will be such that average incentive earnings will equal x per cent."
   1.2 Physical definitions:
       The standard may be represented by an observable physical activity such as: "Walking at three miles an hour," or "dealing 52 cards in four equal piles four inches apart in one-half minute."
   1.3 Film definitions:
       Sometimes films are made of the physical activities which were mentioned in the preceding paragraph or of typical jobs being performed at what is considered a standard rate. The Society for the Advancement of Management has prepared a number of such films. Once the over-all definition of standard has been established for the organization, each of the remaining steps which will be discussed must be performed every time a standard is established for an operation.
2. *Determine the work unit which will be used to record production.*
   It should be repeated that the time value actually applies only to a given sequence of motions, sub-operations, or operations, and the work unit is used to record the completion of these. It is very important that the work unit be such that it will be affected directly by the amount of effort expended by the persons performing the operation —that is, the harder they work, the greater the number of work units they will produce.

3. *Analyze and record the method.*
To begin with, the method is studied to determine the possible degree of improvement. The operation is then divided into elements or sub-operations for further analysis.

3.1 Each element is recorded in sufficient detail to permit the operation to be reproduced from the written description. A sketch is made of the workplace, and all machine feeds, speeds, horsepower, welding currents, and so on, are recorded.

3.2 Any other conditions such as noise, heat, or light are recorded when considered an important factor.

4. *Record the time.*
The time to perform the operation is recorded by means of a stop-watch or a constant speed movie camera.

4.1 The times are recorded by elements within the operation.

4.2 Sufficient number of cycles are recorded to achieve statistically reliable results.

5. *Rate performance.*
At the *same time* as the operation is being timed, the performance of the person being studied is compared with the concept of standard. It is with this step that opinion enters into the time study process. This comparison between the observed performance and the standard concept is known as rating and is the source of many labor disputes.

5.1 The rating is indicated by a percentage, such as 110 if the time study man feels that the operator is doing 10 per cent better than the concept of standard.

5.2 When the average element time is multiplied by the rating factor for that element, the result is called the *base time.* For example, if the average time for the element is .50 minutes and the rating factor was 110 per cent, the base time would be .55 minutes.

6. *Additional allowances.*
The base time represents the time required to perform the element at standard pace. Individuals are not expected to spend their entire working day on specific operations.

6.1 Adjustments must be made for personal time and for unavoidable delays such as waiting for an inspector to check out a part, consulting with the supervisor, and so on. These adjustments are called allowances.

6.2 Allowances should only be made for delays so short that it is impractical to use "down time" cost codes to record them.

6.3 The allowances are stated as a percentage of the working day.

6.4 The base time is increased by the allowance percentage to establish the standard time. For example, if the base time is .55 minutes and the allowances equal 15 per cent, the standard time would be equal to .55 plus .0825, or .6325.

If we review these steps in stop-watch time study, we will also be listing the essential elements of an "engineered" standard.

1. Standard time or standard performance is defined for the organization.
2. The work unit is established to record the performance of the operation.
3. The method of performing the operation is analyzed and recorded.
4. A person performing the operation is observed, and the time he takes to perform each element is recorded.

5. Concurrently with the recording of the time, the performance of the individual being observed is compared with standard and a rating factor established. The average time is multiplied by the rating factor to establish the base time.

6. Allowances for personal time and short unavoidable delays are added to the base time, thus establishing the standard time.

OTHER FORMS OF WORK MEASUREMENT

There are several other approaches to work measurement, and we will discuss each one briefly and compare it with the steps required for an "engineered" standard.

1. *Predetermined Time Systems.* There are many predetermined time systems being used in industry. Some of the best known ones are:

"Methods Time Measurement"—MTM

"Work Factor"

"Basic Motion Time"—BMT

"Motion Time Analysis"—MTA

"Dimensional Motion Time"—DMT

"Ordnance Predetermined Time System"—OPTS

All of these systems are similar in that a detailed analysis is made of many jobs to determine time values for basic motions such as reach, move, turn, release, and the like. These times are then recorded in tables according to the variables that affect the time required for their performance such as distance in the basic motion reach. To illustrate, a table from the Ordnance System developed by Dr. I. P. Lazarus is reproduced in Table 14-1.

To establish the standard for an operation, the operation is broken into elements as in stop-watch time study. The elements in turn are divided into the basic motions of which they are composed. Reference is made to the table of predetermined times to obtain the base times. Then all the base times involved in the element are added to determine the element base time, and allowances are added to establish the standard time.

The predetermined time system approach to work measurement is especially important when considering Production Control because this technique can be used to compare alternate methods with only a written description of these methods, before tooling for production is committed to one of the methods.

This approach is limited, however, in that very detailed analysis is required, and it takes considerable effort to develop an individual standard. Usually predetermined time systems are not used for establishing standards for individual jobs, but rather for the development of standard data which will be discussed later in this chapter.

If we compare the predetermined time system method of establishing standards with stop-watch time study, we find that in fact the steps required for an "engineered" standard have been performed:

1. There is a definition of standard which has been incorporated in the original data.
2. The work unit is established with the same considerations as in stop-watch time study.
3. There is a very detailed methods description, and usually great attention is paid to methods improvement, which is one of the strongest points in favor of the predetermined time technique.
4. The timing and rating steps have been performed by the developers of the system.
5. Allowances are added as a separate step.

2. *Work Sampling.* Work sampling will be discussed in detail in Chapter 24. The prime use of this technique is to determine the time distribution of men or equipment, but it may also be used to develop standards. With the work sampling technique, the work situation is sampled to determine the amount of time spent in producing different work units. The work units can be as fine as is desired. Methods may be described and possibly improved prior to making the work sampling study, and it is even possible to rate performance at the same time the activity of an individual is analyzed and assigned to one of the categories which has been defined.

Depending upon the degree of methods analysis and improvement, the precision with which the work unit is established, and whether or not the activity is rated, standards developed through work sampling may be considered to be "engineered." The prime advantage of using work sampling for establishing standards is that broad coverage can be obtained in a limited amount of time and standards developed through work sampling are usually superior to the other gross techniques such as historical data. The main disadvantages in the use of this technique are:

1. It requires the unbiased cooperation of the persons being studied.
2. Usually the methods descriptions are incomplete, and there is much less attention paid to methods improvement than there would be if the more precise techniques were used.
3. The rating accuracy is very questionable.

3. *Technical Estimates.* When the technical estimate approach is used for work measurement, the job for which the standard is being developed is broken down into operations and sometimes even elements by the process planner. The finer the breakdown made, the more accurate the time estimates. A person who is familiar with the technical aspects of the work estimates the time required

TABLE 14-1 BASIC TIME TABLES IN 1/100,000 MINUTES (.00001 MIN.)*

Time for TRANSPORT LOADED, TRANSPORT EMPTY, and WALKING

A	B	C	D	E	F	G	H	I	J	K	L	M	N	O	P	Q	R	S	T	U
Distance (inches)	Basic Time	Bx 1.01	Bx 1.02	Bx 1.03	Bx 1.04	Bx 1.05	Bx 1.06	Bx 1.07	Bx 1.08	Bx 1.09	Bx 1.10	Bx .10	Bx .20	Bx .30	Bx .40	Bx .50	Bx .60	Bx .70	Bx .80	Bx .90
1	172	174	175	177	179	181	182	184	186	187	189	17	34	52	69	86	103	120	138	155
2	243	245	248	250	253	255	258	260	262	265	267	24	49	73	97	122	146	170	194	219
3	297	300	303	306	309	312	315	318	321	324	327	30	59	89	119	149	178	208	238	267
4	343	346	350	353	357	360	364	367	370	374	377	34	69	103	137	172	206	240	274	309
5	384	388	392	396	399	403	407	411	415	419	422	38	77	115	154	192	230	269	307	346
6	420	424	428	433	437	441	445	449	454	458	462	42	84	126	168	210	252	294	336	378
7	454	459	463	468	472	477	481	486	490	495	499	45	91	136	182	227	272	318	363	409
8	485	490	495	500	504	509	514	519	524	529	534	49	97	146	194	243	291	340	388	437
9	515	520	525	530	536	541	546	551	556	561	567	52	103	155	206	258	309	361	412	464
10	543	548	554	559	565	570	576	581	586	592	597	54	109	163	217	272	326	380	434	489
11	569	575	580	586	592	597	603	609	615	620	626	57	114	171	228	285	341	398	455	512
12	594	600	606	612	618	624	630	636	642	647	653	59	119	178	238	297	356	416	475	535
13	619	625	631	638	644	650	656	662	669	675	681	62	124	186	248	310	371	433	495	557
14	642	648	655	661	668	674	681	687	693	700	706	64	128	193	257	321	385	449	514	578
15	665	672	678	685	692	698	705	712	718	725	732	67	133	200	266	333	399	466	532	599
16	686	693	700	707	713	720	727	734	741	748	755	69	137	206	274	343	412	480	549	617
17	707	714	721	728	735	742	749	756	764	771	778	71	141	212	283	354	424	495	566	636
18	728	735	743	750	757	764	772	779	786	794	801	73	146	218	291	364	437	510	582	655
19	748	755	763	770	778	785	793	800	808	815	823	75	150	224	299	374	449	524	598	673
20	767	775	782	790	798	805	813	821	828	836	844	77	153	230	307	384	460	537	614	690
21	786	794	802	810	817	825	833	841	849	857	865	79	157	236	314	393	472	550	629	707
22	805	813	821	829	837	845	853	861	869	877	886	81	161	242	322	403	483	564	644	725
23	823	831	839	848	856	864	872	881	889	897	905	82	165	247	329	412	494	576	658	741
24	841	849	858	866	875	883	891	900	908	917	925	84	168	252	336	421	505	589	673	757
25	858	867	875	884	892	901	909	918	927	935	944	86	172	257	343	429	515	601	686	772
26	875	884	893	901	910	919	928	936	945	954	963	88	175	263	350	438	525	613	700	788
27+	892	901	910	919	928	937	946	954	963	972	981	89	178	268	357	446	535	624	714	803
28	908	917	926	935	944	953	962	972	981	990	999	91	182	272	363	454	545	636	726	817
29	924	933	942	952	961	970	979	989	998	1007	1016	92	185	277	370	462	554	647	739	832
30	940	949	959	968	978	987	996	1006	1015	1025	1034	94	188	282	376	470	564	658	752	846
31	955	965	974	984	993	1003	1012	1022	1031	1041	1051	96	191	287	382	478	573	669	764	860
32	971	981	990	1000	1010	1020	1029	1039	1049	1058	1068	97	194	291	388	486	583	680	777	874
33	986	996	1006	1016	1025	1035	1045	1055	1065	1075	1085	99	197	296	394	493	592	690	789	887
34	1001	1011	1021	1031	1041	1051	1061	1071	1081	1091	1101	100	200	300	400	501	601	701	801	901
35	1015	1025	1035	1045	1056	1066	1076	1086	1096	1106	1117	102	203	305	406	508	609	711	812	914
36	1029	1039	1050	1060	1070	1080	1091	1101	1111	1122	1132	103	206	309	412	515	617	720	823	926
37	1044	1054	1065	1075	1086	1096	1107	1117	1128	1138	1148	104	209	313	418	522	626	731	835	940
38	1058	1069	1079	1090	1100	1111	1121	1132	1143	1153	1164	106	212	317	423	529	635	741	846	952
39	1072	1083	1093	1104	1115	1126	1136	1147	1158	1168	1179	107	214	322	429	536	643	750	858	965
40	1085	1096	1107	1118	1128	1139	1150	1161	1172	1183	1194	109	217	326	434	543	651	760	868	977
41	1099	1110	1121	1132	1143	1154	1165	1176	1187	1198	1209	110	220	330	440	550	659	769	879	989
42	1112	1123	1134	1145	1156	1168	1179	1190	1201	1212	1223	111	222	334	445	556	667	778	890	1001
43	1125	1136	1148	1159	1170	1181	1193	1204	1215	1226	1238	113	225	338	450	563	675	788	900	1013
44	1138	1149	1161	1172	1184	1195	1206	1218	1229	1240	1252	114	228	341	455	569	683	797	910	1024

Time for all elements given in ten-thousandths of a minute.

† Use 27″ for WALKING, per standard pace. For WALKING by distance, per foot, use line below.

Columns 1–11

Element	1	2	3	4	5	6	7	8	9	10	11
WALKING, per foot	396	400	404	408	412	416	420	424	428	432	436
WALKING 45	1151	1163	1174	1186	1197	1209	1220	1232	1243	1255	1266
WALKING 46	1164	1176	1187	1199	1211	1222	1234	1245	1257	1269	1280
WALKING 47	1176	1188	1200	1211	1223	1235	1247	1258	1270	1282	1294
WALKING 48	1189	1201	1213	1225	1237	1248	1260	1272	1284	1296	1308
GRASP Contact	174	176	177	179	181	183	184	186	188	190	191
Contact-Pinch	278	281	284	286	289	292	295	297	300	303	306
Pinch	357	361	364	368	371	375	378	382	386	389	393
Wrap	566	572	577	583	589	594	600	606	611	617	623
POSITION 1st Degree	200	202	204	206	208	210	212	214	216	218	220
2nd Degree	223	225	227	230	232	234	236	239	241	243	245
3rd Degree	311	314	317	320	323	327	330	333	336	339	342
4th Degree	412	416	420	424	428	433	437	441	445	449	453
5th Degree	556	562	567	573	578	584	589	595	600	606	612
ASSEMBLE ¼ inch‡	00	00	00	00	00	00	00	00	00	00	00
3/8	24	24	24	25	25	25	25	26	26	26	26
1/2	89	90	91	92	93	93	94	95	96	97	98
5/8	154	156	157	159	160	162	163	165	166	168	169
3/4	219	221	223	226	228	230	232	234	237	239	241
7/8	284	287	290	293	295	298	301	304	307	310	312
1	349	352	356	359	363	366	370	373	377	380	384
1⅛	414	418	422	426	431	435	439	443	447	451	455
1¼	479	484	489	493	498	503	508	513	517	522	527
1⅜	544	549	555	560	566	571	577	582	588	593	598
1½	609	615	621	627	633	639	646	652	658	664	670
1⅝	674	681	687	694	701	708	714	721	728	735	741
1¾	740	747	755	762	770	777	784	792	799	807	814
1⅞	805	813	821	829	837	845	853	861	869	877	886
2	870	879	887	896	905	914	922	931	940	948	957
2⅛	935	944	954	963	972	982	991	1000	1010	1019	1029
2¼	1000	1010	1020	1030	1040	1050	1060	1070	1080	1090	1100
2⅜	1065	1076	1086	1097	1108	1118	1129	1140	1150	1161	1172
2½	1130	1141	1153	1164	1175	1187	1198	1209	1220	1232	1243
2⅝	1195	1207	1219	1231	1243	1255	1267	1279	1291	1303	1315
2¾	1260	1273	1285	1298	1310	1323	1336	1348	1361	1373	1386
2⅞	1325	1338	1352	1365	1378	1391	1405	1418	1431	1444	1458
3	1391	1405	1419	1433	1447	1461	1474	1488	1502	1516	1530
DISASSEMBLE All	476	481	486	490	495	500	505	509	514	519	524
RELEASE LOAD All	189	191	193	195	197	198	200	202	204	206	208

Columns 12–20

Element	12	13	14	15	16	17	18	19	20
WALKING, per foot	356	317	277	238	198	158	119	79	40
WALKING 45	1036	921	806	691	576	460	345	230	115
WALKING 46	1048	931	815	698	582	466	349	233	116
WALKING 47	1058	941	823	706	588	470	353	235	118
WALKING 48	1070	951	832	713	595	476	357	238	119
GRASP Contact	157	139	122	104	87	70	52	35	17
Contact-Pinch	250	222	195	167	139	111	83	56	28
Pinch	321	286	250	214	179	143	107	71	36
Wrap	509	453	396	340	283	226	170	113	57
POSITION 1st Degree	180	160	140	120	100	80	60	40	20
2nd Degree	201	178	156	134	112	89	67	45	22
3rd Degree	280	249	218	187	156	124	93	62	31
4th Degree	371	330	288	247	206	165	124	82	41
5th Degree	500	445	389	334	278	222	167	111	56
ASSEMBLE ¼ inch‡	00	00	00	00	00	00	00	00	00
3/8	22	19	17	14	12	10	7	5	2
1/2	80	71	62	53	45	36	27	18	9
5/8	139	123	108	92	77	62	46	31	15
3/4	197	175	153	131	110	88	66	44	22
7/8	256	227	199	170	142	114	85	57	28
1	314	279	244	209	175	140	105	70	35
1⅛	373	331	290	248	207	166	124	83	41
1¼	431	383	335	287	240	192	144	96	48
1⅜	490	435	381	326	272	218	163	109	54
1½	548	487	426	365	305	244	183	122	61
1⅝	607	539	472	404	337	270	202	135	67
1¾	666	592	518	444	370	296	222	148	74
1⅞	725	644	564	483	403	322	242	161	81
2	783	696	609	522	435	348	261	174	87
2⅛	842	748	655	561	468	374	281	187	94
2¼	900	800	700	600	500	400	300	200	100
2⅜	959	852	745	639	533	426	320	213	107
2½	1017	904	791	678	565	452	339	226	113
2⅝	1076	956	837	717	598	478	359	239	120
2¾	1134	1008	882	756	630	504	378	252	126
2⅞	1193	1060	928	795	653	530	398	265	133
3	1252	1113	974	835	696	556	417	278	139
DISASSEMBLE All	428	381	333	286	238	190	143	95	48
RELEASE LOAD All	170	151	132	113	95	76	57	38	19

‡ ¼ inch and less of ASSEMBLE is contained in POSITION; at least it could not be evaluated as a separate value.

* Adapted from Lazarus, Irwin P., "A System of Predetermined Human Work Times," Ph.D. Thesis, Purdue Univ., 1952. (Reprinted by permission of Dr. Lazarus.)

to perform each element or operation. Usually this person is a planner (perhaps the same one who prepares the process plan) or the supervisor of the activity. In some organizations, the persons doing the estimating and the supervisors are rotated so that the estimator has knowledge of current shop situations and the supervisors appreciate the estimator's problems.

Technical estimates cannot be considered a source of "engineered" standards since most of the necessary steps for an engineered standard are not performed. Such estimates, however, are a frequent source of time values for planning purposes and are very valuable.

4. *Historical Data.* In many organizations, the number of manhours expended and the number of units produced are recorded and an average time per unit is established. These historical data are then used as the standard if the job is repeated. The principal limitations of this technique are:

1. Lost time due to excessive avoidable and unavoidable delays is included in the historical data and hence becomes part of the "standard."
2. "Shifting" of time charges in some shops makes the data unrealistic.
3. There is no positive concept of what is meant by standard.
4. Constant time values such as set-up time are mixed with the variable time used in producing each unit.

The historical data method definitely cannot be considered a source of "engineered" standards. There is no method improvement or description. Frequently the work units are too gross to reflect work input correctly. There is no concept of standard performance. There is no separate determination of allowances.

In spite of these limitations, given good accounting procedures and accurate reporting, this approach does create data usable for planning purposes. As a matter of fact, this technique is one of the main sources of time data used in planning.

Staffing Pattern. Thus far in our discussion, we have been referring to the measurement of work in terms of relating time to the actual completion of a particular end unit of production. Not all production procedures however culminate in a particular end unit. On the other hand, some production procedures culminate in end units too numerous or too varied to allow for convenient or accurate measurement. In such cases, the staffing pattern is used as a substitute factor in work measurement.

Staffing patterns are used to establish personnel requirements for such jobs as secretaries, receptionists, material handlers, supervisors, and tool crib attendants. The number of people required to satisfactorily carry out the particular function is based upon the estimated or observed number of people to be serviced or "staffed."

STANDARD DATA

All of the techniques discussed up to this time are limited as far as work planning and control are concerned. The job must have been performed previously or must be in process for time value to be developed through either historical data or stop-watch time studies. Historical data are frequently inaccurate, and in many work situations the jobs do not last long enough to merit the cost of a time study. Technical estimates at best are still only estimates. The predetermined time approach to establishing standards is very time consuming and too costly for any but very long-run jobs. However, the standard data concept overcomes many of the disadvantages of the other techniques.

The standard data approach is based on the concept that although many different end products may be created at a given work station, the elements of work which are performed at the work station are limited in number. For example, a work station consisting of a turret lathe may produce thousands of different parts or end products, yet there are only about 135 different elements of work involved in operating a turret lathe. Once the time values have been derived for each of these elements, the standard for any particular end product can be derived very quickly by simply adding together the times for the elements to be performed. This may be done by the process planner or an individual in a standards section. Actually the predetermined time systems are very refined forms of standard data.

The standard element times may be developed through the application of any of the basic work measurement techniques. They may be established by analyzing many time studies of operations performed at the work stations and establishing average times for each of the elements encountered. Frequently, they are established by building up times for the elements with a predetermined time system. Sometimes technical estimates or historical data can be used, but when this is done, the recorded data are usually on a gross basis such as operations.

Time values derived from a standard data system may or may not be considered engineered standards, depending upon the type of basic work measurement technique upon which they are based.

The standard data technique is of particular importance to persons involved in Production Control. This approach provides a fairly accurate and reliable estimate of times for operations prior to the actual performance of the operations and thus serves as an excellent source of time values for the scheduling and loading functions. Standard data can also be used to establish standards in work situations where there are many short-run jobs, and thus management control can be greatly improved.

ANALYZING THE WORK MEASUREMENT FUNCTION

Although work measurement is not considered one of the basic functions of Production Control, the Production Control system designer depends on it for the development of an effective Production Control system. The lack of any work measurement data or procedures for establishing standards is one of the greatest obstacles to implementing Production Control systems in nonmanufacturing activities. Even in manufacturing activities, the designer frequently finds that the weakest link in his system is the work measurement data upon which all scheduling and loading must be based.

Not only are scheduling and loading dependent upon work measurement, but the work authorization function also relies to a large extent upon a performance evaluation system based on accurate time standards to insure that only authorized work is performed.

When establishing a system or analyzing an existing system from the design viewpoint, the designer must investigate the adequacy of the work measurement data and the means by which they are obtained. For this reason, the systems designer should be familiar with the techniques of work measurement and the advantages and limitations of each. The designer should determine whether engineered standards are being developed and maintained by a regular time study department as the basis of performance evaluations or wage incentives. If such standards are in existence in the organization, they should be the source of the time data used in the scheduling and loading functions.

In some organizations, work is only measured after production has begun. In such cases, the designer should investigate the applicability of a standard data system for providing the necessary planning data and making the work measurement function more effective for short-run jobs.

If the work is extremely varied, or if the management does not wish to invest in an "engineered" standards program, the designer must do his best to develop the most reliable work measurement data possible through technical estimates or historical data. If technical estimates are to be used, the responsibility for estimating must be assigned to a person who is familiar with current shop practice. Also the degree of detail in which the job will be broken down must be established. This breakdown should be performed in the process planning function, and, as noted previously, the greater the refinement, the more accurate the time estimate. Historical data may be used if the repeat business is sufficiently consistent to permit the use of data generated during previous jobs. Sometimes when there are only slight modifications from job to job, historical data can be combined with technical estimates to produce very satisfactory data. When historical data are to be used, a procedure must be established for collecting time and production data so that the procedure can be applied in future planning. It is extremely important that the fixed and variable times be separated and that production time data do not contain "overhead" time.

It should be mentioned that all work can be measured. Extreme variability in the end product or the fact that a limited number of people are involved in producing the end product may make it impracticable to measure the work. In such cases, a staffing pattern should be established, and control should be maintained on a strictly budgetary basis.

The designer must recognize that to maintain adequate control over a work situation, there must be some means of measuring productive effort. This means is provided by the time standard.

Summary   While work measurement is not considered one of the basic functions of the Production Control System, the time data generated by this activity are essential in the performance of scheduling and loading, the subject of the next chapter.

Some of the most important uses of work measurement are:

1. Planning activities
2. Establishing management objectives
3. Comparing alternative methods
4. Paying wage incentives

The steps involved in the establishment of stop-watch time standards or "engineered" standards are as follows:

1. Defining standard
2. Establishing the work unit
3. Analyzing and recording the method
4. Timing the operation
5. Rating the performance of the person being timed
6. Applying allowances

Other forms of work measurement may also be used in certain situations. The value of stop-watch time study compared to other forms of work measurement should be understood by anyone requiring work measurement.

These forms are:

1. Predetermined time systems
2. Work sampling
3. Technical estimates

15

Loading and Scheduling

In the preceding chapters, we have considered the problems encountered in the planning and control of the individual elements of production—materials, tools, product design, and so forth. The importance of all of these elements cannot be overemphasized, but unless there is effective over-all coordination among these elements and the production facilities, it will be impossible to obtain the desired goods and services efficiently. The principles and methodologies of loading and scheduling are designed to obtain this necessary coordination.

Loading is defined as the assignment of work to a facility. The facility may be people, equipment, work groups, or an entire plant or group of plants. Loading these facilities would consist essentially of assigning work tasks *without* specifying when the work is to be done or the sequence in which it is to be done.

Scheduling is defined as the assignment of work to a facility and the specification of the time and the sequence in which the work is to be done. Scheduling is actually the time-phasing of loading. This refinement of loading generally leads to better coordination, but it must also be pointed out that it cannot always be employed.

There are two good general guides in establishing an effective loading and scheduling program:

1. The loading function should be done at the higher echelons of the organization, and the scheduling function should be done at the lower echelons.
2. Whenever an informal loading and scheduling system can effectively coordinate a facility, it should be used.

The rationale behind these general guides will be considered separately.

Since both the loading and the scheduling functions are responsible for coordinating the activities of multiple facilities, it is imperative that the functions be centralized if full effectiveness is to be realized. A rather trite, but nevertheless illustrative, example of the need for centralization of coordination can be drawn from the game of football. If a fully coordinated team action is to be attained, the assignment of work tasks to the individual team members must originate from one source: the quarterback. To decentralize the function of work assignment into any type of subgrouping will obviously lead to disorganization. The same thing is true in the production "team." In order to attain coordinated action from the individual production components, it is essential that a centralized planning group be established. Unfortunately, it is often felt by management that centralization of the loading and scheduling functions is either unnecessary or too difficult. The result is that an attempt is made to decentralize. At best, this approach will lead to mediocrity in the operation. The importance of centralized coordination does not imply that every detail of work assignment has to originate from a single source. On the contrary, the over-all coordination is obtained by establishing broad goals of accomplishments at the centralized point and delegating to a lower level of the organization the detailed planning necessary to realize the goal.

Quite often, and especially in small organizations, management will argue that it does not load or schedule its operations. This is not true. All organizations *must* load and schedule, but in the case of many smaller activities the functions are performed on an informal basis. For example, an office manager might schedule work in the office simply from his knowledge of what is expected of his group, assigning the work to the individuals in such a way as to assure its completion. A similar situation might arise in a small manufacturing shop. The shop manager may know the approximate number of hours of work in the shop and when he can expect the completion of various jobs. When new orders arrive, the shop manager will often estimate the completion date with remarkable accuracy. The functions of loading and scheduling in cases such as these are being performed with little or no formalized system. The coordinator of the facilities is so familiar with the situation that he can retain all of the pertinent information in his head. There is nothing wrong with informal systems such as these, *if* they are effectively coordinating the operation of the production facilities. The major trouble with informal systems is that as the facility's activity expands the coordinator will not, or does not want to recognize the need for introducing some type of formal system of loading and scheduling until the facility's effectiveness has fallen far below an acceptable level. Then, all too often, when the coordinator does recognize the need for a formal system, a complex one is introduced which would work well in a large organization but fails in the relatively small one.

Whenever an informal system can effectively coordinate a facility, it should be used, but some type of check on effectiveness should be periodically applied in order to assure the adequacy of the system.[1] As the need for the introduction of a formal system of loading and scheduling is apparent, the entire operation should be analyzed and formal systems introduced on an evolutionary rather than a revolutionary basis.

INFORMATION REQUIRED FOR LOADING AND SCHEDULING

Before loading and scheduling can be accomplished successfully, it is necessary to have complete information about the individual elements

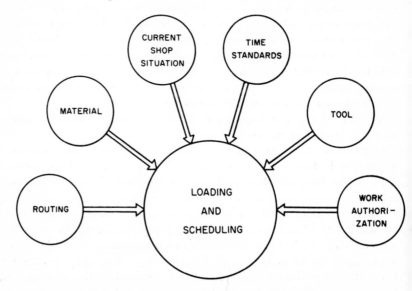

Fig. 15-1. Loading and scheduling information. In order to plan the work of a facility, information from many parts of the organization must be available.

---

[1] See Chapter 25, "Evaluating the Production Control System."

of production. In Figure 15-1, the type of information that is required is shown. The operation sequence for the item must be known in order to assign the job to the proper facilities in the correct succession. In addition to the operation sequence, the time standards are needed to determine the total time at each operation. The availability of tools and material must be taken into consideration. The proper work authorization must be available. Finally, the current load position of the production facilities must be known. This multiplicity of information can cause tremendous problems in communications. Many basically well conceived loading and scheduling programs fail because one or more of these areas of information is inadequate.

DEVELOPING A LOADING AND SCHEDULING SYSTEM

There are three major steps in establishing a loading and scheduling system:

1. The selection of the unit of measure for the work to be done
2. The selection of the facilities or operations to be controlled
3. The selection of the type of loading or scheduling technique to be used

Selection of the Unit of Measure   There are many different units that could conceivably be selected to measure work: manhours, machine-hours, dollars of sales or standard production units such as tons, gallons, gross, and so on. The final selection of the unit of measure will depend on the type of activity. For example, in an office it might be advisable to use the unit of measure of manhours. In a computer laboratory it would be better to use machine-hours. In a high volume output situation such as in a highly specialized office sending out invoices, or in the mass production of radios, the number of *units of production* would probably be sufficient. Often in the industries in which the output is sold by weight (candy factories, basic steel mills, flour mills, and the like), pounds or tons of production can be used for loading and scheduling purposes. The general rule in the selection of the measure of output is to select that unit which is "natural" to the activity. The preceding list should help in determining the type of measure that would be the best for a given activity.

Selection of the Facilities or Operations To Be Controlled   This is the second step in establishing a loading and scheduling system. While it may be true that a work stoppage at any operation could result in the complete disruption of the coordinating plans, it should *not* be assumed that every operation must be formally controlled. In many activities a large number of the facilities can be informally controlled simply because there is ample capacity at these points in the process for adequately servicing the remaining critical operations of the activity. The facilities selected to be formally controlled must offer sufficient savings in time and money to warrant the cost of formal control. The "non-bottleneck" facilities will still be controlled but in an informal manner. For example, in the activities of a maintenance crew, it might be advisable to schedule the activities of the individual workers but not the workers' tools and equipment. The potential saving in scheduling the workers' activity is sufficient to warrant the expense of formal control but is not sufficient in scheduling the tools and equipment. A good general rule to use in the selection of facilities to be formally controlled is: *Control only those facilities in which the cost of control can be clearly justified.* If at a later date it becomes evident that the initial selection was incorrect, additional facilities can be added to the formal system. Application of this general rule will result in the greatest return for the least cost.

Selection of the   After the unit of work measurement and the
Techniques To Be Used   facilities to be controlled have been de-
                             termined, the system of loading or scheduling must be selected. There are several basic systems available along with an infinite number of variations. The final selection, however, is again dependent upon the type of activity. In the following discussion the basic systems will be described separately and some of the variations will be pointed out.

TYPES OF LOADING AND SCHEDULING SYSTEMS

The four basic types of loading and scheduling to be considered are:

1. Master scheduling
2. Perpetual loading
3. Order scheduling
4. Loading by schedule period

Master scheduling   Although the control offered by the master sched-
                         uling technique is relatively crude, it often provides all that is necessary to coordinate an activity. As an illustration, we will consider a computer center consisting of a large electronic computer and its auxiliary equipment consisting of key punches, printers, collators, and so forth. From past experience it is known that the maximum number of hours that the equipment can be operated is twenty

Max. Production – 100 Hours
Min. Production –   8 Hours

WEEK 1	WEEK 2	WEEK 3	WEEK 4	WEEK 5	WEEK 6
35	18	8	10	12	
15	25	12	6		
25	30	15			
12		5			

Fig. 15-2. An example of a master schedule form. As jobs are accepted into the facility, the estimated hours of work are posted to the week in which the jobs are to be completed.

hours per day. The remaining portion of the day must be available for routine maintenance and repair work. Thus, in a five-day week the maximum number of computer-hours (machine-hours) that can be planned will be one hundred. There is also a minimum production. If the work load is not sufficiently large it would not normally pay even to start this expensive equipment. Assume that the minimum number of hours will be eight hours in one day. For control purposes it is necessary to plan the work into the computer center in such a way that the maximum is never exceeded in any one week and the work load never drops below the minimum. In addition, it must always be possible to determine accurately, within one week, the time of completion for any new job. In Figure 15-2, a diagram of the master schedule for the computer center is shown. In the heading at the top of the sheet both the maximum available and the minimum required computer-hours per week are recorded—in this case 100 and 8 hours respectively. The lower portion or body of the master schedule has six columns representing the next six-week period. As each job arrives at the center, the person maintaining the master schedule estimates the number of hours the job will require. This estimate may be based upon a detailed breakdown of the elements of the job and the application of established elemental time standards, or it may be simply an estimate based upon judgment. The former would probably be more accurate, but the latter might be adequate.

The numbers in the weekly columns of the master schedule in Figure 15-2 indicate the number of hours for the various jobs already assigned to the computer center. For the first week there are four jobs totaling 87 hours, for the second week there are three jobs totaling 73 hours, while for the third week there are four jobs totaling 40 hours of work assigned to the center.

Assume that a new job arrives at the center which requires completion during the second week. If the number of hours required by the job does not exceed 27 hours it can be assigned directly to the second week's work schedule. If the number of hours required by this job is greater than 27 hours but less than 40 hours, it can be assigned to both the first and the second weeks' work schedule and still be completed before the due date. If the job requires more than 40 hours, it cannot be promised for completion by the end of the second week unless some previously scheduled jobs can be postponed. Through the application of this simple posting system (master scheduling) an effective, although elementary, method for analyzing a work load can be obtained.

The usefulness of master scheduling can be extended by relatively minor modifications. Frequently companies are faced with the problem of maintaining enough flexibility to take care of rush orders, but they still wish to schedule the routine work. In such a situation it is possible to allocate a certain percentage of the total weekly capacity for the rush jobs and schedule the routine jobs into the remaining capacity. If the rush orders do not materialize, the next period's routine jobs can be

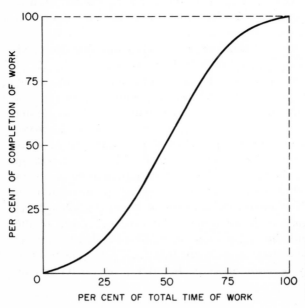

Fig. 15-3. Time pattern for long-run jobs. When a long-run job is scheduled into a facility, the actual pattern of completion must be taken into consideration.

completed early. If the rush orders do materialize, there will be sufficient capacity to take care of them. By analysis of the records of past rush jobs the proper percentage of capacity to be reserved can be quickly determined.

In cases where the jobs scheduled for a facility extend over two or three periods, it is usually unrealistic to assign an equal amount of work to each period. In most long-term jobs a definite pattern, similar to Figure 15-3, will be observed. At the start of a job the work accomplishment is relatively slow, then speeds up considerably, only to slow down appreciably toward the finish. The poor work performance at the beginning of the job is usually attributed to the many problems encountered in "getting the job going." Similarly, the poor work performance at the end of the job is caused by "clearing up the loose ends." This pattern is so common that it can be identified in almost every type of activity at one time or another. By taking this pattern into account when assigning a long-run job to a master schedule, a much more realistic time apportionment will be realized.

Master scheduling has several advantages which makes it applicable to many situations. Among these advantages are:

1. The system is simple and understandable.
2. The system can be easily kept current.
3. The system can be maintained by clerical workers.
4. The operation of the system can be quickly taught to clerical personnel.
5. The over-all cost of operation of this system is less than any other formal loading or scheduling device.

Master scheduling, however, has one major limitation. The information obtained from a master schedule does not provide detailed job information. It will provide the user with an over-all picture of the work in a facility but, in general, the information is too gross for detailed job planning. This is especially true in activities where the time values are based upon estimates. However, this disadvantage does not greatly reduce the usefulness of this technique.

There are many situations where master schedules are commonly used. Frequently in smaller operations the only formal loading and scheduling technique will be a master schedule. Larger organizations will use the master schedule as a basis for loading entire plants, with the more detailed scheduling being performed at the lower organizational levels. Master scheduling lends itself to the control of research and development laboratories. The best time estimates available for research and development projects are usually gross, and therefore a more refined load or schedule device is not warranted. The over-all planning in foundries, steel mills, repair shops, and many other production facilities often employs master schedules very effectively.

LOAD ANALYSIS SHEET						
ORDER NUMBER	DEPARTMENT A	DEPARTMENT B	DEPARTMENT C		Y	DEPARTMENT Z
2897	27	16				12
2898		18	22			
2907	51		12			18
2909	42	7	66			15
2917	19		14			
2927	7	12				10
2928		60	18			12
2929	16	47				12
2932	17		17			28
2936	38	53	28			7
2943	51					
23		16				8
3032	23	87	10			
3040	61	42				12
TOTAL HOURS STILL TO BE COMPLETED	1870	2589	1403			1376

Fig. 15-4. Load analysis sheet. As each order in the open file is analyzed, the number of hours still to be completed is recorded on the analysis sheet.

Perpetual loading   It is common practice to place a copy of every order going to the facility into an open order file. The copy remains in the file until the job is reported completed. During the time the job progresses through the work cycle, progress is reported by such media as completed job tickets, inspection tickets, material requisitions, rework tickets, and move tickets. As these various papers are received they are filed along with the original copy of the work order in the open order file. Since the original work order specifies both the facility assigned to do each operation and the estimated operation time, it is possible to determine the total amount of work remaining ahead of each facility simply by analyzing the information in the open order file. Typically, a clerk determines the status of each job by going through the entire open order file once a week and posting the load information on a form similar to Figure 15-4. Addition of the figures in each column provides the total estimated hours of work ahead of the various departments. This might appear to be a laborious and time consuming job, especially if there were 1000 or 2000 orders in the shop at one time, but in practice this is not the case, for in most plants not all of the jobs are worked on during any single week. This means that the load information on many of the orders would not change from the preceding week's report. By analyzing only the orders with new reports and duplicating the load information on the inactive jobs, the task of filling out the Load Analysis Sheet is greatly simplified.

Once the total number of hours of work ahead of each department has been determined, it is possible to convert it into a common denominator such as weeks of work. If Department A in Figure 15-4 had an estimated 800 manhours available each week, it would mean that there is 1870/800 or 2.3 weeks of work ahead of this department. Similarly, if Department B has an estimated 650 manhours available each week, it would have 2589/650 or 4.0 weeks of work ahead of it. By placing all of the departments' work on the same basis, it is possible to compare the relative loads of the entire facility.

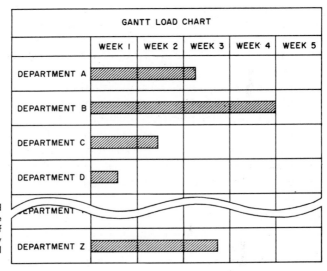

Fig. 15-5. Gantt-type load chart. In a load chart, the amount of work ahead of each facility is graphically illustrated by a horizontal bar.

It is often desirable to represent graphically the load figures to simplify the evaluation of the relative work loads. In Figure 15-5, a typical Gantt-type load chart is shown. Each row of the chart represents one of the departments. The horizontal distance represents time. In this particular example, each vertical column represents one day. The load of each department is shown by a horizontal bar. With a chart such as this, it is possible to determine quickly the over-all load condition in the shop. As an example, Department B has considerably more work ahead of it than any other department. This fact is clearly shown in the chart. On the other hand, Department D has an extremely small work load. Both departmental situations should be investigated since overload and underload are both undesirable. Sometimes it is advantageous to show

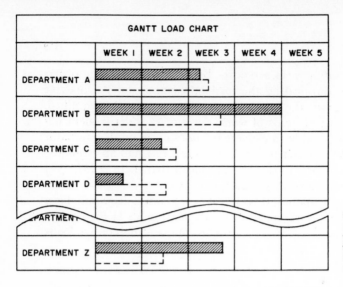

Fig. 15-6. Modified load chart. The solid line indicates the current load against the facility, whereas the dotted line shows the load situation the week before.

additional information on the Gantt load chart. In Figure 15-6, a typical modification is shown. The dark line, as before, indicates the actual work load ahead of each department. The dotted line indicates the amount of work ahead of the departments at the end of the preceding week. The Gantt chart now takes on new meaning. Management can not only evaluate the current situation, but can also quickly determine whether the situation is becoming better or worse.

Still other modifications may be used. Four or five different-colored bars may be used to show such things as last week's load, forecast work load, and so forth. As the amount of information shown on the load chart increases, the advantage of clear, easily understood representation of data diminishes. As a general rule it is wise to minimize the number of bars used.

Perpetual loading has the following advantages:

1. The system is simple and understandable.
2. The system can be maintained by clerical workers.
3. The time necessary to obtain the load information from the open order file is relatively short.
4. The operation of the system can be quickly taught to the clerical workers.
5. The over-all cost of operating the system is relatively small.
6. The numerous modifications of the system make it applicable to many situations.

The major disadvantage of the perpetual work loading technique is that the information is very gross, and can therefore be misleading at times. As an example, Figure 15-5 indicates that Department B has a four-week load ahead of it. The load chart does not tell *when* the work will occur. It is possible, although unlikely, that Department B is currently idle because the four-week work load may not occur for several more weeks.

The number of possible places for applying the perpetual work load technique is almost unlimited. It has been successfully applied in manufacturing shops, in offices, for maintenance crews, and for research and development activities, to name but a few examples. Wherever time estimates can be established for the work being assigned to a facility, the load type of analysis can be used. The number of applications is limited only by the imagination of the Production Control system designer.

Order scheduling   The most elaborate method of planning the activity of a facility is the order scheduling technique. For each individual operation assigned to a facility the set-up time, starting time, completion time and in many cases even the material move time are specified. Because of the great amount of detailed planning, it is theoretically possible to know the exact status of every job in the shop and predict the various completion times with a high degree of accuracy. Before order scheduling can be successfully applied to an activity, it is necessary to have a well developed program for setting accurate time standards, and an adequate system of communications.

An order scheduling Gantt chart is shown in Figure 15-7. Each horizontal row denotes a particular facility or group of facilities. As in all Gantt charts, the horizontal scale represents time intervals—hours, work days, or work weeks. The selection of the time interval depends upon the individual situation. For illustrative purposes, Figure 15-7 is established on a daily basis.

The following minimum information is required for each job to be scheduled into production by means of an order scheduling system:

1. The number of items to be produced.
2. The name of the machine or work center assigned to do the work of each operation.

Fig. 15-7. Order schedule chart. The bars represent the jobs to be done at various times by the different facilities.

ORDER SCHEDULE CHART								
DAY	1	2	3	4	5	6	7	8
FACILITY NO. 1	▨		▨	▨				
FACILITY NO. 2	▨	▨		▨			▨	
FACILITY NO. 3	▨			▨				
FACILITY NO. 4	▨	▨		▨				
FACILITY NO. 5	▨	▨			▨	▨		
FACILITY NO. 6	▨	▨						

3. The time necessary to set up for each operation.
4. The total production time necessary to do each operation.
5. The required date of completion for the job.

In addition it is necessary to know the exact status of every job already scheduled into the plant.

The scheduling of a new job can begin once the basic information is known. A considerable amount of judgment must be used when arranging the new operations on the schedule board. In general, schedulers will begin with the last operation placed on the board at the required completion date and then work backwards to the first operation. The reason for this is to keep as much current time available as possible in order to take care of rush jobs with a minimum amount of disruption. If the rush jobs do not materialize, the starting times of the jobs can be advanced as required. Order scheduling is basically a "cut-and-try" process in which the various operations are arranged and rearranged until an acceptable over-all plan has been devised. A few general rules of procedure can be developed which will aid in reducing the technique to a relatively simple clerical job.

Order scheduling has one major *advantage* over all other loading and scheduling techniques: *each individual phase of operation is planned in detail.* Theoretically, there should never be a question about what is expected of each production facility. It is possible to estimate the earliest possible completion date of any new order with assurance that the promised delivery can be met.

There are a number of restricting *disadvantages* to order scheduling:

1. It is the most costly method of planning the activity of a facility.
2. It is difficult to "track" the process when operation times are short.
3. It is difficult to maintain when there are many active orders.
4. It is difficult to maintain when there are many operations on each order.
5. Accurate time standards must be available.
6. It is difficult to maintain when frequent changes are necessary.
7. Communications within the organization must be highly developed.

Loading by schedule periods   The loading-by-schedule-period technique for planning the activity of a facility incorporates many desirable features. The Gantt chart for this procedure is shown in Figure 15-8. As usual, each horizontal row is devoted to an individual facility or group of similar facilities. The time interval, shown as horizontal distance, can be any appropriate value depending upon the situation; weeks were selected in the illustration. Each order to be scheduled must be prepared in the same manner as for the order scheduling technique. The job is broken down into its separate set-up and operation times. Instead of attempting to do the de-

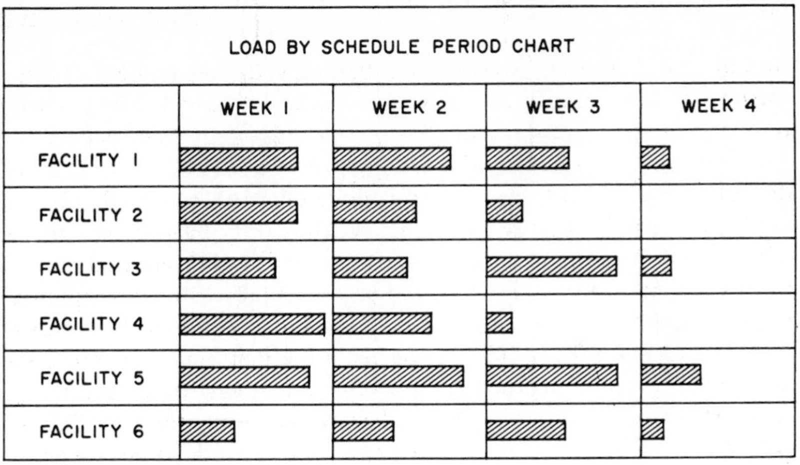

Fig. 15-8. Gantt chart for loading by schedule periods. The bars indicate the work ahead of the facilities for the different weeks.

tailed planning normally associated with order scheduling, each operation is assigned to its appropriate facility in a different time period. The first operation would be loaded into its facility in time period one. The second operation would be loaded into time period two, the third operation into time period three and so on. No attempt is made to specify when, within the time period, the work is supposed to be started. The loading simply indicates the time period in which the operation must be completed.

At the beginning of every time period, each facility is provided a list of work which it must complete during the period. It is the responsibility of the foreman in charge of the facility to determine the best sequence for doing the work. He must take into consideration similar set-ups,

Fig. 15-9. Weekly work list. Every facility being scheduled receives a list of the jobs on which work is needed.

WORK LIST FOR FACILITY NUMBER   1

WEEK ENDING   July 20

FACTORY ORDER NUMBER	OPERATION NUMBER	% COMPLETE	REMARKS	
REQUIRED WORK				
5876	20			80% OF ANTICIPATED CAPACITY
5893	40			
5975	10			
5981	20			
6003	30			
EXTRA WORK				
5890	30			30% OF ANTICIPATED CAPACITY
5985	20			
6000	20			

similar raw material, individual differences between operators, and so forth. In short, the foreman is responsible for the detailed scheduling but is required to stay within the framework of the load period.

Figure 15-9 shows a typical work list provided to a facility. It is usually desirable *not* to load the facility completely with required jobs during any time period. The work list shows that only 80 per cent of the facility's theoretical capacity has been used up by the required jobs. This means that it is very likely that the required jobs will be completed during the time period unless a 20 per cent cumulative error has been made. The remaining jobs shown in the list in Figure 15-9 are not required during this period, but they are available to the facility in order to avoid running out of work. The "excess" work is frequently a portion of the jobs required at the end of the next time period. The advantage of establishing the work load in this manner is that it provides a high degree of flexibility in the activity. Whenever a rush job must be loaded into the facility, it can be handled by using the "extra" 20 per cent capacity. By establishing the work load in this manner the rush job can be accomplished without the disruption of the regularly scheduled jobs.

A minimum of progress reporting has to be made under the "schedule period" system. At the end of each period, the foreman in charge of each facility can report the completion of the work simply by using the available space on the work list (Figure 15-9). The per cent completion is indicated in the appropriate column, and the list is returned to the central coordinating group. With this information the load plans can be developed for the subsequent period. The only other progress report-

Fig. 15-10. Assembly chart. The interrelationship between the individual parts and sub-assemblies going into a final assembly can be shown on a chart such as this. Notice that the time scale was reversed for ease of development.

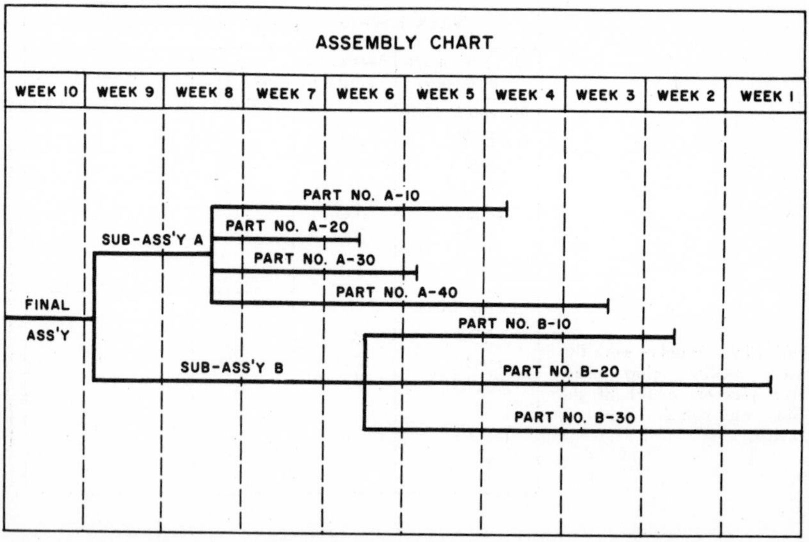

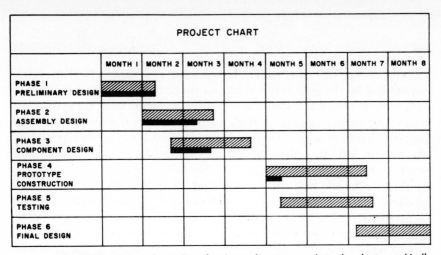

PROJECT CHART									
	MONTH 1	MONTH 2	MONTH 3	MONTH 4	MONTH 5	MONTH 6	MONTH 7	MONTH 8
PHASE 1 PRELIMINARY DESIGN	▨▨							
PHASE 2 ASSEMBLY DESIGN		▨▨						
PHASE 3 COMPONENT DESIGN			▨▨					
PHASE 4 PROTOTYPE CONSTRUCTION					▨▨			
PHASE 5 TESTING					▨▨			
PHASE 6 FINAL DESIGN								▨▨

Fig. 15-11. Project chart. For planning a long-term project, the chart graphically shows the interrelationship of all the phases. Progress can be shown by the heavy line under the bars.

ing required in this system is the notification of major production breakdowns or other major disruptions.

The salient *disadvantage* of the technique of loading by schedule periods is that *the total in-process time of the individual orders tends to be relatively long.* For the system to function correctly, it is necessary to avoid scheduling any more than one operation during each individual time period, since there is no assurance of the completion of the operation until the end of the period. If a typical order has ten operations in different work areas, it will require ten time periods for completion. The over-all in-process time can naturally be reduced by keeping the time periods as small as possible. If the individual operation times are short, a schedule period of one or two days can be used successfully. If the operation times are long, the schedule period can be correspondingly increased. However, it is desirable from the over-all planning point of view to keep the schedule periods as long as possible. This provides the maximum amount of flexibility to the foreman in establishing the most desirable job sequence and reduces the progress reporting to a minimum.

There are two other graphical aids available to the Production Control group which justify mentioning. They are the assembly chart and the project chart. Both of these charts are useful only in highly specialized situations and therefore will not be considered in detail. The reader should become familiar with the basic format of the charts by studying Figures 15-10 and 15-11.

DETERMINATION OF THE AVAILABLE STANDARD HOURS

One of the problems in loading and scheduling that has not been considered thus far is the establishment of the relative capacity of the facility. Frequently, the orders to be scheduled into a facility are accompanied by the established standard times for each operation. Unfortunately, there is not a direct relationship between these standard times on the job orders and the available clock times of the facility. If a job requires forty standard hours to be completed on a particular facility, there is no assurance that the job can be finished during a forty-hour work week. On the contrary, it is highly unlikely that the job would be finished by the end of the forty-hour week. The reason for this is that the standard times do not take into consideration all of the variables involved in production. For example, the standard time does not include allowances for absenteeism of the workers, unexpected production stoppages due to lack of material and orders, the efficiency of the production unit, and the like. These variables must be taken into consideration if effective scheduling is to be accomplished.

It is possible either to modify the standard hours of the individual job or to modify the facility's clock hours in order to get the two of them on a comparable basis. Since the former would require considerably more work than the latter, it is common practice to establish the number of standard hours of work available on each facility. The conversion of a facility's available clock time into available standard hours is illustrated in the following example:

Step 1. *Determination of the maximum gross hours:* The "work week" of an employee is defined as the total number of hours that the employee is supposed to work. This is frequently the standard 40-hour week less the morning and afternoon rest breaks. For scheduling purposes, it is necessary to consider the actual number of hours in the "work week"; therefore, if a facility has 20 people working a "40-hour" week, the maximum gross hours would be:

Maximum gross hours = 20 men × [40 hrs. − 10 (10 minute breaks)]
= 20 men × 38.3 hours
= 760 manhours

Step 2. *Determination of the available gross hours:* A certain percentage of the maximum gross hours will not be available because of absenteeism. In all likelihood this percentage will vary during the year. Around the major holidays and seasons such as New Year's, Fourth of July, Labor Day, Thanksgiving, Christmas and the first day of the fishing and hunting season the absentee rate generally rises. An investigation of the personnel or payroll records will reveal such conditions. Assume that the facility under consideration can expect a 5 per cent absentee rate, then the available gross hours will be:

Available gross hours = 760 manhours − [(0.05) (760 manhours)]
= 760 manhours − 38 manhours
= 722 manhours

Step 3. *Determination of the actual gross hours:* The 722 manhours are the total number of hours we can expect to have people available to work. People cannot and will not work every minute of the working day. At the beginning of the day a certain amount of time is lost getting started, and similarly at the end of the day, time is lost when the worker prepares to leave. The same type of situation occurs at the beginning and end of the coffee breaks and lunch period. In addition to this, the workers may be idle because of lack of tools, lack of work, lack of material, lack of instructions, and so forth. This means that there will not be a total of 722 clock hours of work available to the facility. The amount of actual lost time can be determined by a properly designed work sampling study which is described in Chapter 24. Assume that a study reveals that 8 per cent of the time is lost because of the above mentioned factors. This means that the actual gross hours would be:

Actual gross hours = 722 manhours − [(0.08) (722 manhours)]
= 722 manhours − 58 manhours
= 664 manhours

Step 4. *Determination of the available standard hours:* During the 664 hours that the people are physically working, their efficiency might be either under or over 100 per cent of standard. This also will have to be taken into account. The past efficiency of the facility can be determined by analyzing the records of the cost accounting section, Production Control section or the payroll section. In the example problem it will be assumed that the efficiency of the facility was determined to be 105 per cent. The available standard hours would therefore be:

Available standard hours = 1.05 (664 manhours)
= 697 manhours

The 697 standard hours available in the facility is exactly comparable to the standard hours indicated on the production order. If an order requires 697 production hours, the facility should be able to complete it during a one-week period. It is very important in scheduling to understand the difference between the original 760 maximum gross hours and the final 697 standard hours. The calculations are simple and the data can be obtained fairly easily, yielding an estimate of capacity which is accurate enough for practical purposes.

Summary   Loading and scheduling is one of the most important phases of any Production Control system. It is the technique by which the over-all coordination of the various functions and facilities within an organization is accomplished. Whenever a loading or scheduling system must be initiated into an organization, the various problems discussed in this chapter should be carefully studied. The proper selection of a specific technique along with its necessary modifications will result in an effective, trouble free system. Improper selection will result in a costly, ineffective system.

16

Mathematical Loading and Scheduling

Once a basic system for loading and scheduling has been installed and has proven to be adequate, it is possible to improve its effectiveness by the use of certain mathematical techniques. In this chapter three simple and understandable mathematical aids will be considered:

1. Statistical Load Control
2. Index Method of Scheduling
3. Linear Programming Method of Scheduling

In order to avoid an excessive amount of mathematical procedure, only the basic concept and procedure of each technique will be discussed in this chapter. The reader should refer to the appendix in the back of the book if he desires to develop a better understanding of the mechanism and the theory of the techniques.

STATISTICAL LOAD CONTROL

In most organizations it is desirable to have a perpetual backlog of work in order to have a sufficient lead

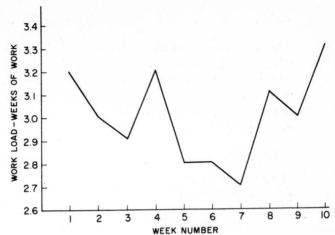

Fig. 16-1. Graphical load chart for a typical manufacturing department.

time to plan the work schedule for the facility. This backlog work load will vary from day to day and week to week; however, it is important that the load does not become excessively low or high. Whenever the work load changes significantly, it is then necessary to increase or decrease the facility's capacity. Graphical reports, similar to Figure 16-1, are a means of presenting to management significant changes in the work load. From this management can subjectively evaluate the load pattern and decide whether or not the production rate should be changed. The experienced manager can often ascertain what action should be taken, but there is a strong tendency to either over-control (make changes in the production capacity when none are warranted) or under-control (delay changes in the production capacity when they are needed). Statistical load control is a means of reducing the guess work in such decisions by indicating significant work load deviations.[1] Such significant deviations require corrective action.

Statistical load control is illustrated by the following example. An organization found, by analyzing the open order file for a ten-week period, that one of its facilities had the following load conditions:

Week Number	Load (weeks)
1	3.2
2	3.0
3	2.9
4	3.2
5	2.8
6	2.8
7	2.7
8	3.1
9	3.0
10	3.3

[1] See Appendix I for a discussion of the principles of control charting. Particular attention should be paid to the assumption of a normal population.

The management of the organization feels that the situation in the facility during this ten-week period is representative of the typical and desirable load condition. It now wants to have a device which is capable of indicating any significant shift in the average load size. The first step to be taken is to calculate $\overline{X}$ and σ for the system. This calculation is shown in Table 16-1.

TABLE 16-1

DETERMINATION OF $\overline{X}$ AND σ FOR
STATISTICAL LOAD CONTROL

X_i	$(X_i - \overline{X})$	$(X_i - \overline{X})^2$
$X_1 = 3.2$	3.2 — 3.0	0.04
$X_2 = 3.0$	3.0 — 3.0	0.00
$X_3 = 2.9$	2.9 — 3.0	0.01
$X_4 = 3.2$	3.2 — 3.0	0.04
$X_5 = 2.8$	2.8 — 3.0	0.04
$X_6 = 2.8$	2.8 — 3.0	0.04
$X_7 = 2.7$	2.7 — 3.0	0.09
$X_8 = 3.1$	3.1 — 3.0	0.01
$X_9 = 3.0$	3.0 — 3.0	0.00
$X_{10} = 3.3$	3.3 — 3.0	0.09
$\Sigma X_i = 30.0$	$\Sigma(X_i - \overline{X})^2 =$	0.36

$$\overline{X} = \frac{\Sigma X_i}{n} = \frac{30.0}{10} = 3.0 \text{ weeks}$$

$$\sigma = \sqrt{\frac{\Sigma(X_i - \overline{X})^2}{n}} = \sqrt{\frac{0.36}{10}} = 0.19 \text{ weeks}$$

Next, it is necesary to decide how reliable to make the testing device. As with any statistical control system, the establishment of the control limits is dependent upon the number of times we are willing to be in error. Whenever a point falls outside of the 3 σ limits, we will be correct approximately 99.7 per cent of the time in assuming a new load pattern exists. Since in this example a weekly evaluation is used, we would expect, by chance alone, to assume a new load pattern exists when it does not really exist, only once every six and one-half years. Obviously, this is somewhat overcautious, and therefore the 2 σ limits are usually used. The 2 σ limits would permit a chance error approximately once every five months. This is considered to be tolerable in most situations.

Figure 16-2 shows the graphical representation of the control system for this problem. The 2 σ limits are shown along with the $\overline{X}$ line. The sample week number is shown horizontally. As long as the future weekly load values remain within the control limits, it is assumed that business

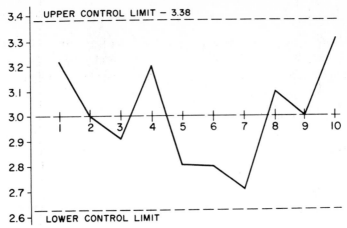

Fig. 16-2. Statistical load control Chart.

is going as usual. If the load value should fall outside of the control limits, it is assumed that the nature or pattern of the business activity has changed. This assumption will be correct approximately 95 times out of 100. A conservative manager may consider the first point which falls out of control as a warning signal only. When this signal occurs, he will take the necessary preliminary steps to increase or decrease production. If the load conditions of the following week are still out of control, the plans for the production-rate changes are put into action. This double check reduces the probability of error to less than 1 per cent and reflects a very conservative policy.

The preceding weekly load control charting technique is applicable to situations in which the seasonality of activity is relatively unimportant. In cases where seasonal fluctuations are relatively large in magnitude the procedure must be slightly modified. It has been found in practice that the standard deviation of the load changes with the average load size. The larger the average load size, the greater the variability of the loads. The smaller the average load size, the smaller the variability of the loads. The relationship of σ and $\overline{X}$ expressed as a percentage is called the coefficient of variation. Algebraically it is expressed as:

$$\text{Coefficient of variation} = K = \frac{\sigma}{\overline{X}}$$

Since for any given work load situation the coefficient of variation remains relatively constant regardless of the level of activity, the coefficient can be used to determine control limits in cases where seasonal fluctuations are large.

In the ten-week illustrative load problem discussed previously, the value of $\overline{X}$ was determined to be 3.0 weeks and σ turned out to be 0.19 weeks. If during the rush season this department had an average work load $(\overline{X})$ of 4.0 weeks it is logical that the σ of the system would change, but the coefficient of variation would remain constant. It is possible to determine the value of the coefficient from the original $\overline{X}$ and σ in the following manner:

$$K = \frac{\sigma}{\overline{X}} = \frac{0.19}{3.0} = 0.063$$

Using the same relationship for the rush season and solving for the new σ:

$$K = 0.063 = \frac{\sigma}{4.0}$$
$$\sigma = 4.0 \ (0.063) = 0.25 \text{ weeks}$$

During the rush season the control limits would be computed using the σ value of 0.25 weeks, while during the slower weeks the control limits of the load chart would be based on a σ value of 0.19 weeks.

An organization with the problem of seasonality would use two different control charts: one for the slow season and one for the busy season. This type of approach is shown graphically in Figure 16-3. By using two charts, the control is greatly improved, yet it does not require an undue amount of extra work.

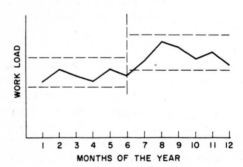

Fig. 16-3. Load control charts for seasonal activities.

INDEX METHOD OF SCHEDULING

The index method of scheduling is not entirely a mathematical procedure in that some subjective evaluation is required of the analysts. However, it is easy to learn, simple to apply, and provides considerable improvement over the conventional scheduling techniques. It can be successfully applied to situations in which a large number of work assignments must be scheduled to a large number of production facilities. The technique does not require expensive electronic computers or an excessive amount of clerical hours to solve a relatively large schedule. Finally, it can be used for problems for which no precise mathematical procedure has been developed.

Generally speaking, the standard method of assignment of work tasks
to different facilities is on a "first come, first served" basis. As each order
arrives in the scheduling department, it is assigned to the best machine
available at the time. In cases where the backlog of work at the machines
is very small, this type of scheduling is usually necessary. In most cases,
however, it is possible to accumulate at least a small group of jobs and
then assign them to the facilities in a better fashion by use of the index
method. This is illustrated in Table 16-2 where ten orders must be assigned to the four machines.

In this particular situation there are four similar machines, each capable of doing any of the ten jobs but with the differences noted in the
table. In the column for each machine are the hours required to do a
given job. Machine *A*, being a new piece of equipment, can do all of the
jobs faster than the rest of the machines. Machine *D* is too old and slow
for most of the production work and is generally considered standby
equipment for emergency use only. The orders are received by the
schedule clerk in a sequential basis. If he used the first come—first served
method of assignment, the jobs would be assigned to the machines as
noted by the circled numbers in Table 16-2. This method of scheduling
results in a total of 195 machine hours being required to complete the
ten jobs.

TABLE 16-2

JOB ASSIGNMENTS SCHEDULED BY CONVENTIONAL MEANS

Job Number	Machine A	Machine B	Machine C	Machine D	
1	10 hours	15 hours	14 hours	12 hours	
2	18 hours	20 hours	22 hours	27 hours	
3	17 hours	21 hours	25 hours	28 hours	
4	16 hours	17 hours	24 hours	25 hours	
5	12 hours	20 hours	17 hours	impossible	
6	16 hours	22 hours	19 hours	28 hours	
7	12 hours	impossible	18 hours	22 hours	
8	15 hours	18 hours	16 hours	20 hours	
9	25 hours	30 hours	27 hours	35 hours	
10	18 hours	25 hours	29 hours	32 hours	**Totals**
Total Hours Available	65 hours	65 hours	65 hours	65 hours	260
Total Hours Required	61 hours	48 hours	54 hours	32 hours	195

An inspection of Table 16-2 shows that a reduction in the total number of machine hours is possible. For example, by shifting job number
10 from Machine *D* to Machine *A* and job number 2 from *A* to *D*, a net

savings of five hours would be realized. Remember, however, that the schedule clerk did not have the information about job 10 when he scheduled job 2. Therefore, it isn't surprising that some obvious improvements are possible. Trying to make improvements in the job assignment by such haphazard inspection methods is not very efficient or effective when there is a large number of jobs and machines. The index method, on the other hand, provides a systematic procedure for developing a better schedule.

In order to apply the index method to scheduling, *it is first necessary to accumulate a backlog of jobs for the facility to be scheduled.* Generally, the larger the number of jobs the greater the savings that can be expected by such scheduling. The jobs are arranged in a table such as Table 16-3 showing the production times for the various job-machine combinations. The bottom of the columns shows the available machine times.

When the table is completed, the next step is to *assign all of the jobs to the best facility regardless of the machine time availability.* In this case the circled numbers under Machine *A* are the initial assignments. If it were possible to use this machine exclusively, it would require 159 hours to complete the ten orders. This is the optimal answer, although not feasible, because Machine *A* has only 65 working hours available. Since the initial first come—first served solution required 195 hours, it is known that the final index solution will fall somewhere between 195 and 159 hours.

TABLE 16-3

FIRST JOB ASSIGNMENT SCHEDULED BY THE INDEX METHOD

Job Number	Machine A	Machine B	Machine C	Machine D
1	10	15(1.50)	14(1.40)	12(1.20)
2	18	20(1.11)	22(1.22)	27(1.50)
3	17	21(1.25)	25(1.47)	28(1.65)
4	16	17(1.06)	24(1.50)	25(1.56)
5	12	20(1.67)	17(1.42)	impossible
6	16	22(1.38)	19(1.18)	28(1.75)
7	12	impossible	18(1.50)	22(1.83)
8	15	18(1.20)	16(1.06)	20(1.33)
9	25	30(1.20)	27(1.08)	35(1.40)
10	18	25(1.39)	29(1.62)	32(1.78)
Total Hours Available	65	65	65	65
Total Hours Required	159	0	0	0

The next step is to develop the index numbers for all of the non-optimal combinations. This will be used as a guide in rearranging the assignments in order to secure the best feasible schedule. This is done with each job separately by dividing the number of hours for the best machine time into the number of hours for the non-optimal machines. These index numbers are shown in Table 16-3 written in brackets next to the machine times. The following examples are given to illustrate the calculation of the index numbers.

Job Number 1
Machine A $= 10$ hours optimal

Machine B $= \dfrac{15}{10} = 1.50$

Machine C $= \dfrac{14}{10} = 1.40$

Machine D $= \dfrac{12}{10} = 1.20$

Job Number 2
Machine A $= 18$ hours optimal

Machine B $= \dfrac{20}{18} = 1.11$

Machine C $= \dfrac{22}{18} = 1.22$

Machine D $= \dfrac{27}{18} = 1.50$

Since the index numbers are actually ratios between the nonoptimal machine hours and the optimal machine-hours, it would seem logical that if a reassignment of jobs is necessary, the shift should be made to those machines with the smallest index number. Following this procedure would cause the least costly shifts to be made first followed by the more costly reassignments. This logic is the basis for the next step in the index method: *find the smallest index number in the table for an alternate process, remove the work task from the optimal machine, reassign it to the alternate process with the smallest index number.* As the orders are shifted from one machine to another, it is necessary to keep track of the reduction and increase in the various work loads in order to ascertain when the best feasible solution is reached. This can be conveniently done by using the data shown in Table 16-4.

In Table 16-4 the various shifts in job assignments and the changes in the total hours required on the individual machines are shown on the left side and the bottom of the table respectively. Job number 8, with an index number of 1.06 on Machine *C,* was reassigned first. This reduced the total hours required on Machine *A* by 15 and increased the hours on Machine *C* by 16. Next, job number 9, with an index number of 1.08 on Machine *C,* was moved from Machine *A* to Machine *C.* This reduced the hours on Machine *A* by 25 and increased the hours on

Machine *C* to 43. Job number 4 was moved next to Machine *B* and so forth until the sixth reassignment (job number 1) provided a feasible solution.

The final work assignment solution shown in Table 16-4 requires a total of 169 machine-hours to complete the same jobs which required 195 machine-hours under conventional scheduling. This is a saving of 24 hours, or approximately 10 per cent. This is the magnitude of the typical savings realized using the index method for scheduling purposes.

It should be noted that the final work assignment established by the index method is not necessarily the best possible solution. All this

TABLE 16-4

JOB ASSIGNMENT BY THE INDEX METHOD

	Job Number	Machine A	Machine B	Machine C	Machine D
(6th Shift)	1	10	15(1.50)	14(1.40)	12(1.20)
(4th Shift)	2	18	20(1.11)	22(1.22)	27(1.50)
	3	17	21(1.25)	25(1.47)	28(1.65)
(2nd Shift)	4	16	17(1.06)	24(1.50)	25(1.56)
	5	12	20(1.67)	17(1.42)	impossible
(5th Shift)	6	16	22(1.38)	19(1.18)	28(1.75)
	7	12	impossible	18(1.50)	22(1.83)
(1st Shift)	8	15	18(1.20)	16(1.06)	20(1.33)
(3rd Shift)	9	25	30(1.20)	27(1.08)	35(1.40)
	10	18	25(1.39)	29(1.62)	32(1.78)
	Total Hours Available	65	65	65	65
	Total Hours Required	159	0	0	0
	1st Shift	—15	0	+16	0
	2nd Shift	144 / —16	0 / +17	16 / 0	0 / 0
	3rd Shift	128 / —25	17 / 0	16 / +27	0 / 0
	4th Shift	103 / —18	17 / +20	43 / 0	0 / 0
	5th Shift	85 / —16	37 / 0	43 / +19	0 / 0
	6th Shift	69 / —10	37 / 0	62 / 0	0 / +12
	Final Total Hours Required	59	37	62	12

technique is capable of doing is to provide a systematic procedure for arranging the work tasks in a better fashion than is normally achieved with the first come—first served basis. It is entirely possible, in fact highly likely in the case of a large number of jobs and facilities, that a better solution might be obtained if a more highly developed mathematical procedure were available. Unfortunately, more refined methods are not readily available, and the index method must be relied upon in many cases.

LINEAR PROGRAMMING METHOD OF SCHEDULING

Unfortunately, it is commonly believed that linear programming is too difficult to be used by the average Production Control group. This is certainly not true; in Appendix III a description of the mechanics of the Modi method of linear programming is provided. This method can be successfully used in establishing long-range scheduling plans without resorting to the use of electronic computers or highly trained mathematicians. In conjunction with the following example of scheduling analysis, it will provide the reader with a highly useful and workable tool. To illustrate the technique, a relatively simple example will be considered, but the exact procedure can be applied to a much more complex problem.[1]

A manufacturer is faced with a highly fluctuating sales demand for his product. The sales forecast for the first eight months of next year is as follows:

Month:	Anticipated Sales:
January	200 units
February	150 units
March	250 units
April	250 units
May	350 units
June	300 units
July	150 units
August	200 units

There are two different methods commonly used in situations such as this. The first, illustrated in Figure 16-4a, is to fluctuate the manufacturing rate in the same pattern as the sales demand. This would assure the manufacturer of keeping up with sales while maintaining a minimum inventory. However, the maximum production capacity would have to

---

[1] The numbers used in this example are not intended to be realistic. They were selected so that only whole numbers would appear in the problem. In an actual situation, the addition and subtraction would be complicated by the decimals which would have to be used, but the basic procedure would be *exactly* the same.

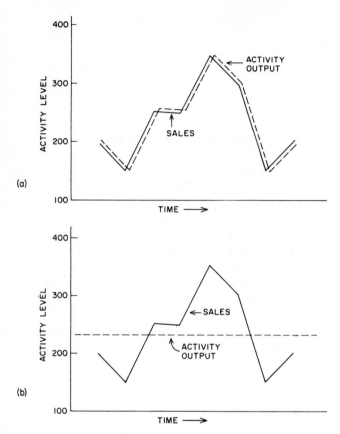

(a)

(b)

Fig. 16-4. Methods of producing for variable sales.

be adequate to meet the highest monthly demand. This naturally leads to manufacturing below capacity most of the year in addition to creating the problem of expansion and contraction of the labor supply.

The second common approach to this problem is to manufacture at a constant rate and allow the finished goods inventory to take care of the fluctuation in demands, as illustrated in Figure 16-4b. This type of policy will stabilize production, but it requires a considerable build-up of inventory during the slow periods. In many cases the cost of storage is prohibitively high, making this policy impracticable.

Applying linear programming to the same problem will result in the best compromise policy, minimizing the total of the cost of fluctuating the production level and the cost of inventory storage. However, it is first necessary to have additional information about the cost of manufacturing and storing the product. The following simplified information will be assumed for illustrative purposes:

Month:	Number of Work Days:	Straight-Time Production Capacity:	Overtime Production Capacity:
January	22	200 units	44 units
February	19	190 units	38 units
March	21	210 units	42 units
April	20	200 units	40 units
May	22	220 units	44 units
June	21	210 units	42 units
July	17	170 units	34 units
August	22	220 units	44 units

Current inventory level as of January 1 $= 100$ units
Required inventory level on August 31 $= 125$ units
Straight time production cost $\qquad = \$10$ per unit
Overtime production cost $\qquad = \$14$ per unit
Monthly storage cost:
    Items produced on straight time $\qquad = 10$ per cent per unit per month
    Items produced on overtime $\qquad = 15$ per cent per unit per month

These data must be placed in a Modi table as shown in Figure 16-5. The columns represent the straight-time and overtime capacities for the various months. The first eight rows represent the monthly sales demands. It should be noted that January has a sales estimate of 200 units of which 100 units will be supplied out of the current inventory.

Fig. 16-5. The initial modi table. If this initial solution is actually used, the total cost of manufacturing and storing the items will be $21,316.00.

The remaining 100 units must be manufactured as indicated at the end of the January row. The final inventory of 125 units is shown as a separate row. Actually these 125 units could be added to the August demand and the row for the final inventory eliminated if it is desired. Since the total production capacity for the next eight months exceeds the sales demand, an extra "slack" row must be introduced. This row is simply a mechanism to meet the mathematical restriction of the Modi method that the supply must equal the demand. The values appearing in this row in the final table will have little practical significance. The small numbers in the upper left hand corner of each square represent the total manufacturing and storage costs. For example, the January straight-time column costs are determined as follows:

January sales — $10 production cost   = $10
February sales — $10 production cost plus $1 storage charge = $11
March sales   — $10 production cost plus $2 storage charge = $12
April sales   — $10 production cost plus $3 storage charge = $13

and so forth.

The initial Modi table is shown in Figure 16-5. The squares above the heavy line are impossible situations, e.g., the February production cannot be allocated to January sales. If the initial solution to the problem were adopted by this manufacturer, the total cost of production and storage would be $21,316 for the eight-month period. The actual steps of the Modi method are not shown because of space limitations. Using Appendix III as a guide, the reader can calculate the optimal solution shown in Figure 16-6.

The total cost of producing and storing the items using the plan shown in Figure 16-6 is $20,434. This is a saving of approximately 5 per cent, or $892. It could be argued that through observation alone, the cost of the initial solution could be reduced. This is true only because the example was deliberately simplified for illustration purposes. The trial-and-error method can reduce the initial cost, but it would never be known when the *best* answer had been obtained. The Modi solution shown in Figure 16-6 is *optimal*—there is *no* better solution.

The Modi method shows [2] that there are many other plans which will result in the same cost. One of these optimal solutions is shown in Figure 16-7. The total cost of any optimal solution is exactly the same—$20,434. In this case the alternate has little practical significance since it simply earmarks the same amount of inventory for different months in the future. The actual monthly inventory levels in both solutions are exactly the same. In actual application of this technique, the alternate optimal solutions may have real meaning and should be considered.

_____

[2] Whenever the $R_i + K_j = C_{ij}$ for the cells not in the optimal solution, an alternate optimal solution exists. The reader should refer to Appendix III for a more complete explanation of this property of the Modi table.

Fig. 16-6.

	JAN ST	JAN OT	FEB ST	FEB OT	MAR ST	MAR OT	APL ST	APL OT	MAY ST	MAY OT	JUNE ST	JUNE OT	JULY ST	JULY OT	AUG ST	AUG OT	DEMAND
JAN	100																100
FEB			141	19													150
MAR					208	42											250
APL			8		2		200	40									250
MAY	45		41						220	44							350
JUNE	48										210	42					300
JULY													116	34			150
AUG															200		200
FINAL INVENTORY	7													54	20	44	125
SLACK		44		29													73
PRODUCTION CAPACITY	200	44	190	38	210	42	200	40	220	44	210	42	170	34	220	44	1948

Fig. 16-6. Optimal solution to the modi problem. If this plan is used, the total cost of manufacturing and storing the items will be $20,434.00.

Fig. 16-7.

	JAN ST	JAN OT	FEB ST	FEB OT	MAR ST	MAR OT	APL ST	APL OT	MAY ST	MAY OT	JUNE ST	JUNE OT	JULY ST	JULY OT	AUG ST	AUG OT	DEMAND
	100																100
FEB	93		48	9													150
MAR					208	42											250
APL			8		2		200	40									250
MAY			134						172	44							350
JUNE									48		210	42					300
JULY													116	34			150
AUG															200		200
FINAL INVENTORY	7													54	20	44	125
SLACK		44		29													73
PRODUCTION CAPACITY	200	44	190	38	210	42	200	40	220	44	210	42	170	34	220	44	1948

Fig. 16-7. Alternate optimal solution to the modi problem. This plan will cost $20,434.00—exactly the same as the solution shown in Figure 16-6.

A word of warning should be given at this point. The mathematical techniques introduced in this chapter are relatively simple and easy to apply. They are useful in many places and should be considered a "tool" just as the schedule board is a "tool." But, simply because the technique is quantitative in nature and mathematically correct is no justification for accepting the final answers without question. All of the numbers used in these techniques are based upon estimates, and therefore the final answer is no better than the original estimates. Judgment must still be used in interpreting the results.

Summary   Once the basic loading and scheduling procedure of the production control system has been established, consideration should be made of applying some of the mathematical aids discussed in this chapter. In many cases, the use of quantitative procedures of loading and scheduling can appreciably reduce the time necessary for analysis and at the same time improve the ultimate program.

The techniques discussed in this chapter have all been successfully applied by various companies with considerable savings. In most cases the actual details can be delegated to clerical workers after the basic procedure has been set up. As was mentioned previously, the numerical results must be carefully scrutinized since there are many intangible variables which the techniques do not take into account. This limitation should not be considered insurmountable. The techniques must be considered as aids, not panaceas.

17

Commercial Loading and Scheduling Devices

It is important to be able to recognize and understand the basic problems encountered in loading and scheduling (as we have done in the preceding chapters) *before* becoming concerned with the available commercial loading and scheduling devices. Once the system has been designed, then the commercial device which best fills the needs of the system can be selected. Many times loading and scheduling systems are designed around a specific commercial device with little thought given to the requirements of the system. This practice can only lead to marginal effectiveness. The fact cannot be overemphasized that the system must be designed *first* and the commercial aid selected *second*. This is the only way that the most effective loading and scheduling system can be established.

As the activity of a small organization increases, a point is reached, sooner or later, where the relatively simple informal loading and scheduling systems are no longer adequate to control the facilities. Some type of formal control system is needed. It is a common practice of the managements of these concerns to look to outside aid, in the form of consultants, for revamping their control systems. Unfortunately, the consultants selected are frequently the representatives of one of the manufacturers of commercial scheduling devices since their services are available at no charge to the customer. Regardless of how well qualified these consultants may be, their principal function is to sell some specific control system. Because of this, the client is literally "sold" the basic "hardware" for a production control system even if some competitive commercial device is more desirable. The ultimate outcome of this type of approach to system design is generally a closet full of hardware items and an office full of disgruntled people. It must be kept in mind that the commercial devices are aids, and not systems. The commercial device should be incorporated into the system; the system should not be built around the device.

There are many different types of commercial loading and scheduling devices available to the designer of a production control system. Each device has both advantages and disadvantages which must be carefully weighed before the final selection is made. The best device in the world will fail if it is applied to a situation for which it was not designed. In the following discussion the basic operation of the devices is considered, and a few of the typical applications are indicated. When using these devices in a production control system, some operational modifications are usually required. Although it's impossible to discuss all these modifications, they become apparent when the device is used in an actual system.

Sched-U-Graph   One of the oldest and best known commercial scheduling devices is the Remington Rand Sched-U-Graph. Basically, the Sched-U-Graph board is an enlarged Kardex file made up of a series of heavy kraft paper flaps with clear plastic margins at the bottom (see Figure 17-1). The flaps are attached to the boards so that they overlap each other in the same manner as the shingles on a house, and only their plastic margins are visible. Since the flaps are hinged at the top, any individual flap can be viewed simply by lifting the flaps covering it. A wide selection of board sizes is available.

To illustrate the use of this equipment, the steps for setting up and operating a simple schedule board will be considered.

> Step 1. Assuming an organization is interested in the order scheduling of 50 work facilities, a Sched-U-Graph board with 50 flaps will be selected. Each facility is assigned to a specific flap, and cards with the equipment information are inserted into the plastic margins on the left hand side of the board. (See Figure 17-2C.)
>
> Step 2. As with any Gantt type chart, the horizontal distance on the Sched-U-Graph depicts time. The long strips of paper, shown in

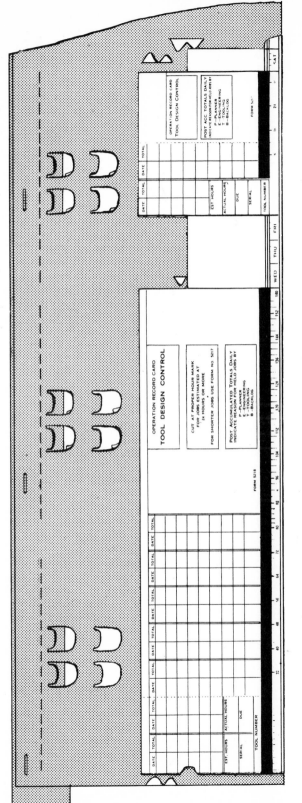

Fig. 17-1. The Sched-U-Graph Flap with load card inserted in the plastic margin.
(Reprinted by permission of the Remington Rand Division of Sperry Rand Corp., New York.)

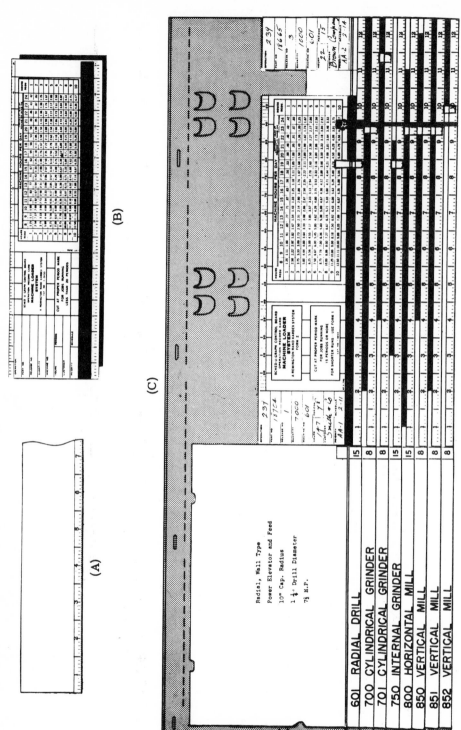

Fig. 17-2. The basic parts of the Sched-U-Graph equipment.

Figure 17-2A, with the appropriate time scale printed along the bottom edge, are inserted across the board in each plastic margin.

Step 3. The special load cards, shown in Figure 17-2B are used to represent the operations required on each order. The time scale along the bottom edge of these cards is the same as the time strips used in Step 2. If an operation requires three time units for completion, the load card is cut to the length of three time units. The colored bar at the bottom of the card now represents the time necessary to complete the operation.

Step 4. The load cards are inserted in the appropriate plastic margins at the time location at which the operation is to be performed. The color bars on the load cards are always visible as shown in Figure 17-2C. These bars take the place of the heavy lines drawn on the paper Gantt charts.

Step 5. Special colored signals can be moved along the plastic margins to indicate the progress of the work. The heavy vertical line shown at the beginning of the tenth day in Figure 17-2C represents the "today" line. The status of the work on each machine can be determined by observing the position of the progress signals in relationship to the "today" line.

Produc-Trol Board   Another very common type of mechanical Gantt chart is the Wassell Company's Produc-Trol board. There are three major parts of the Produc-Trol as shown schematically in Figure 17-3: the index record panel, the tape pegs and the peg board. The description and purpose of each part will be considered separately.

Fig. 17-3.   Schematic sketch of the Produc-Trol board.

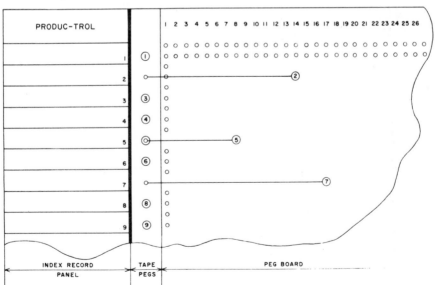

1. *Index Record Panel.* The index record panel is made up of a series of heavy kraft paper flaps with clear plastic margins at the bottoms. The construction of these flaps or pockets is very similar to the ones on the Sched-U-Graph. The pockets in the panel are numbered consecutively from top to bottom. Each machine or facility to be incorporated into the system is assigned a specific pocket, and an appropriate title card is inserted into the margin.

2. *Tape Pegs.* For every pocket in the index record panel there is a corresponding numbered tape peg. The tape peg, as its name implies, is a special peg to which is attached a tape or string. The string threads into a hole between the index record panel and the peg board section and is attached to a hidden mechanism which keeps it under tension at all times.

3. *Peg Board.* For every pocket in the index record panel there are two rows of holes in the peg board. The upper row of holes is used for inserting colored signal pegs to indicate special information. The lower row of holes is used for the tape peg.

To illustrate the use of the Produc-Trol board, the steps for setting up and operating a simple work-load chart are described in the following paragraphs:

Step 1. There is a large selection of different sizes of boards available to the user of Produc-Trol. The first step is to determine the right board size for the number of facilities included on the load chart.

Step 2. Each machine or facility incorporated on the load chart is assigned to a specific pocket in the index record panel. An appropriate title card is inserted into the pocket so that the card's bottom edge is visible through the plastic margin at all times. The machine identification is printed on the bottom edge of the card, while the detailed information about the machine is written on the covered portion of the card.

Fig. 17-4. The Produc-Trol board used as a load chart. Two columns of holes represent one work day. The position of the tape peg indicates the load on the facility, whereas the square peg represents the load one week before.

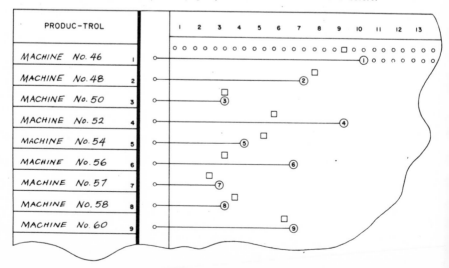

Step 3. The horizontal distance between the peg holes represents the time interval. As an example, if the selected Produc-Trol board has 200 columns of peg holes, the distance between adjacent columns might represent four hours of work. It would, therefore, be possible to show up to 100 days of load against a facility.

Step 4. The actual work load assigned to a facility can be shown with the tape peg. The white string attached to the peg would graphically show the load against the facility. In Figure 17-4 the relative loads against the nine facilities can be quickly seen.

Step 5. By inserting special colored pegs in the upper row of holes, the historical load pattern against the various facilities can be shown. As an example, last week's load might be denoted by a yellow peg, while the month's average load might be indicated with a red peg.

Boardmaster   Another commercial aid for the loading and scheduling of facilities is the Boardmaster, made by the Graphic Systems Company. This device, shown in Figure 17-5, is made up of a large number of paper cards mounted in special grooves on an aluminum board. When all of the paper cards are mounted on the board, they form essentially a large sheet of paper. The vertical and horizontal spaces between the cards rule the board off into rows and columns. If a heavy line were drawn across the face of each card in a single row, it would appear, from a few feet away, to be one continuous line across a solid piece of scored paper. If it were desired to reduce the length of the apparent line, plain cards could be substituted for the marked cards at one end of the row. This operation of changing cards, shown in Figure 17-6, becomes very rapid with experience.

Fig. 17-5. Boardmaster's basic panel.

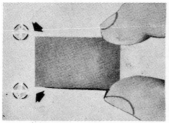

1. INSERT LEFT END

Insert both corners of left side in grooves at same time. Do not try to insert corners separately.

2. INSERT RIGHT END

Hold left end of card in groove opening. Bend right side of card. Allow card to enter right openings

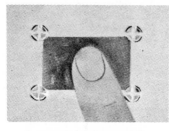

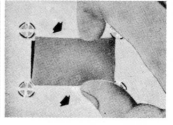

3. TAP TO SETTLE

Light tap of finger helps to settle card in grooves. Card is now anchored at 4 corners.

4. TO PULL OUT

Grasp card with tips of fingers at center buckle. Pull away from board. Card pops out of grooves.

Fig. 17-6.   Paper cards on a Boardmaster Panel.

To set up the Boardmaster as a load chart for a group of facilities, the following steps would be taken:

Step 1. Each facility to be included on the load chart would be assigned to a specific row. The facility's identification could be type-written or handwritten on the small cards, and then the cards could be inserted into the grooves as row headings.

Step 2. As with any Gantt-type chart, the horizontal distance would denote time. Once the time scales have been selected, intervals could be written on the cards and used as column headings.

Step 3. The Graphic Systems Company sells special "Gantt Chart Cards" with heavy black lines printed on them. The bar depicting the load on a facility could be "drawn" on the board by using enough of the special cards to get the length of the bar desired. The remaining portion of the row would then be filled in with blank cards.

After all of the cards had been placed on the Boardmaster, it would look like a regular load chart drawn on paper. When a revision in the chart was needed, it would not be necessary to replace all of the cards. Only a few cards at the right-hand end would have to be changed. The changes would be limited to the substitution of the "Gantt Chart Cards" for blank cards or vice-versa in order to modify the bar lengths to reflect the new load conditions.

A. PLASTIC TUBE

B. OVERRIDING SLIDE

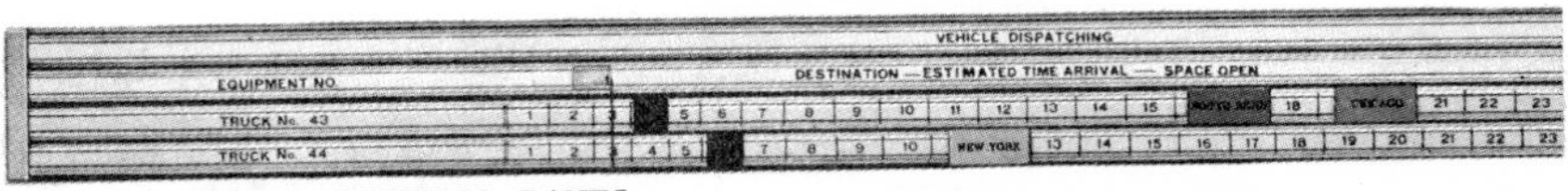

C. VISUAL CONTROL PANEL

Fig. 17-7. Acme Visual Control Panel equipment. (Reprinted from "Visual Control Panels" [No. 1140], by permission of the Acme Visible Records, Inc., Crozet, Va.)

Visual Control Panels A relatively new commercial device for loading and scheduling purposes has been introduced by the Acme Visible Records Company under the name Visual Control Panels. These boards are made up of a number of plastic channels called *tubes,* shown in Figure 17-7A, which are held in a special metal panel. Because of the flexibility of the plastic, the tubes can be removed and reinserted into the panel very rapidly. Special strips of paper, which can be placed inside the tubes, are available with different time intervals printed on them. The left hand portions of these strips are blank so the user of the panels can write in any desired facility identification. When the tubes, with the inserted time strips, are placed into the panel, they form a conventional Gantt type chart.

By using the overriding signals, shown in Figure 17-7B, on top of the plastic tubes, the various work to be completed by the facilities can be scheduled. Changes in plans can be made very quickly by sliding the overriding signal along the tube to the new schedule date. The visual Control Panel used for truck scheduling is shown in Figure 17-7C as an illustration of an operating board.

Summary In this chapter four of the common commercial loading and scheduling devices were considered. Only the basic operation of the equipment was discussed to enable the reader to understand what "hardware" items are available. There are many variations in the use of all of these devices, and these are covered in descriptions available from the manufacturers. It must be remembered, however, that the needs of the Production Control system should be established *before* trying to select the commercial equipment to be used. The *equipment is an aid, not a system.*

18

Dispatching

In the list of functions in Table 3-1, page 18, dispatching is shown as the action phase of the Production Control system. After the prior-planning and action planning phases have been completed, it is necessary to put the plans into action. This is the function of dispatching.

Dispatching is defined as the physical release of the work authorization to the operating facility in accordance with a previously established plan of activity developed by the scheduling function. This generally accepted definition of dispatching does not convey the importance or the many congruent duties of the individual doing the dispatching. In addition to releasing the work authorization, the dispatcher is commonly expected to:

1. Authorize the stores department to release the required raw material.
2. Authorize the tool department to release the required tools, inspection devices, and so on.
3. Provide the worker with the appropriate time tickets.
4. Notify the inspection department of necessary inspections.
5. Authorize the movement of material from one facility to another.

It can be seen that the dispatcher's job is an important key to effective operation of the facility; without close control of it there is little assurance that the work will be accomplished in accordance with the established plan. Many organizations fail to realize the importance of controlling the dispatching function, and when they experience difficulty in securing the anticipated output from the facility, they unjustly criticize the planning phases of the production control system.

One of the major reasons that control of the dispatching function is difficult to establish is that it is the point at which the plans developed by the production control staff must be introduced into the line portion of the organization. This brings up the question of line and staff prerogatives. Although this book is primarily interested in the principles of a production control system, it is necessary at this point to consider some of the problems in the organizational design. It would be foolish for anyone to attempt to design a dispatching system without a knowledge of the problems which will inevitably arise.

In the organization chart, shown in Figure 18-1, the line portion of the organization is indicated with dark lines while the staff portion is shown with lighter lines. The production control section originates the schedule plans for the organization and is faced with the problem of where the schedule should be introduced into the line. In the case of the company shown in Figure 18-1, there are four levels at which the schedule could be introduced:

1. Plant Manager Level
2. Plant Superintendent Level
3. Foreman Level
4. Operator Level

These levels are indicated by the arrows on the right hand side. Since

Fig. 18-1. Possible organizational levels for introducing the work schedule into the line.

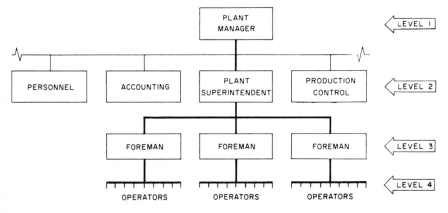

there are both advantages and disadvantages in choosing any one of these levels, each choice will be considered separately.

Plant Manager Level   The Production Control section is a staff reporting to the Plant Manager. It would be consistent with good organization principles to have the schedules submitted directly to him for review and approval. Once the Plant Manager approves the schedule and forwards it to the Plant Superintendent, it would become a mandatory directive to the line portion of the organization and the question of line prerogatives would never arise. The advantages of introducing the schedule at the Plant Manager level are:

1. There is no violation of line prerogatives.
2. The Manager is kept informed of the expected activity in the plant.
3. The function of dispatching is done by the line portion of the organization.

The disadvantages of introducing the schedule at the Plant Manager level are:

1. The Manager should not be concerned with the detailed planning and probably does not have the time or the inclination actually to review the schedule.
2. Considerable time is spent as the schedule is "passed down the line." This would lead to inflexibility in the control system.

When an organization establishes the policy of submitting the detailed schedule to the Plant Manager level, it is only a matter of time before the Manager becomes aware that his approval of the schedule serves no real purpose. He cannot devote much time to analyzing the information and soon his approval becomes automatic. In order to avoid wasting the Manager's time, a company policy can be established which will automatically give the necessary approval to all schedules developed by the Production Control group. The schedules could now be submitted directly to the Plant Superintendent, bypassing the Manager Level.

Plant Superintendent Level   Submitting the schedules to the Plant Superintendent has the same general advantages and disadvantages as submitting them to the Plant Manager. The Plant Superintendent's responsibilities normally do not include the analysis of the detailed schedule. He is responsible to make sure that the operating organization performs the scheduled work at an acceptable level of performance. He cannot be expected to determine if the scheduled work best fulfills the need of the overall plant operation. If the schedule is introduced into the line at this level, approval of the Plant Superintendent soon becomes automatic and, therefore, meaningless.

Foreman Level   Many organizations, realizing that the higher level
                      supervisor should not be concerned with the detail
planning found in work schedule, have the production control group
submit the schedule directly to the foreman. In theory, the foreman has
the prerogative of accepting or rejecting the staff's "suggested" work plans.
In practice, the foreman usually welcomes the staff planning since it
reduces the responsibility of his job. As long as he follows the schedule
provided by the production control group he cannot be criticized for not
doing the work in the best sequence.

The advantages of introducing the schedule at the foreman level are:

1. Only the foreman level receives the detailed schedule.
2. Higher-level supervision is not needlessly burdened with details.
3. The function of dispatching is carried out by the foreman—a traditionally acceptable practice.
4. The time interval between the establishment of the schedule and its
   execution is less than when several levels of supervision have to review and approve each schedule.

The main disadvantage of introducing the schedule at the foreman
level is that the foreman must perform all of the clerical work associated
with the dispatching function. (See page 192.)

By introducing the schedule into the line at the foreman level the
actual dispatching function, the physical release of the work authorization, is performed by the foreman. The assignment of work duties
traditionally is the responsibility of an immediate supervisor. This concept is so deeply ingrained in many organizations that it would be
impossible to design and operate a production control system in which
the dispatching is done by anyone other than the foreman. Whenever this
situation is present, there is very little choice for the designer of the
Production Control system. The function of dispatching must be done
by the foreman.

Operator Level   Many foremen continually complain about the amount
                       of paper work that is expected of them. The typical
complaint is, "How do they expect me to run my department if I have to
spend all day making out paper forms?" This question is well taken because the direct supervision of the operators is a full time job, yet it is
essential that *all* phases of the dispatching function be adequately covered.
This problem has led many organizations into making dispatching a staff
function. Under this type of system a staff member from Production
Control, and not the foreman, physically releases the work authorization.
This practice theoretically violates a foreman's prerogative, but it is frequently the only practical solution to a difficult problem.

Under the staff dispatching system, a member of the central Production

Control group, usually titled the dispatcher, is physically located within the shop area. The number of dispatchers necessary to service the entire shop will vary from organization to organization. Sometimes one dispatcher, working a portion of the day in each department, can take care of all of the dispatching, while in other organizations it is necessary to assign a full time dispatcher to each separate department. The number of required dispatchers depends upon the quantity of orders being processed, the number of employees involved and so forth. The departmental dispatchers receive the schedule of the jobs along with the authorization to begin manufacturing in the form of production orders. After he has established the tentative work assignment for each member of the department, the dispatcher will frequently consult with the departmental foreman to obtain his suggestions and approval. This informal discussion about the tentative work assignment may lead to improvements in the plan, but, more important, it allows the foreman to be aware of the planned work assignment. Although the dispatcher is not a staff member reporting to the foreman in the strictest organizational sense, he is a staff aid and must have the approval and support of the foreman. This informal association must be amiable or else the effectiveness of the dispatcher will be seriously hampered.

The advantages of introducing the schedule at the operator level by the use of staff dispatchers are:

1. The foreman is relieved of clerical duties.
2. The dispatching function is performed by specialists.
3. Closer coordination is possible by having *all* functions of production control under staff control.

The disadvantage of introducing the schedule at the operator level by the use of staff dispatchers is that the foreman's traditional prerogative of assigning work to the people he supervises is lost.

CENTRALIZED AND DECENTRALIZED CONTROL OF DISPATCHING

The amount of control held by the centralized production planning and control group over the individual facilities can be varied considerably by the quantity of work released to the shop for dispatching at one time. It is sometimes necessary for a centralized group to plan and control closely in order to get the required over-all coordination. In other cases in which the individual facilities are more autonomous, a large portion of the detailed planning can be decentralized to the various facilities. We will consider the operation of both systems.

Centralized Control of Dispatching   Whenever it is necessary to coordinate the activities of the various facilities closely, centralized control of dispatching is mandatory. As an example, the work in the individual departments of an automobile factory, such as the body shop, paint room, engine assembly department, and so forth, must be controlled from a centralized section to be sure that the proper components are available for the final assembly line. The dispatching function can be accomplished by teletyping the list of work to be done in the *immediate* future. The foreman of the department has little or nothing to say about the work to be done. If a staff dispatcher is located in the department the nature of his job is clerical rather than planning.

The advantages of centralized control of dispatching are:

1. There is effective coordination between individual facilities.
2. Schedule changes can be made rapidly.
3. The foreman is relieved of the duties of dispatching.

The disadvantages of centralized control of dispatching are:

1. The communication system must be rapid and dependable.
2. There is very little flexibility in assigning work on the basis of the individual operator's capacity, limitations, and so on.

Decentralized Control of Dispatching   It is frequently desirable to decentralize the control of dispatching in order to avoid a complicated centralized scheduling system. Decentralization can be accomplished by releasing relatively large groups of orders to the facilities at one time. As an example, under this system the dispatcher may receive all of the work to be assigned during the week and it is his responsibility to make the assignments in such a manner as to meet most effectively the schedule.

The advantages of decentralized control of dispatching are:

1. The work assignments are made taking into consideration the capabilities of the people and the equipment.
2. There is a large amount of flexibility in work assignments.
3. The communication system is not as critical as in the centralized control.
4. The foreman has an opportunity to review the work assignments.
5. Savings in set-up cost can be made by knowing the best work sequence.

The disadvantages of decentralized control of dispatching are:

1. It is difficult to secure coordination between facilities.
2. Changes in the schedule cannot be made rapidly.

Dispatching is one of the most important functions of the production control system; therefore it must be carefully planned and established. It is at this point of the system that the plans established in the earlier phases of production control are released to the working facility. In order to be assured that the established schedule will be followed, it is necessary to have close control over dispatching. It is a little more difficult to obtain the necessary control if the dispatching function is performed by the foreman, but many well-run systems are in operation despite this problem. From the production control staff viewpoint, staff dispatchers aid immeasurably in obtaining the necessary over-all control.

Summary In this chapter considerable time was spent in discussing the relationship of the dispatching function and the typical line-staff organization. It must be understood that the formal concepts of good organizational design must be unavoidably violated when the production plans are introduced into the line. This problem can usually be resolved by informal agreement between the line and staff.

The two basic methods of control of dispatching were considered. These methods are centralized control and decentralized control of dispatching. Both of them have advantages and disadvantages. The selection of the method to be used in any particular organization must be based on a careful analysis of the problems which exist in that organization. Only by this analysis can the most effective production control system be designed.

19

Progress Reporting — Principles

The first portion of the follow-up phase of Production Control shown in Table 3-1 is progress reporting. *Progress reporting is defined as the transmission of the information of accomplishment by the operating facility to the planning group and the interpretation of the information for necessary corrective action.*

A basically well conceived Production Control system may be ineffective simply because the progress reporting system is not adequately designed. One of the principles of a good Production Control system (see Chapter 3) is: "The system must furnish timely, adequate, and accurate information." This is another way of saying that there must be good progress reporting.

In most situations at least two major groups of people receive work progress information:

1. *The production control group*—for evaluating the actual performance against the anticipated performance and subsequently for replanning the facilities activity.
2. *The accounting group,* normally cost accounting—for recording the labor and material expenditures for the various jobs.

In general the information required by the planning

group and the accounting group is very similar, and consequently it is desirable to have one progress reporting system which will be adequate for both groups. Table 19-1 lists the *major* items of information required by both groups. The reader should notice the similarity in the lists. Although we will primarily discuss the specific requirements of the Production Control group, the requirements of cost accounting as well as other organizational functions must be considered when designing the progress reporting system.

TABLE 19-1

PROGRESS REPORTING INFORMATION

Work Planning and Control Group	Accounting Records and Control Group
1. Job identification	1. Job identification
A. Order number	A. Order number
B. Operation number	B. Operation number
2. Time of report	2. Time of report
3. Work accomplishment	3. Work accomplishment
	4. Time used for accomplishment
	5. Reasons for time variations
	6. Operator identification
	7. Spoilage

Fig. 19-1. The trouble with most progress reporting systems.

The most common cause for failure in securing an acceptable progress reporting system is dramatized in Figure 19-1: "Too much—too late."

WHY "TOO MUCH"?

There are understandable reasons for having "too much" information. When initially establishing a progress reporting system, it is common practice to enumerate *all* the information that could possibly be reported and then attempt to incorporate the entire list into the system. This procedure will naturally reduce the chances of overlooking some item which *may* become important in the future. It also effectively avoids many criticisms when initially presenting the system to the people involved. In general, people will be more critical about insufficient information than they will about too much information.

The "illegal" files found in many organizations frequently cause over-accumulation of information. Since many progress reporting systems have both the Production Control and the accounting information on the same form, there is a considerable amount of extraneous information available to both groups. Using the excuse of tighter and more effective control, either or both groups may establish files on potentially useful information. All too often, the potential usefulness never materializes, but the argument is, "Maybe someday I'll need back-up material." These illegal files detract from the real needs of the progress reporting system to the extent that the system is not able effectively to fulfill its objectives.

A general rule can be concluded from the problem of too much information: *Only the absolute minimum of information should be included in the progress reporting system.* Each bit of informatiton required from the operating facility should be evaluated before it is incorporated into the system. Since it is impossible to anticipate and plan for all the unusual situations that may occur, the system should be designed to accommodate only the general situation. The special cases should be handled as special. As an example, if the date on which a job is started and completed is normally needed by the production control group it should be included in the progress reporting system. However, if the clock time at which the job is started and completed is not useful, it should not be introduced into the reporting form. Reporting this needless information is too costly, both in time and money, to tolerate the argument that it *might* be useful *sometime*.

Another major cause of too much information is the duplication of reports found in many progress reporting systems. There are a number of different paper forms which are commonly used by an activity to report

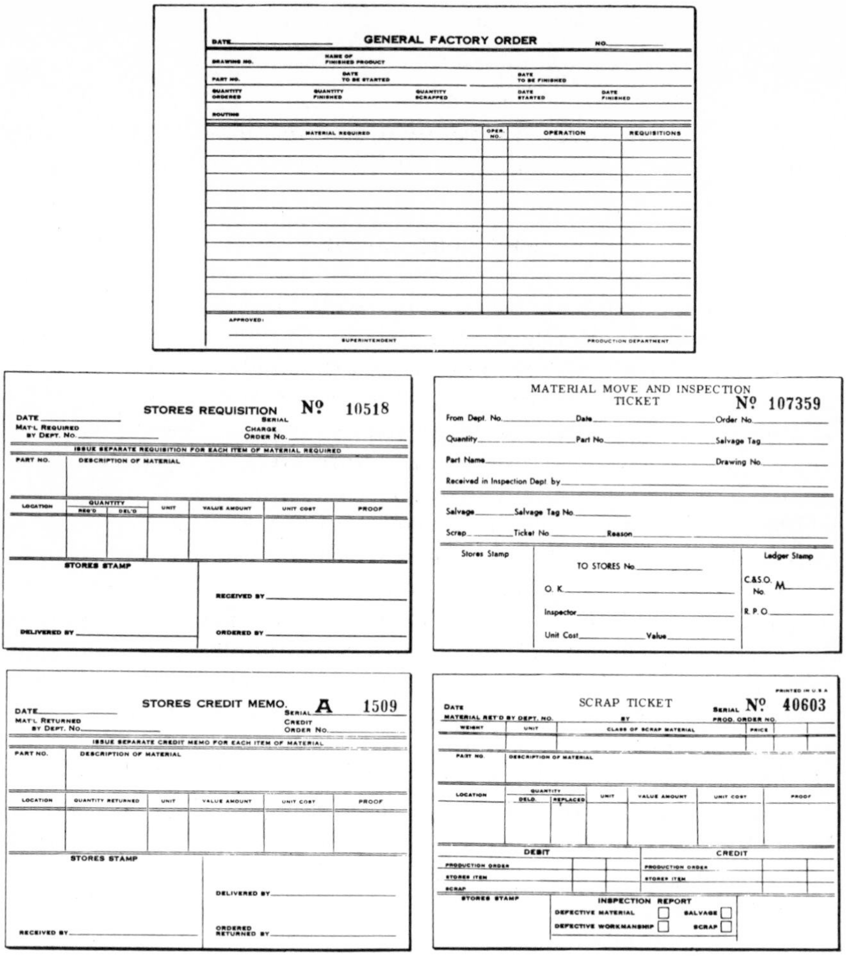

GENERAL FACTORY ORDER

DATE ______ NO. ______

DRAWING NO. | NAME OF FINISHED PRODUCT

PART NO. | DATE TO BE STARTED | DATE TO BE FINISHED

QUANTITY ORDERED | QUANTITY FINISHED | QUANTITY SCRAPPED | DATE STARTED | DATE FINISHED

ROUTING

MATERIAL REQUIRED	OPER. NO.	OPERATION	REQUISITIONS

APPROVED:

SUPERINTENDENT ______ PRODUCTION DEPARTMENT ______

STORES REQUISITION Nº 10518

DATE ______ SERIAL

MAT'L REQUIRED BY DEPT. NO. ______ CHARGE ORDER NO. ______

ISSUE SEPARATE REQUISITION FOR EACH ITEM OF MATERIAL REQUIRED

PART NO. | DESCRIPTION OF MATERIAL

LOCATION	QUANTITY REQ'D	QUANTITY DEL'D	UNIT	VALUE AMOUNT	UNIT COST	PROOF

STORES STAMP

RECEIVED BY ______

DELIVERED BY ______ ORDERED BY ______

MATERIAL MOVE AND INSPECTION TICKET Nº 107359

From Dept. No. ______ Date ______ Order No. ______

Quantity ______ Part No. ______ Salvage Tag ______

Part Name ______ Drawing No. ______

Received in Inspection Dept. by ______

Salvage ______ Salvage Tag No. ______

Scrap ______ Ticket No. ______ Reason ______

Stores Stamp | TO STORES No. ______ | Ledger Stamp

O. K. ______ | C.&S.O. No. M ______

Inspector ______ | R. P. O. ______

Unit Cost ______ Value ______

STORES CREDIT MEMO. SERIAL A 1509

DATE ______

MAT'L RETURNED BY DEPT. NO. ______ CREDIT ORDER NO. ______

ISSUE SEPARATE CREDIT MEMO FOR EACH ITEM OF MATERIAL

PART NO. | DESCRIPTION OF MATERIAL

LOCATION	QUANTITY RETURNED	UNIT	VALUE AMOUNT	UNIT COST	PROOF

STORES STAMP

DELIVERED BY ______

RECEIVED BY ______ ORDERED RETURNED BY ______

PRINTED IN U.S.A.

SCRAP TICKET SERIAL Nº 40603

DATE

MATERIAL RET'D BY DEPT. NO. ______ BY ______ PROD. ORDER NO. ______

WEIGHT	UNIT	CLASS OF SCRAP MATERIAL	PRICE		

PART NO. | DESCRIPTION OF MATERIAL

LOCATION	QUANTITY DELD.	QUANTITY REPLACED	UNIT	VALUE AMOUNT	UNIT COST	PROOF

DEBIT			CREDIT		
PRODUCTION ORDER			PRODUCTION ORDER		
STORES ITEM			STORES ITEM		
SCRAP					

STORES STAMP

INSPECTION REPORT

DEFECTIVE MATERIAL ☐ SALVAGE ☐

DEFECTIVE WORKMANSHIP ☐ SCRAP ☐

Fig. 19-2. Sample forms used by one company for progress reporting.

work accomplishment. These include job tickets, move tickets, inspection tickets, and material requisitions. All of these forms have special purposes, but the information on them is frequently exactly what is required for progress reporting. A generally accepted principle of systems design is to use the various paper forms for multiple purposes, so it is desirable to use the job tickets, inspection ticket, and the like for progress reporting. The problem is that too many times *all* of the forms are used. This means that the same information is reported two or more times. The idea that this practice provides convenient checks on the validity of the reported information is not sound. Accurate information, reported one time, is all that a system requires. Multiple reporting is certain to create too much information. An example of this type of situation was found in the production shop of a major industrial equipment manufacturer. The paper forms for this particular organization, shown in Figure 19-2, had the following distribution and usage:

1. *General Factory Order:* A heavy card-stock copy of this form was used as the official authorization to manufacture. Additional copies of the form were sent to every department involved in the processing of the particular order to be used as job tickets. At the end of each day or at the completion of the operation, whichever occurred first, a copy was sent to the production control department to report the progress on the order. If an operation required three days to be completed, at *least* three copies of the factory order were sent to the production control office for each operation.

2. *Stores Requisition:* Multiple copies of this form were sent to the shop as authorization to withdraw the needed material. One copy of the form was returned to the production control department to report the issuance of material. The major problem here was that a separate stores requisition was needed for every item required by the job. If an order required ten different items, ten different material requisitions were sent to production control.

3. *Material Move and Inspection Ticket:* As work was completed on an order, an inspector would check the quality of the product and fill out a material move and inspection ticket. This form authorized the movement of the material to another operation. A copy of the form was sent to Production Control to report the move. If a job was broken up into four lots, four reports were made out for each operation.

4. *Stores Credit Memo:* If any item was returned to the storeroom for any cause whatsoever, a stores credit memo was made out. A copy of this form was sent to Production Control.

5. *Scrap Ticket:* Whenever material had to be scrapped, a copy of the scrap ticket was sent to Production Control so that it would know the status of the order.

The total number of pieces of paper needed for progress reporting under this system was appalling. A typical order would have as a minimum:

$$\frac{10 \text{ operations}}{\text{order}} \times \frac{2 \text{ General Factory Order Forms}}{\text{operation}} = 20 \text{ forms/order}$$

$$\frac{10 \text{ items}}{\text{orders}} \times \frac{1 \text{ Stores Requisition Form}}{\text{item}} = 10 \text{ forms/order}$$

$$\frac{4 \text{ lots}}{\text{operation}} \times \frac{10 \text{ operations}}{\text{order}} \times \frac{1 \text{ Move Ticket}}{\text{lot}} = 40 \text{ forms/order}$$

$$\text{Total} = 70 \text{ forms/order}$$

Considering the fact that there were approximately 1000 active orders in the shop at a time, it can be seen that the progress reporting "system" of this organization had to be replaced. There was clearly too much information arriving at the production control office.

The frequency of progress reporting directly determines the quantity of information received by production control. There are only two basically different ways of establishing when the progress reporting will take place:

1. According to a fixed time interval
2. According to the work accomplishment

Both ways, or a combination of the two, are used in various organizations. The best method is entirely contingent upon the circumstances.

Progress Reporting at   Progress reporting at fixed time intervals is very
Fixed Time Intervals   common in all types of activities. The report
                               of the accomplishment of a whole organization
is normally made on an annual, quarterly, or monthly basis. This type
of reporting is exemplified by the familiar financial reports such as balance
sheets and profit and loss statements. To a certain extent the time intervals are determined by legal requirements and business traditions, but
their selection is based upon a sound principle. Since the over-all goals
of most organizations are for the long run, progress reporting on a more
frequent basis would be meaningless. The same thing is also true in activities such as building construction, research and development, and
many office activities. Because these are long-term in nature, the time interval for progress reporting is also relatively long.

In most manufacturing plants, service organizations, maintenance
shops, and so forth, the production work tasks are accomplished in a
relatively short period of time. Because of this, the time interval between
progress reports may be considerably shorter than a week or a month.
Daily or even hourly reports are fairly common in such activities. Frequent reporting theoretically allows the planning group to maintain the
shop work schedules on a current basis. It should be carefully noted,
however, that as the time interval between progress reports decreases,
the number of reports to be analyzed increases. This can become a major
cause for too much information in many organizations.

Progress Reporting by   Because of the inherent difficulty in establish-
Work Accomplishment   ing the proper time interval for progress re-
                               porting in some types of activities, reports
are made at the completion of a specified work task. In a shop which
has a large number of relatively short operations, progress reporting may
be made at the end of each operation. Similarly, maintenance workers
frequently report only at the completion of each assigned task. Research
and development projects can be broken down into sub-stages and a report is made as each stage is finished. This procedure of progress reporting has the advantage that only pertinent information is reported. It is
practical only if the time required to complete the work task is relatively
short.

Many, if not most, systems use a combination of reporting at fixed time
intervals and at the completion of a work task. In a typical manufactur-

A. CONVENTIONAL PROGRESS REPORTING

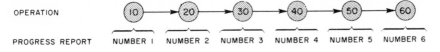

B. AUDIT POINT PROGRESS REPORTING

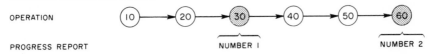

Fig. 19-3. Two progress reporting systems.

ing organization a report is made whenever an operation has been completed. At the end of the day another report is made on the status of the jobs still in operation. In this manner, the planning group is aware of the exact work accomplishment on each factory order. If a combination such as this is properly established and maintained, the problems of progress reporting are usually minimized. However, it is commonly found that organizations will report the completion of each operation via a job ticket or move ticket and then have a "recap" report made out by the foreman at the end of the work day. This is usually an attempt to check the accuracy of the information and is a highly questionable practice.

The principle of "management by exception"[1] is one of the most important principles applicable to progress reporting. In Figure 19-3, a six operation job is shown. Upon the completion of each operation a progress report is sent to Production Control. This means that as the job moves through the plant *at least* six separate reports have to be processed. If the data used to schedule the work in the plant are reasonably accurate, the actual accomplishments of the shop should be close to the planned assignments. If this is true, there is no reason to report the concurrence of performance. Theoretically it should be possible for production control to assume that the established plans are being followed unless a non-concurrence report is made by the shop.

Adapting the principle of management by exception to progress reporting requires some modification of the reporting system. This is usually in some system of audit points such as illustrated in Figure 19-3b. Management by exception is practiced for the first two operations and then a check report is made at the completion of the third operation. The completion of the next two operations does not necessitate a report

---

[1] See Chapter 3 for a discussion of the principle of management by exception.

while the end of the last operation does. The number of reports is reduced 66 per cent, while the actual control has not been diminished appreciably. When initially establishing the audit-point system it is usually
advantageous to eliminate reports only on a limited number of operations that have a history of dependability. As the system proves its value,
a larger number of operations can be "reported" on the exception principle basis.

There are many possible places where audit points can be logically established. If a specific operation has a history of poor performance it
would be unwise to use the principle of exception on it since the "exception" is the general rule. This type of undependable operation lends
itself to audit points. In some cases all of the operations performed in
one department may be reported on the exception principle. In this
case the audit point might be the movement of the material from one
department to another. Inspection operations of the semi-finished items
frequently can be used as audit points. In some cases the audit points
are arbitrarily established at the end of every second, third, or fourth
operation. Regardless of how the audit points are assigned, the basic
concept is to alleviate the problem of too much information flowing into
the Production Control department.

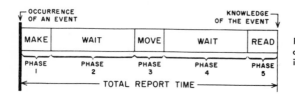

Fig. 19-4. The typical time scale and phases of the progress reporting cycle.

WHY "TOO LATE"?

There are many conditions which may cause the progress report to be
too late. These can be categorized after analyzing the progress reporting
cycle. Figure 19-4 shows that there is a minimum of five phases needed
to complete any progress report. The cycle is not finished until *all* five
phases have been completed. This means that an excessive delay in *any*
phase could cause the progress report to be too late. In order to define
the five phases fully, a typical paper form of progress reporting will be
used as an illustration:

Phase I—Make—This phase of the progress reporting cycle occurs immediately after the event to be reported. It consists of the time necessary to determine if the work task has been completed and to record
the pertinent information on the appropriate forms.

Phase II—Wait #1—Once the information has been recorded a certain
amount of time will elapse before the information is transmitted to

the planning group. In the case of reporting via a paper system, Phase II consists of the time the paper form is waiting to be carried to the planning office.

*Phase III—Move—*This phase consists of the time consumed in the physical transmission of the information. In the case of the paper system, Phase III consists of the actual movement of the paper forms.

*Phase IV—Wait #2—*Once the information has been physically moved to the planning group a certain amount of time elapses before the information is interpreted. In the case of a paper reporting system, it is the time spent waiting until someone finally reaches into the pile of reports and obtains the specific report under consideration.

*Phase V—Read—*Once the specific report form is obtained by the planning group for consideration, a certain amount of time is spent in segregating the pertinent information and interpreting its contents.

If a progress reporting system has a history of providing late information to the production control group, the cause can be located by analyzing each phase of the reporting cycle. The concept of dividing the cycle into phases is useful for developing a systematic method of analyzing and correcting the problem. In many cases where late reporting is a problem the first action usually taken is that of trying to reduce the time for the "move" phase. This is rarely the cause of serious trouble as will be seen in the next chapter. In the majority of the cases where information is too late, the basic trouble is found in one of the other phases.

BASIC METHODS OF PROGRESS REPORTING

The designer of a progress reporting system has a large variety of commercial devices available to him. All of these devices have certain advantages and disadvantages but they should never be considered as basic methods of progress reporting. They are simply the "hardware" for the system. The actual design of the reporting system should begin with the determination of which of the three basic *methods* of progress reporting best suits the needs of the particular organization. After this question has been answered, the specific commercial device can be selected.

The three basic methods of progress reporting are:
1. Written systems
2. Oral systems
3. Electronic systems

Written Systems of Progress Reporting   The most common system of progress reporting is the written report. Under this system, the information to be transmitted by the facility is commonly recorded on a paper form by the dispatcher or the foreman and then the paper is physically transported to the production control group.

The advantages of the written systems are:

1. A written record of the information is available for future reference.
2. Errors in interpreting the information are reduced to a minimum.
3. A large amount of data can be recorded at one time on a single sheet of paper.

The disadvantages of a written system are:

1. It takes a relatively long time to complete the progress reporting cycle.
2. Important information may be lost because of misplaced paper.
3. Too much information is frequently incorporated on the forms.
4. It is too easy to establish unnecessary files—again, too much information.

Oral Systems of   The oral system of progress reporting is a fairly
Progress Reporting   popular one; under this system, phones and radios
                          are the common media for transmitting the neces-
                          sary information.

The advantages of the oral systems are:

1. The progress reporting cycle is relatively short.
2. It is difficult, if not impossible, to establish unnecessary files.
3. Questions about the progress report can be answered immediately.

The disadvantages of the oral systems are:

1. Errors in interpretation of information are likely.
2. A record is not generally available for future reference.
3. It is difficult to transmit large quantities of information.

Electronic Systems of   Electronic data transmission, data storage, and
Progress Reporting   data processing are widely used in organizations
                             in which large amounts of data must be
handled and will become more common in all types of business and industry in the near future. The advantages of electronic systems of progress
reporting are:

1. Large quantities of information can be processed at one time.
2. The progress reporting cycle is greatly shortened.
3. Errors in data processing are greatly reduced and may be virtually eliminated in some systems.
4. A permanent written record may be quickly obtained.
5. The first four phases and a portion of the fifth phase of progress reporting can be completely machine-controlled.

The disadvantages of electronic systems of progress reporting are:

1. The equipment cost is relatively high and is therefore economically limited to applications where large volumes of data are processed.
2. Specialized training of personnel is required.
3. Special physical facilities are usually required.

Summary   Before an effective Production Control system can be de-
               signed, it is absolutely essential that the underlying diffi-
culties involved in progress reporting be thoroughly understood. This
is the only way that duplication of reporting information can be avoided.

20

Progress Reporting — Systems

When designing or redesigning the progress reporting system of any organization, it is important first to identify the specific phase or phases of the cycle that offer the greatest potential savings. After this has been done, the designer is in a position to evaluate and select intelligently the commercial equipment that will provide the best service to the system.

This chapter is devoted to the description of some of the commercial equipment available for reducing the time of the various progress reporting phases. Although there are a multitude of patented devices on the market, the discussion in this chapter will be restricted to:

1. Prewritten Papers
2. Pneumatic Tube Equipment
3. TelAutograph Equipment
4. Teletype Equipment
5. Telephone and Intercom Equipment
6. Radio and Closed Circuit Television Equipment
7. Telecontrol Equipment

Since the segregation and interpretation of data—the

"read" phase of progress reporting—is a highly specialized problem, it will be considered separately in Chapter 21.

PREWRITTEN PAPERS

The major reason for adopting the use of prewritten papers is that it provides a rapid, accurate method of generating multiple copies of the various forms needed by the production control system. The reduction in the "make" phase of the progress reporting cycle is an important auxiliary benefit realized by the use of prewritten papers. This is illustrated in Figure 20-1.

Fig. 20-1. The potential savings in time by using prewritten papers.

└AREA OF POTENTIAL SAVINGS IN TIME

It is axiomatic that operating personnel have a strong objection to, and perhaps disdain for, record keeping. The usual argument is, "We are here to produce, not to shuffle papers." There is no denying the truth of this statement, yet it is important that the progress reporting function be adequately administered. Prewritten papers provide a mechanism for appreciably reducing the amount of time spent by the operating personnel on clerical work. In addition, prewritten papers aid in obtaining more accurate reports and reducing the likelihood of errors in interpretation.

In order to illustrate the use of prewritten paper forms, an assembly order system for a manufacturing organization will be taken as an example. In Figure 20-2, the major paper forms for the organization are shown.

These forms are used as follows:

Form 1. A spirit-duplicating master is made out for each standard assembly and filed by model number for future use. Notice that the information on this master pertains to one unit of the assembly. All of this information remains the same regardless of the number of units to be assembled on a specific order.

Form 2. Whenever an order is received for a particular assembly model, the master for the assembly (Form 1) is taken out of the file and a hectographic variable master (Form 2) is typed. The information on Form 2 pertains to the specific order and includes factory order number, quantity to be assembled, starting dates, and so on. Forms 1 and 2 are placed in a reproduction machine and standard order forms (Forms 3, 4, 5, 6, and 7) are printed. After the standard set has been printed, the variable master is destroyed and the assembly master is returned to file.

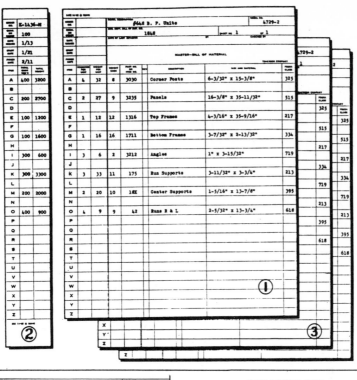

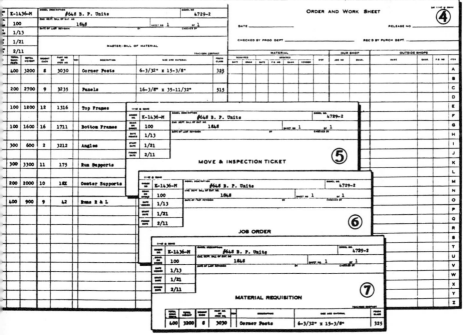

Fig. 20-2. An example of a prewritten paper system. (Reprinted from "Ditto Assembly Order System," Ditto System Release No. 3, by permission of Ditto, Inc., Chicago, Ill.)

The following forms make up a standard set of papers to be used in the Production Control system. The number of copies of each form can be varied as required. In most prewritten paper systems all of the forms are enclosed in a special envelope which is sent to the shop. The forms can then be taken from the envelope by the shop personnel as the need arises.

Form 3. There are two copies of the assembly order reproduced in the standard order set. One copy is used by the Production Control section as a source of information for loading and scheduling purposes. The second copy is used by the shop as the basic authorization to assemble the required number of units.

Form 4. The order and work sheet form is used by the Production Control department for analyzing the component parts status. If it is necessary to order additional component parts for the required assemblies, the ordering information can be recorded on this form. By using a work sheet such as this, all of the information pertaining to the assembly order and its component parts is conveniently consolidated in one place.

Form 5. The move and inspection tickets are printed with only the information appearing at the top of the hectographic master reproduced on them. Since the list of component parts is not required on these tickets, the list appearing on the master is masked out when it is placed in the reproducing machine. The lower portion of the move and inspection tickets has space provided for the shop to record the progress reporting information.

Form 6. The job order tickets are reproduced in the same manner as the move and inspection tickets.

Form 7. A copy of the material requisition form is made for each component part going into the assembly. The top portion of the requisition is a duplicate of Forms 5 and 6. At the bottom of the requisition the information about the specific component part is printed.

The Production Control system of this organization uses Forms 5, 6, and 7 as media for progress reporting. Whenever a report has to be made by the shop, the information is recorded on one of the appropriate prewritten forms. Using these forms relieves the shop personnel of the task of reproducing the fixed information (order number, item identification, and so on), but more important, the fixed information is in a legible typewritten form. Only the actual report must be hand written by the foreman. This arrangement reduces the chance of misinterpretation and eliminate a considerable amount of clerical work on the part of the foreman.

The advantages of the prewritten paper systems are:

1. The clerical work at. the operating levels is reduced.
2. Errors in the basic typewritten information are impossible once the material has been proofread.
3. The cost of most of the available systems is moderate.

The disadvantages of the prewritten paper systems are:

1. There is a tendency to reproduce too many forms.
2. Unnecessary files are frequently established because of the availability of extra copies.

PNEUMATIC TUBES

The introduction of a pneumatic tube system for the movement of papers in an organization can essentially eliminate the first "wait" phase and substantially reduce the "move" phase of the progress reporting cycle. The system is simply a series of tubes going from the Production Control section to the various operating departments in the shop. When a progress report is to be made by one of the facilities, the information is written on a standard report form and the paper is put inside a special capsule. The capsule is then placed in the tube and shot by air from the shop to the Production Control department. Normally the pneumatic tube systems are constructed with pairs of tubes: one tube for sending the capsule in one direction, the second tube for sending it in the opposite direction. This provides a rapid two-way written communication system between the shop and production control.

Pneumatic tubes are particularly beneficial whenever the production facilities are spread out over a large area. They have been successfully used in large single-story buildings, multiple-story buildings and in multi-building operations. In such situations as these, much time would be needed to hand carry the progress reports. Pneumatic tubes can save both time and money.

Advantages of the pneumatic tube system:

1. It provides a method for rapid, two-way communication.
2. Written reports are used.
3. Operating and maintenance costs are low.

Disadvantages of the pneumatic tube system:

1. The number of paths for information flow is limited.
2. Information flow can go only over fixed paths.
3. The initial equipment and installation costs are relatively high.

TELAUTOGRAPH TELESCRIBER

Figure 20-3 shows the potential time savings in the progress reporting cycle when the TelAutograph Telescriber is used. When this device is

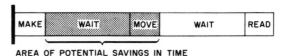

Fig. 20-3. The potential savings in time by using the TelAutograph Telescriber.

MAKE | WAIT | MOVE | WAIT | READ

AREA OF POTENTIAL SAVINGS IN TIME

Fig. 20-4. TelAutograph Telescriber sending and receiving units. (Reprinted from "The Nerve System of Controlled Manufacturing is Communciation," by permission of the TelAutograph Corp., New York.)

incorporated into a Production Control system, the first "wait" phase and the "move" phase of the reporting cycle are completely eliminated. This is due to the fact that the report is transmitted over elecric wires at the same instant it is written.

There are two units to a Telescriber system—the sending or transmitting unit and the receiving unit; these are shown in Figure 20-4. The sending unit has a special stylus which is used to write the message. As the stylus is moved, the small pens, appearing in the windows of both units, record the exact stylus movement on paper. It is therefore possible to transmit written messages, sketches, figures, and the like with this equipment and have a permanent record on paper at both the sending and receiving units.

If the Telescriber is to be used for progress reporting, the various work centers in the shop have sending units located in them. The receiving units are in the Production Control office. The foreman or dispatcher using the transmitter writes out the progress report and the report is duplicated at the same time on the receiving unit. The major difference in the end result between the conventional paper system of progress reporting and the Telescriber system is that the written message is instantaneously available at the Production Control office.

Points to be considered in using the TelAutograph Telescriber are:

1. Written messages can be transmitted between the shop and Production Control instantaneously.
2. A permanent record of all messages is available at both the sending and the receiving units.
3. The equipment is relatively easy to operate.
4. The sender must have a legible handwriting to avoid the possibility of misunderstanding the information.

5. Although the initial investment for equipment is relatively high, Tel-Autograph offers equipment on a lease basis, which includes service, for less than a dollar a day for a transceiver unit and half that amount for a receiver unit.

TELETYPE EQUIPMENT

The use of teletype equipment for progress reporting has the same potential time reduction in the reporting cycle as the TelAutograph Telescriber shown in Figure 20-4. The use of the equipment will eliminate the first "wait" phase and the "move" phase of the cycle. A basic teletype machine has a keyboard very similar to the one found on an ordinary typewriter. As the operator actuates the keys, each character is converted into an electrical signal. These signals are transmitted by wire to one or more receiving machines where the message is recorded, character by character, exactly as it was sent. An important feature of the teletype equipment is its ability to produce a punched tape or operate from a punched tape. In the integrated data processing systems, discussed in Chapter 21, this feature is very important.

Teletype equipment is frequently used for Production Control communications in the mass production industries. In such systems the machines are located in both the Production Control office and the various work centers in the shop. The teletype is used for the centralized dispatching system discussed in Chapter 19—and the same equipment can be used by the shop for reporting the work accomplishment.

The advantages of the teletype equipment are:

1. Information can be rapidly sent to and from the Production Control office.
2. Written records of the messages are available at both the sending and the receiving machines.
3. The equipment can be incorporated into an integrated data processing system.

The disadvantages of the teletype equipment are:

1. The equipment is costly to install and operate.
2. Trained operators must be used to utilize the maximum speed of transmission.
3. Only the standard alphabet and numerical symbols can be transmitted, not pictorial matter as on the Telescriber.

TELEPHONE AND INTERCOM EQUIPMENT

All of the oral methods of progress reporting have the same potential for time reduction in the reporting cycle—they remove the three middle phases of "wait—move—wait" as shown in Figure 20-5, and span directly between "make" and "read." It would appear from this that the oral systems offer the greatest reduction in the reporting cycle and should therefore be used whenever possible, but this conclusion is hasty. The

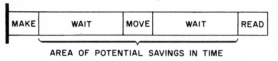

Fig. 20-5. The potential savings in time by using oral methods of progress reporting.

oral methods of progress reporting have a special problem which does not occur with the written methods and does not show on the reporting cycle graph. Before an oral report can be made, it is necessary to have both the sender and the receiver of the information available at the same time. The "wait" phases of the reporting cycle cannot occur with oral reporting but there is a potentially significant time delay while the necessary connections are made. To explain this problem, consider the reporting of the work accomplishment at the end of a day. In a written reporting system all of the facilities could make out the necessary reports at the same time without interfering with one another. In an oral reporting system, only a restricted number of facilities could make their reports simultaneously. Because of this problem, most organizations limit oral reporting to emergency or exceptional situations and consider it an auxiliary to some other type of progress reporting system.

The advantages of telephone and intercom equipment are:

1. The equipment is easy to use.
2. Emergency reports can usually be made very rapidly.
3. The equipment cost is relatively small.

The disadvantages of telephone and intercom equipment are:

1. There is no record of the information transmitted.
2. Errors in interpretation of information are likely.
3. Only a relatively small amount of information can be sent at one time.
4. It is necessary to have both the sender and the receiver of the information available at the same time.

RADIO AND CLOSED-CIRCUIT TELEVISION

Many companies have adopted the use of radio as the means for communicating with mobile type of equipment. As an example, there are many uses of radio as a means for controlling the activity of material handling equipment. In such a system there is a centrally located dispatcher in charge of the operators of the material handling equipment. As the need for the services of the equipment is reported by phone, the dispatcher establishes the best schedule to be followed and then radios instructions to the equipment operators. After the job has been completed, the equipment operator radios this information to the dispatcher and receives the next instruction. In order to have this type of communication economically feasible, one or more of the following would have to be the situation:

1. There are a large number of units to be controlled.
2. The equipment to be controlled is expensive to operate.
3. The equipment must be used over a large area.

The industrial use of closed-circuit television is a rapidly expanding field. The use of television has been mainly in collecting instantaneous data from greatly inaccessible locations. As an example, television has been used on large machinery to allow the operator to see what is actually occurring at various parts of the machine. Television has also been used extensively in controlling hazardous processes such as those in which radiation is an important factor. It is possible to use television to transmit pictures of schedule and report boards. Pioneering work is still required to develop these uses extensively in the area of Production Control.

TELECONTROL EQUIPMENT

The Telecontrol equipment manufactured by Hancock Industries uses several unique communication features to secure a centralized dispatching system along with semi-automatic progress reporting. Although the equipment is relatively expensive, it can, when used in an appropriate situation, be amortized in a reasonable length of time.

There are two major elements in the Telecontrol system:

(A) Control box

Fig. 20-6. The equipment for the Telecontrol system. (Reprinted by permission of the Hancock Telecontrol Co., Mich.)

1. Control Box (see Figure 20-6A). Control boxes are located at each of the work centers incorporated into the system. The box is so wired that the completion of every unit of production is automatically recorded on the centralized control board. The position of the toggle switch, under the control of the machine operator, indicates if the job is running as planned. If the machine operator needs to communicate with the foreman or the Production Control section, he can throw the toggle to "alarm." This lights a signal both on the control box and in the Production Control office.

The foreman can indicate "downtime" by turning the key-operated switch. A telephone hand set, carried by the foreman, can be plugged into the control box at any time for direct communication with the Production Control office.

2. Control Board (see Figure 20-6B and C). On the control board, located in the Production Control section, there is a separate panel to record the production information from each of the control boxes. On the left hand side of each panel is the identification of the production machine, the job currently on the machine, and the machine operator. To the right is a group of four counters which record the amount of production time chargeable to the job, the amount of downtime, the current total production in units and the number of units still to be produced on the job.

(C) Control board

The operation of the Telecontrol system is very simple. Before each shift, the Production Control section decides what jobs will be performed on each machine. Individual operators are then assigned. This information is attached directly to the control panel for each production unit and the total number of units to be produced on the job is set on the counters. When the master control clock signals the beginning of the shift, the green signal lights up on the control boxes at the various work stations and also on the control panel, indicating that the machine is ready for production. This signal tells the operator that he may start work. The operator at each work station knows that as long as the green light on his control box remains on, his work is being recorded on the counters in the Production Control office. A flashing green light goes on automatically when the order-balance counter reaches zero. The operator turns off his machine and waits for the foreman, who has been notified over the public address system, to give him new work.

When an operator notices that an emergency situation is developing, such as running out of material, machine malfunction, and so forth, he throws the toggle switch on the control box. This starts the red light flashing on both his control box and the control board. The Production Control section notifies the foreman over the public address system which work station requires his assistance. When the foreman gets to the machine he can communicate directly to the Production Control section by using his telephone hand set. If the machine is to be down for a short period of time, the key-operated downtime switch is turned. If the machine is going to be down for an extended period of time, the worker can be assigned to another machine or job.

The advantages of the Telecontrol equipment are:

1. Strong centralized control is possible.
2. Continuous progress reporting is provided.
3. Oral communication between the shop and the Production Control section is maintained by use of the telephones and public address systems.

The disadvantages of the Telecontrol equipment are:

1. The equipment is expensive.
2. The system is usable only for certain types of facilities.

Summary   A number of different types of commercial devices, designed to reduce the first four phases of the progress reporting cycle, have been considered in this chapter. In Chapter 21, data processing equipment will be discussed separately. It is difficult, however, to separate these two classes of equipment because in an actual work situation they must be integrated into an effective Production Control system. The reader should be aware of the fact that the separation made in this book is but a convenience for discussion purposes.

21

Data Processing

Up to this point, the commercial equipment designed to reduce the first four phases of the progress reporting cycle has been considered. Attention will now be turned to the last portion of the cycle—the "read" phase. It will be recalled that the "read" phase of the progress reporting cycle is concerned with the segregation and interpretation of the data. (See Figure 21-1.) Because of the importance and the relative complexity of the data processing equipment, this entire chapter will be devoted to it. Before an understanding of the operation of the various devices can be achieved, it is necessary to digress to a certain extent and discuss the way in which data can be processed and stored by using a binary system.

BINARY NUMBERING SYSTEM

The decimal numbering system is so engrained in our minds that we seldom realize that the original selection of the number 10 as the basis for counting and performing arithmetic operations was purely arbitrary. The only apparent reason for using this number is the fact that man has 10 digits on his 2 hands. If we had 12 fingers, it is very likely that our numbering system would have 12 as the basis. Actually, it doesn't matter if we use 2 (binary), 5, 10, 12, or any other number as a base, the mechanics of addition, multiplication, and so on are equally applicable to *all* systems.

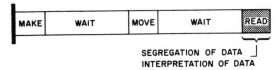

Fig. 21-1. The portion of the progress reporting cycle affected by data processing equipment.

SEGREGATION OF DATA
INTERPRETATION OF DATA

It so happens that most of our efficient electro-mechanical devices and systems for receiving, processing, and storing data do receive and process these data, and transmit or store their results, in binary form. This is because the components are built to react to data inputs by taking one of two states. Each unit of information on a standard punched card, for example, is conveyed through the presence or the absence of a hole. Each basic unit of an electronic computer is built either to block or to permit the flow of current through it to another unit. Therefore all information "fed" into these devices must be expressed in terms of the fundamental yes-no form of these basic binary components. All data, and their verbal interpretations, must be reduced to numbers, and the numbers have, not 10 different digits, but 2. It behooves us to learn a little more about binary numbers, then, before we consider the equipment based on them.

The first problem encountered in developing any numbering system is to obtain a set of symbols to be used to denote different numerical magnitudes. There are ten different symbols in our familiar decimal system. By using different combinations of these symbols any numerical magnitude can be expressed.[1] The single symbol "3" denotes the magnitude of three. The symbol combination "25" denotes the magnitude of twenty-five. We know this because we have been taught it, we have used it, and we are accustomed to it.

In the binary system there are only two symbols to learn: 0 and 1. By using different combinations of these two symbols any numerical magnitude can be expressed. The reader may not recognize these combinations because the binary system is not used in everyday work. To overcome

---

[1] It is beyond the scope of this discussion to introduce some of the special numbering problems such as the negative numbers, infinitely large numbers, and so on; only the positive, whole, finite numbers will be considered.

this difficulty, the combinations of symbols for the first two numbers are given in Table 21-1.

In order to illustrate the fact that the arithmetic rules learned in school for the decimal system are equally applicable to the binary system, a few simple examples will be given. The basic addition rule, which most

TABLE 21-1

THE SYMBOLS USED IN THE DECIMAL AND BINARY
SYSTEMS OF NUMBERING

Decimal System Symbols	Binary System Symbols
1	1
2	10
3	11
4	100
5	101
6	110
7	111
8	1000
9	1001
10	1010

of us do not even think of any more because it has become so routine, says: "Add the numbers in the right hand column and record the total. If the total should turn out to be a two digit number, record the right hand digit and 'carry' the left-hand digit to the next column to the left. Consider this 'carried' digit as another number to be added in with the rest of the column numbers." In an actual addition using the decimal system the rule would appear as:

$$\begin{array}{r} 19 \\ +\ 5 \\ \hline 2^14 \end{array} \quad \textit{Answer 24}$$

Now let us add numbers in the binary system *using the same rule:*

Decimal	Binary	
2	10	
+1	+ 1	
3	11	*Answer* 11 (check this answer against Table 21-1)
3	11	
+1	+ 1	
4	1¹0¹0	*Answer* 100 (check this answer against the Table)
5	101	
+4	+ 100	
9	1¹001	*Answer* 1001 (check this answer against the Table)

DATA PROCESSING

223

Going back to the basic multiplication rules, you will find that the operation of multiplying in the binary system is *exactly* the same as in the decimal system:

Decimal	Binary
2	10
×2	× 10
4	00
	10
	100

Answer 100 (check this answer against the Table)

Decimal	Binary
3	11
×2	× 10
6	00
	11
	110

Answer 110 (check this answer against the Table)

Decimal	Binary
5	101
×2	× 10
10	000
	101
	1010

Answer 1010 (check this answer against the Table)

These six simple problems should serve to illustrate the fact that the *rules* of arithmetic can be used in either the decimal or the binary numbering system. The reader can verify that the subtraction and division rules work, by setting up and solving a few simple problems. When doing these problems remember you have only the symbols "0" and "1" in the system. You must use *only* these two symbols.

As was mentioned earlier, the reason for discussing the binary numbering system is the fact that data processing and storage by machines must be based on a two-state condition. This will be brought out as the various data processing systems are discussed.

MARGINALLY PUNCHED CARDS

Many times it is desirable to have a mechanical means for segregating and arranging progress reporting data where the size of the system does not warrant the adoption of expensive data processing equipment. Under such circumstances marginally-punched data cards—the "poor man's punch-card system"—provide an economical solution to a difficult problem.

To explain the operation of the marginally-punched cards, a simplified card is shown in Figure 21-2. It will be assumed that this is one card out of a deck of 10 cards upon which is printed information about 10 different orders. The orders are numbered from 1 to 10.

The upper right-hand corner of the card is cut off diagonally so that it is possible to tell at a glance if all of the cards in the deck are oriented in the same direction. An upside-down or backward card can be detected immediately because it breaks the symmetry of the pattern by showing a square instead of a diagonal upper right-hand corner.

Next to the diagonal corner is a group of four holes with the numbers 1, 2, 4, and 7 printed under them in that order from right to left. This group of holes is called a "field" of the card and it is in this field that information can be stored. The storage of information in the field of the card is accomplished by notching out different combinations of holes using a special notching punch. (Notice that the marginally-punched cards us a binary system—a notch, or no-notch.) These combinations are shown in Figure 21-3. Notice that it is not necessary to memorize the combinations. They can be determined by summing the small numbers under the notched-out holes.

The separation of the cards into numerical groups can be accomplished by using a special sorting needle. The deck of cards is first arranged into a squared stack so that the holes in the field are in alignment. The sorting needle is inserted into one of the holes and the deck is picked up. The cards with notches will drop out of the deck while the cards with no notches will stay on the needle. By inserting the needle in the different holes in the proper sequence, the cards can be arranged in numerical order. The steps necessary to sort numerically are:

Step 1: Insert the sorting needle in the number 1 hole and allow the notched cards to drop free.
Step 2: Place the dropped cards behind the cards remaining on the needle.
Step 3: Repeat the procedure for the number 2 hole, then the number 4 hole and finally the number 7 hole.

After the needle has been inserted in the four holes, the cards will be in numerical sequence. Cards with the same number, i.e., the same factory order number, will be together. Figure 21-4 illustrates the complete operation of sorting a group of cards into numerical order following the steps given above.

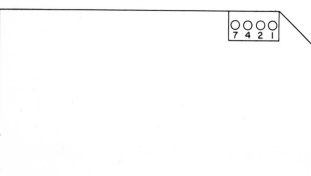

Fig. 21-2. An example of the marginally-punched card. In the upper right-hand corner one information field is shown.

For illustrative purposes, only one field was placed on the cards in Figure 21-4. In practice, it is possible on a 3⅜" x 7½" card to have approximately 20 different fields arranged around the periphery of the card. This provides ample space to store the information necessary to analyze the progress reports. If the marginally-punched cards are used as the paper media for progress reporting, the deck of reports could be sorted by:

a. Factory Order Number
b. Work Center Number
c. Operation Number
d. Employee Number
e. Date
f. Part Number
g. Straight Time and Overtime Work
and so forth

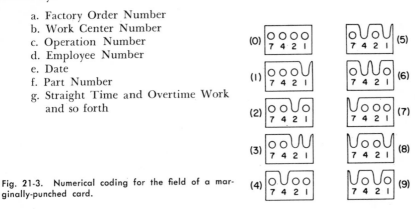

Fig. 21-3. Numerical coding for the field of a marginally-punched card.

The marginally-punched card manufacturers have developed numerous coding systems, hand-punching devices, and electric punching machines, to make the cards applicable to many situations.

The advantages of incorporating these cards into a Production Control system are:

a. The cards provide a method of rapid sorting of coded data.
b. The system is economical to install and use.
c. The cards do not have to be filed in sequence.
d. The cards can be corrected for mispunched ones by using a special mask.

The disadvantages of the marginally-punched cards are:

a. The system provides no means for machine computation.
b. The system provides no means for data summary.
c. The punching operation can become a time consuming task, especially if hand punches are used.

PUNCHED CARDS

In many Production Control systems a tremendous amount of data must be segregated and analyzed in order to maintain the control necessary for the effective operation of the facility. There is a practical limitation to the number of people who can be assigned to the manual processing of data. When this limit is reached, data processing by ma-

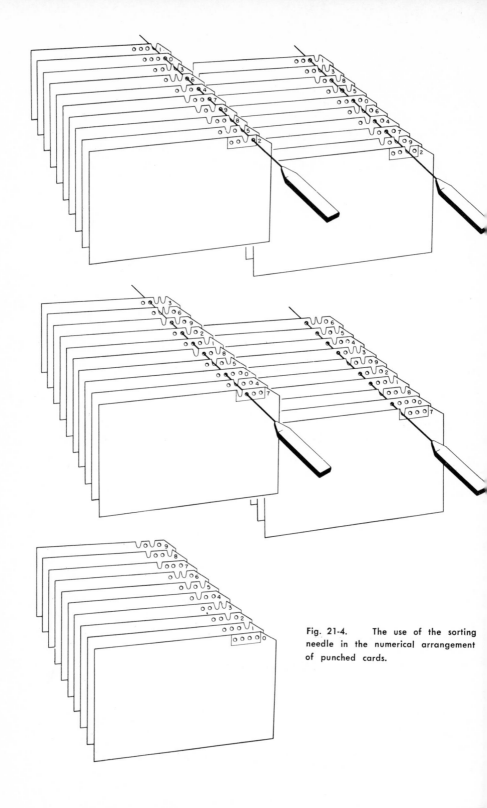

Fig. 21-4. The use of the sorting needle in the numerical arrangement of punched cards.

chines becomes mandatory. The use of punched-card equipment in Production Control systems has provided many organizations with better control at a lower cost than they had using manual data processing systems. It must not be assumed however that punched cards will always improve the effectiveness of the Production Control function. In many cases the manual systems are better. Careful analysis of the data processing system and its problems must always be made before adopting punched cards. This fact cannot be over-emphasized.

The punched-card systems appear complicated to many people because it is difficult to understand how the various machines can perform so many different operations. There is no question that the punched-card equipment is complex, but the operations which the machines can perform are very simple. The designer of a production control system does not have to know *how* the machines operate but rather *what* the machines can do. The basic idea behind the punched-card systems will be discussed in this section along with an illustrative example of the use of punched cards. In Appendix IV the various machine methods of data processing are described. Installing a punched-card system requires the services of specially trained engineers, who are available from the manufacturers of the equipment.

In Figure 21-5 a sample punched card is shown. The corner of the card is cut off diagonally in order to tell at a glance if all of the cards in a deck are oriented properly. There are 80 columns on the card and a single letter or number can be stored in each column. This means, that the basic document can accommodate a record made up of 80 digits. This may appear to be a very limited size record, but for most Production Control systems the capacity is adequate.

In order to code the 10 digits in the decimal numbering system and the twenty-six letters of the alphabet, the card must use twelve spaces or

Fig. 21-5. An illustration of a punched card.

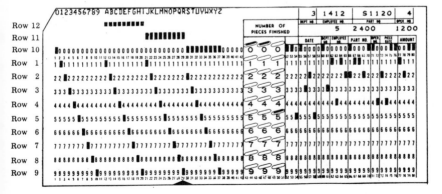

rows in each column. The standard numbering of these rows is shown to the left of the example card. The coding of data into the card is done on the following basis:

Character:	Hole Punched in Row Number:	Character:	Hole Punched in Row Number:
0	10 only	I	12 and 9
1	1 only	J	11 and 1
2	2 only	K	11 and 2
3	3 only	L	11 and 3
4	4 only	M	11 and 4
5	5 only	N	11 and 5
6	6 only	O	11 and 6
7	7 only	P	11 and 7
8	8 only	Q	11 and 8
9	9 only	R	11 and 9
A	12 and 1	S	10 and 2
B	12 and 2	T	10 and 3
C	12 and 3	U	10 and 4
D	12 and 4	V	10 and 5
E	12 and 5	W	10 and 6
F	12 and 6	X	10 and 7
G	12 and 7	Y	10 and 8
H	12 and 8	Z	10 and 9

Using this code, it is possible to "read" the punched card shown in Figure 21-5 by looking at the location of the holes in each individual column.

The data processing machines read the card by "looking" at the location of the holes in each individual column with small electric brush contacts. For each of the 960 locations on the card (80 columns $\times$ 12 rows) only one of two conditions can exist: there is a hole, or there is no hole. If a hole exists in a location an electric contact is made by the brushes and the machine will perform one operation. If there is no hole, the card will act as an electrical insulator and the machine will perform a different operation. Since the cards are fed into the machine so that all 80 columns are "read" at the same time, the data in the cards can be processed very rapidly.

To illustrate the operation of a punched card system, the step by step procedure for the inventory record routine, shown in Figure 21-6, will be considered.[2] The objective of this routine is to post a large number of daily inventory transactions to the inventory ledger cards.

Step 1. The daily inventory transactions can be broken down into four major classifications:
A. The issuance of material by the storeroom for the various production orders.
B. The allocation of material by the Production Control group for future production orders.
C. The ordering of replenishment material by the purchasing

---

[2] The reader should refer to Appendix IV for a more complete description of the individual machine operations.

department.

D. The receipt of new material into the storeroom.

The information for each individual transaction is first punched into separate cards.

Step 2. Since the transaction-punched cards are not arranged in any kind of order, the first machine operation is to arrange the cards numerically by part number.

Step 3. For every item in the inventory there is a punched card in the "previous balance summary deck." These cards, containing all of the information for the last posting to inventory ledger cards, are arranged numerically by part number. The next machine operation is to collate the current transaction cards into the previous balance summary deck. The cards for the inactive items in the summary deck are arranged in a separate deck.

Step 4. The accounting machine receives the cards from the collator and performs the necessary arithmetic operations for posting. As an example, it will take the previous inventory balance and subtract the number of items issued and determine the new balance.

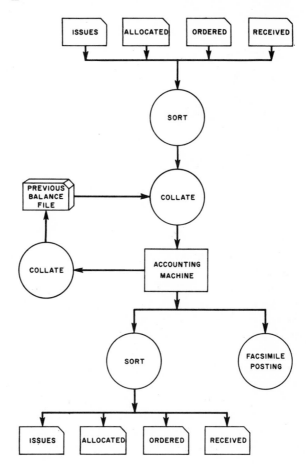

Fig. 21-6. Flow diagram of a simple punched-card inventory control system.

The accounting machine will make these calculations for each of the transactions.

Step 5. As the accounting machine makes the calculations for a transaction, the information is printed on a transaction listing. The final listing is sent to a facsimile posting machine for transfer to the inventory ledger cards.

Step 6. As the accounting machine completes all of the transactions for a particular inventory item, a new balance summary card is punched. All of the new cards are then collated into the summary deck for the inactive items obtained in Step 3. This collation makes current the information in the previous balance summary deck.

Step 7. All of the cards from the accounting machine are sent to a sorting machine where they are separated into the five groups shown in Figure 21-6. The obsolete summary cards are destroyed and the other decks are used for further analysis.

It is apparent by studying the preceding illustration of a punched card system, that the operations performed by the machines are basically simple. In fact, to a large extent, the data processing by machines simulates the exact operations required by the manual system.

The advantages of punched cards are:

   a. The system is capable of handling large amounts of data.

   b. The system can summarize data in many different ways, for control purposes.

   c. The system is very fast and accurate.

The disadvantages of punched cards are:

   a. The equipment is very expensive (either rented or purchased).

   b. The system usually cannot be used economically in small Production Control systems.

   c. The system is difficult to explain to the organization's personnel.

DIGITAL COMPUTERS

The number of installations of digital computers for processing Production Control data has been increasing since 1950. As technological innovations are made in the computer equipment, the potential areas of application are increasing at an even faster rate. It is still questionable, however, if these machines, despite their tremendous ability to process data, will ever become the panacea that some people now believe they are. The usage of computers in this area can be expected to increase in the future and every Production Control systems designer should have a basic understanding of what a computer can do.

It is unfortunate that a lot of grossly misleading information about computers has been circulated in the popularized types of books and periodicals. The terms "giant brain," "thinking machine," and the rest of the clichés have become part of the household vernacular. These terms have tended to put a very practical piece of equipment into the realm of mystery and allowed people's imaginations to run rampant about the machine's ability. The following discussion will explain the type of contribution a digital computer can be expected to make in a Production Control system.

The digital computer is capable of doing three basic things:

    a. It can perform the basic arithmetic operations of addition, subtraction, multiplication, and division.

    b. It can perform comparisons of numbers to determine which one has the greatest magnitude. This is commonly called "logical decisions."

    c. It can store large quantities of information and produce specific items of information whenever they are needed.

It is evident from the above list that the computer is actually rather limited in the number of different operations it is capable of doing. It is not the number of different operations which makes the digital computer unique; it is the speed at which the operations are performed. Using the IBM 705 computer as an example, the processing rates are:

    8,400 additions or subtractions (5-digit numbers) per second.

    1,250 multiplications (5-digit $\times$ 5-digit numbers) per second.

      550 divisions (4-digit into a 6-digit number) per second.

  29,400 logical decisions per second.

  58,800 characters can be "taken out of file" per second.

One of the unique features of a computer, which makes it particularly adaptable to the processing of Production Control information, is its ability to store data and to secure the data out of storage very rapidly. It is possible to provide the computer with a "limitless" storage capacity by using large files of punched cards or punched tapes. The trouble with these systems is the speed of access. Although information can be fed into machines from punched cards at a very rapid rate (up to approximately 200 cards or 16,000 numbers per minute), it is too slow for many computers. Because of the mechanical problems of handling the individual cards and "reading" their information with brush contacts through the holes, it is unlikely that the feed rate of punched cards and punched tapes will ever be increased to the rate of other information feed systems.

Many computers use magnetic tapes as the medium for feeding information into the machine and for storing large quantities of information. As with punched cards, data is coded on the basis of a two-state condition, but with magnetic tapes the conditions are magnetic spot-nonmagnetic spot, instead of the hole or no-hole states. Figure 21-7 shows the coding for a seven channel tape. The lines denote magnetized spots and the blank spaces non-magnetized spots. As the tape passes through the reading head of the computer at the rate of approximately 75 feet per second, all seven channels are read simultaneously. This provides an input rate of approximately 15,000 numbers per second. The storage capacity of one standard 2,400-foot tape is equivalent to over 25,000 punched cards. The problem with magnetic tapes, however, is the time necessary to obtain a *specific* piece of information. If the desired information happens to be at the end of the tape, it would take more than six minutes to reach it.

The magnetic drum found inside many computers provides a convenient method of storing information, but more important, it makes it possible to secure any specific piece of information in a very short period of time. The surface of the drum is divided into many small areas in which information can be stored by coded magnetic spots. As the drum is ro-

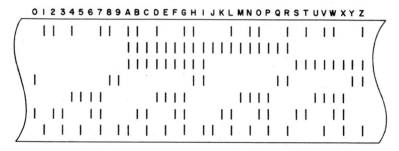

Fig. 21-7. Coding used on a seven-channel magnetic tape.

tated past a set of reading heads, the information in any specified area can be read. To reduce the time required for any area to get under the reading head, the drum is rotated continuously at very high speeds, frequently in excess of 12,000 r.p.m. The maximum time necessary to secure the information in any area on the drum is in the order of 0.005 seconds. This means that any information stored on the magnetic drum can be taken out of "file" in less than 5 micro-seconds. The limitations of the drum storage is the number of digits that can be stored because of the physical dimensions of the drum. Conventional drums can only store approximately 20,000 characters.

The magnetic disc storage system provides a mechanism for storing large quantities of information with relatively fast access to any specified file. This system is made up of a stack of two-sided discs as shown in Figure 21-8. The surface of each disc is divided into area in which information is stored by coded magnetic spots. The stack of discs shown in Figure 21-8 is capable of storing 5 million numbers and by using multiple stacks, the file can be expanded even further. This device provides both the filing capacity and the speed of access required by even the largest types of business applications.

To illustrate the use of a computer in a progress reporting system, we will discuss a portion of a system developed for the Navy Department for use in its manufacturing arsenals. It must be emphasized that this is only one possible use of the computers in the Production Control area. It is beyond the scope of the text to consider even a portion of the many actual computer systems currently operating in American industry.

The Navy system presupposes that four basic types of information must be included in all progress reports:

1. Information about the job order upon which work has been performed.
2. Information about the person performing the work.
3. Information about the machine upon which the work was performed.
4. Information about the work accomplishment.

To get this information into a computer, the system employs some rather unique devices. In order to understand the operation of the equipment, it will be necessary to consider these devices:

1. *Information about the job order:* For every operation on the jobs

Fig. 21-8. Magnetic disc storage unit.

going into the shop, a punched card similar to the one shown in Figure 21-5 is made up. This card contains the job order number and the operation number both in punched form and printed on the face.

2. *Information about the person:* Every employee in the shop is provided with a special identification card. These cards appear to be the usual plastic coated ID cards but in between the plastic laminations of the card is a layer of fine iron filings. By the use of magnetic spots the employee's name and identification number is coded into the card for machine reading.

3. *Information about the machine:* A plastic identification card, similar to the employee's card, is made up for every machine facility in the shop. This card is hung on a hook attached to the machine. The coded magnetic information in the machine card gives the machine and department identification.

Every department in the shop is equipped with the special recorder shown in Figure 21-9. This recorder is a remote input reader for a computer located in the Production Control area. When information is fed into the recorder it is automatically stored in the computer. Progress reporting using this equipment is as follows:

1. Whenever an operation is completed, the worker brings the machine identification card, his own identification card and a regular job ticket to the dispatcher. The dispatcher removes the punched card, containing the information about the job and operation number, out of file. The two identification cards and the punched card are inserted into the slots on the top of the recorder. Inside of the recorder are the necessary reading heads to interpret the coded data.

2. Taking the information from the job ticket, the dispatcher records the work accomplishment by setting the dials located on the front of the recorder. The dispatcher must be careful about the dial settings because this is the only information to be reported which has not been previously verified.

3. By pressing a switch on top of the recorder, all of the information on the cards and the dials is read into the computer. If more than one recorder tries to feed information into the computer at one time, the equipment will automatically sequence the acceptance of the information. This is done so rapidly that the dispatcher would prob-

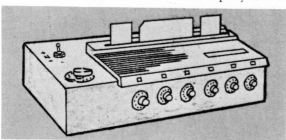

Fig. 21-9. A progress reporting recorder.

ably not be aware of the slight delay. In order to be sure that the computer receives all of the reported information, a small pilot light on the recorder flashes off after the data is properly received.

4. Inside the computer, all of the data reported by the shop is "filed" on a magnetic drum. If the progress report is the second one for a particular operation, the computer will automatically add the additional production data to the production quantity already in file.

5. Whenever a current progress report is needed for production control purposes, the information on the drum is printed out by using a high speed line printer.

The Navy system will ultimately provide the mechanism for producing many of the reports normally compiled by manual methods. As an example, it will be able to make summary cost reports on individual job orders, efficiency reports by departments or individuals, and so forth. In the future the use of this type of equipment will undoubtedly become more general.

At the present time the use of computers in Production Control systems is still in its infancy. New equipment at lower costs is continually being introduced into the market. These technological advancements are making computerized Production Control systems economically feasible in more and more organizations. The influence of computers is only starting. The Production Control systems designer should keep informed about the latest developments in this area.

The advantages of computers in Production Control systems are:

1. Data can be processed at very rapid rate.
2. Direct, swift access to many files is possible.
3. Computers essentially never make an error in computation.
4. Many different reports can be made available for management control.

The disadvantages of computers are:

1. The initial cost and the operating cost is high.
2. It is difficult to explain the operating system to the personnel.
3. It is too easy to generate reports of marginal value.

The use of the various devices discussed in this chapter can frequently reduce the progress reporting cycle appreciably. As with any other commercial device, the needs of the system should be determined first and then the commercial device should be selected. It is especially important to make a careful economic study when installing punched cards or computer equipment. The cost of these systems is very high and the potential savings in time and money must justify the expenditure.

Summary   One of the most difficult tasks in any organization, especially the larger ones, is the processing of the data created by the operation. This problem has received a tremendous amount of attention by the various data processing equipment manufacturers. The result of this is the introduction of many new data processing machines during the recent years. In this chapter the principles behind some of these devices were considered.

Before selecting any particular type of equipment, careful analysis of the requirements of the control system must be made. Every piece of new data processing equipment must be justified on the basis of making monetary savings or creating needed control information. It must always be remembered that this type of equipment is an aid to effective Production Control and does not provide a ready-made system in itself.

22

Corrective Action

The last phase of the Production Control system shown in Table 3-1 is the function of corrective action. Regardless of how accurately and completely the prior-planning and the action-planning phases of the system have been accomplished, it will always be necessary to make some adjustments to compensate for off-standard performance. It is sometimes argued that if the planning phases are carried out properly, there should rarely be a need for corrective action. The reasoning behind this type of thinking is very logical. It should be possible to anticipate the occurrence of any major problems before they occur and make the necessary adjustments while performing the planning functions. The myriad minor problems which may cause the off-standard performance of the facility can, as a group, be "anticipated" by providing sufficient flexibility in the plans to take care of them automatically. As an example, it was suggested in Chapter 15 that when using the loading-by-schedule-period control technique a certain portion of the facility's capacity be reserved for unexpected conditions. If either a rush job should materialize or the facility should not perform at the standard, the basic plans would not have to be changed. The "unanticipated" problems are automatically solved by providing the necessary flexibility in the schedule.

Unfortunately, it is generally not possible or economically feasible to eliminate completely the need for corrective action. There are too many variables entering into any type of production planning and control system. Even if it were possible to isolate and control the majority of the variables, in most cases the cost of doing this would be prohibitively high. It is therefore necessary to recognize that corrective action will have to be taken to some degree and that a mechanism for well-run, systematic corrective action must be established.

It must be remembered that the corrective action function of a Production Control system is *not* a substitute for adequate planning; it is an adjunct to proper planning. The direct and indirect cost of corrective action is considerably greater than the cost of intelligent planning. An organization should have as its goal the elimination of the unexpected work interruptions, but it must realize, that this goal is normally not attainable. Before considering the basic methods of corrective action it will be beneficial to look at some of the problems which may create the need for corrective action.

FACTORS CREATING THE NEED FOR CORRECTIVE ACTION

Although there is an infinite number of possible events that may require corrective action, we may simplify the discussion by considering separately the factors external to the organization from those internal to it.

It is possible to plan for both types of factors if they are important enough to warrant the expenditure of the necessary time and money. In general, the control of the internal factors is somewhat easier than the control of the external factors since it involves the variables which are directly under the jurisdiction of a management group. However, many of the external factors are indirectly controllable and should be investigated if they create the need for an undue amount of corrective action.

External Factors   One of the desirable principles of scheduling, frequently considered as a method of avoiding an excessive amount of corrective action, is the concept of a "firm" schedule. The basic idea behind this concept is to separate the work ahead of a facility into two classifications: the work to be done in the near future and the work to be done at a later date. The schedule of the work to be done in the near future is established on a "firm" basis, and theoretically this schedule will not be modified regardless of the circumstances. This provides the shop with a fixed plan of work and should prevent last-minute changes. Unfortunately, the firm schedule is either non-existent in most organizations or at best covers only a very short future period. The external factors creating the need for corrective action in many cases negate the possibility of establishing an effective firm schedule.

It is difficult to enumerate all of the external factors which may cause corrective action to become necessary, but two of the most common ones are:

1. The changes in the priority of the orders.
2. The delays in receiving the material into the facility.

The unexpected rush order is a major cause for changing the priorities of the orders assigned to a facility. In many organizations the receipt of a rush order leads to the complete replanning of the work schedule. This type of major disruption should be avoided if at all possible.

The careful analysis of the causes of rush orders can frequently lead to their elimination. In one case a machine job shop faced with an excessive number of rush orders began an investigation into the causes of the problem. Their findings typify the situation in many other organizations. The manufacturing facilities of this particular concern had the reputation of missing the promised delivery dates by a considerable margin. The processing of one of their typical orders had the following pattern:

1. The customer, knowing that his order might be delivered late, tells the salesman that he requires it one week before he actually needs it.
2. The salesman, having been disappointed on delivery dates before, writes the order out for delivery several days before the date specified by the customer.
3. The sales office introduces an additional "safety factor."
4. The Production Control department, knowing that the shop is falling behind schedule, subtracts several more days off of the shipping date and places the order into the shop as a "rush" job.

There is little wonder that the manufacturing facilities of this concern had a poor performance reputation. A large percentage of the time was devoted to taking corrective action on the latest rush jobs. Systematic Production Control under such conditions is practically impossible. The elimination of this type of needless rush orders requires a policy decision to be made by the management groups above the Production Control department. In the case of the machine job shop, the introduction of "safety margins" was eliminated by a management edict, the number of rush jobs was automatically reduced drastically, and the shop's delivery performance became considerably better.

Cancelled orders are another cause for changing the priorities of the work assigned to a facility. Whenever a customer cancels an order that has already been semi-processed, it is necessary to make adjustments in the "firm" schedule. The normal business practice in a case such as this is for the customer to compensate the supplier for the cost incurred in processing the order up to that point. It does nevertheless create the need for the Production Control group to take corrective action to adjust the established schedule.

Unexpected delay in the receipt of the incoming material is another external factor which may lead to corrective action. In many cases this is a very difficult factor to anticipate. The delay may be due to the material being "lost" in transit, a strike in the vendor's plant or a breakdown of the common carrier's equipment. Another cause for an unexpected delay which is frequently overlooked is the receipt of substandard material which cannot be used by the purchaser.

Although it is difficult to anticipate future delays in material arrival, there are several ways of avoiding the necessity of extensive corrective action. First, it is generally wise to select and use a *group* of vendors to supply the needs of the organization. If one of the sources of supply should become blocked, another source is readily available. The inventory policy of the organization should be designed to take care of the normal delays. The establishment of an emergency stock discussed in Chapter 12, Material Management, can reduce the number of times corrective action must be taken, but rarely can it completely eliminate it.

Internal Factors   There are many unexpected events occurring within the organization which create the need for corrective action. In Chapter 15 the conversion of the facility's available clock hours into available standard hours was discussed. It will be recalled that the four adjustments that had to be made were:

1. Adjustments for time lost due to tardy starting, coffee breaks, etc.
2. Adjustments for time lost due to absenteeism.
3. Adjustments for time lost due to lack of material, instructions, etc.
4. Adjustment for variations in the facility's work efficiency.

Each of these four adjustments was based upon an analysis of the facility's past performance. This means that the conversion from clock hours to standard hours is an estimation which may or may not be valid in the future. Whenever the facility ceases to perform in the same fashion as it did in the past, corrective action will be necessary. As an example, if the absentee rate should increase or decrease significantly, the number of available standard hours will change. The same situation would arise if the lost time attributable to the other factors should vary or if the facility's efficiency should change significantly. It is very likely that periodically one or more of these factors will change enough to require some kind of corrective action.

Another internal factor that all organizations face at one time or another is an unexpected equipment breakdown. It is possible to anticipate this problem, to a certain degree, by careful analysis of the equipment's past performance in order to estimate the percentage of lost time. This type of approach will usually work only in organizations where there is a large group of similar equipment. Preventive maintenance programs can frequently reduce the number of unexpected machine breakdowns, but these programs can become very costly. Regardless of how the problem of equipment failure is attacked, it is almost certain that some corrective action will have to be taken by the Production Control group because of this factor.

The entire function of Production Control is based on estimates developed to varying degrees of accuracy by many people in the organization. As an example, the time values established as standards of performance are used for scheduling the work into a facility. If these time values are not realistic, the schedule, no matter how cleverly conceived, cannot be realistic. This internal factor will create the need for corrective action. Other internal factors which will cause corrective action to be taken are the estimates for the standard tool requirements, the material reserve quantity, the material usage rates, and so forth. There are bound to be inaccuracies and mistakes in establishing these estimates, with the inevitable result that corrective action is necessary.

As was pointed out previously, it is impossible to anticipate *all* possible conflicting events that will occur in the future. We can, however, isolate those factors which are particularly troublesome and reduce their effect appreciably. For the remaining factors, the only solution available is to establish a systematic corrective action procedure that can be accomplished with a minimum amount of disruption to the facility.

METHODS OF CORRECTIVE ACTION

There are three basic ways for the Production Control organization to take corrective action:

1. Schedule Flexibility: develop a schedule with sufficient flexibility to allow the automatic adjustment to be made.
2. Schedule Modification: adjust the established plans to reflect what is happening in the facility.
3. Capacity Modification: adjust the operation of the facility to conform with the established plans.

It is impossible to establish categorically the most desirable technique to be used in different organizations. It is generally advisable to utilize all three methods of corrective action to varying degrees depending upon the situation.

Schedule Flexibility   When planning the activity of a facility, a certain amount of flexibility can be incorporated into the schedule to compensate for unexpected events. In order to do this, it is necessary to plan for only a portion of the total available standard hours. If a facility has 100 standard hours during a given time period, the scheduler might assign a group of jobs to the facility which should be completed in 85 standard hours. This group of jobs would make up the firm schedule. The scheduler would also provide an additional group of jobs that should be worked on *after* the firm schedule has been completed. The necessary schedule adjustments for off-standard performance or some other internal or external event are made automatically. If the facility requires more than the 85 standard hours to complete the firm schedule, an additional 15 hours are available. This means, that formal corrective action is unnecessary unless the facility is operating at less than 85 per cent effectiveness. If any rush jobs should occur during the time period, the additional 15 hours could be made available to them. This avoids the necessity of a complete revision in the established plans. Should the facility complete the firm schedule in less time than is anticipated, the additional work will be accomplished earlier than planned and the schedule for the next time period can be developed to reflect this condition.

Schedule Modification   Many times when an unexpected event creates the need for corrective action, the most economical adjustment is to modify the established plans to conform to the new set of conditions. This is particularly true if the facility should accomplish more work than was originally anticipated or if the current work load should be temporarily reduced. Schedule modification is such a common practice that many organizations rely on this type of corrective action almost exclusively. Its use avoids the necessity of overtime operations and if all of the assigned work can be accomplished on time, even after rearranging the schedule, this type of corrective action is fairly economical.

The major problem encountered with a schedule modification is making the transition from the original plans to the new plans with a minimum amount of disruption in the operation of the facility. In many organizations one of the major cost items is the time and money spent in setting up the equipment. When schedule modifications are necessary under such circumstances, it is essential that careful consideration be given before breaking any number of set-ups. Changing work before the completion of an operation might make it possible to finish the rush jobs on time but at the same time it reduces the over-all productive capacity of the facility and increases the cost of operation.

Capacity Modification   There are three basic ways in which a facility's capacity can be changed:

1. Change the number of standard hours available.
2. Change the amount of work within the system.
3. Expedite the work.

In most organizations all three methods of capacity modification are available to the Production Control group as means for taking corrective action. There are advantages in each one and they should all be considered before a final choice is made.

Changing the number of standard hours available in a facility is one of the most common ways of obtaining additional capacity. Many times this can be accomplished simply by having enough flexibility in the work force to allow the shifting of manpower from one facility to another. In order to achieve this flexibility it is necessary to have a continuous program of training so that the workers are capable of doing many different jobs. Probably the most common way of changing the number of available standard hours is to have the facility operate on an overtime basis. Changing the number of hours worked provides a considerable amount of flexibility of output, especially on a short-term basis. The work day or the work week can be increased by 25 or 30 per cent without undue hardship upon the employees. Indeed, many workers look forward to overtime operations because of the premium wages they must be paid. Because of these premium wages, the direct labor costs are increased. However, the additional production will absorb a portion of the fixed overhead costs and thus will compensate to a great extent for the wage premium paid to the workers.

Changing the amount of work within the system is a method of corrective action which is often overlooked by Production Control groups. Most organizations have the option of doing the work themselves or "farming out" a portion of the items. The "make or buy" decision should be a function of the current conditions in the facility, but often it is considered an emergency measure. This attitude is understandable. The profit realized by doing the work within the facility is normally higher than the profit associated with "farmed-out" work. The important point that is overlooked, however, is that the policy of subcontracting varying amounts of work as a method of corrective action can appreciably reduce the required capacity of the facility. The work load within the plant can remain relatively constant while the variation in the demand for work capacity can be taken up by the quantity of work released to outside vendors.

Expediting the work through a facility is generally considered to be an undesirable practice which should be avoided if at all possible. However, the old Production Control axiom, "The better the system—the fewer the expediters," is not necessarily true. The proper use of expediters can increase the effectiveness of most Production Control systems and at the same time reduce the cost of operating the system. There are two basic methods of expediting:

1. The "exception" principle.
2. The "fathering" principle.

The selection of the method to be used is naturally dependent upon the type of work being done by the facility. Both methods have their advantages and disadvantages.

Under the exception principle, the expediter's time is devoted to the work which is "off" the established schedule. The scheduling section, using the latest progress reports, lists on a "trouble sheet" the jobs which are falling behind schedule. The expediting section is responsible for investigating and correcting the conditions which are causing the difficulty. The expediters commonly find "lost jobs" in the shop, have material moved from one department to another, secure the needed tools for an operation, have the purchasing department procure additional material, and so forth. After the problems have been investigated and all possible corrections made, a report is made to the scheduling section so that the work can be rescheduled.

The "fathering" system of expediting is based upon an entirely different concept. Each expediter in the system is assigned a group of jobs for which he is responsible. He must make sure that the necessary tools, materials, equipment, and so on are available when needed. If any of the necessary items are missing, the expediter must take the necessary steps to provide them. The "fathering" system is particularly useful for controlling large projects such as construction of buildings, repair of ships, and the manufacturing of custom-made machinery. In such projects it is very difficult to schedule the multitude of relatively small activities.

Summary In any Production Control system, the function of corrective action must be carefully considered. There will always be certain internal and external factors which will create the need for corrective action. It is impossible to eliminate all of these factors. However, analysis of the major trouble areas can often reduce the problem to a minimum. The function of corrective action should not be established on an emergency basis, but should be planned as carefully as material control, scheduling, dispatching, and the rest of the Production Control functions. To a great extent the success of the entire Production Control system depends upon effective corrective action.

23

Procedure Analysis

Basically, procedure analysis has been the object of this text from the beginning. Procedure analysis has been described specifically as it applies to Production Control. *We defined Production Control as "the design and use of a systematic procedure for establishing plans and controlling all the elements of an activity." Procedure analysis is defined as "the scientific method of selecting or designing a preferable data and information processing routine in an activity."*

It is quite obvious from the above that procedure analysis is a useful tool in designing Production Control systems. It was stated in Chapter 1 that one of the objectives of the text was to demonstrate methods of analyzing Production Control problems and designing or evaluating and improving Production Control systems. It is the purpose of this chapter to present the techniques which are used to *analyze* and *develop* systems.

It may seem to the reader at this point that it may have been more logical to place this chapter at the beginning of the text. After all, the objective is to design a system and in order to design a system, it is necessary to make an analysis and go through a "design" procedure. However, designing requires a knowledge of specifications of the product which is being designed. The specifications of the product in this case are the functions and principles of Production Control. Without a thorough knowledge of these, it is virtually impossible for a designer even to begin on the design of the Production Control system.

It is not intended in this chapter to go into a very intensive investigation of procedure analysis. That is left to the texts which are specifically designed for procedure analysis. The intent is only to outline the steps to be used in procedure analysis and to present what is felt to be the best technique to use in procedure analysis.

The basic approach to the problems of procedure analysis is through the use of the "scientific method." A common-sense application of a systematic approach assures the analyst that he will usually produce better procedures than will hit-or-miss ingenuity.

The scientific method consists of the following seven steps:

1. *Objective.* The determination of what is to be studied and changed, the objective sought, and the establishment of a yardstick to evaluate the desirability of any contemplated change.

2. *Analysis.* The breaking down of the procedure into parts or subdivisions so that a comparison can be made to what is known about such parts or subdivisions.

3. *Evaluation.* The application to each of the subdivisions of a checklist to question the best way of doing each part and the possibility of combining parts.

4. *Improvement.* The formulation of a better way, using information brought out in Step 3.

5. *Test.* A preliminary check of whether or not the proposal of Step 4 meets the requirements of the objective. Also, a discussion of the proposal with all personnel concerned.

6. *Trial.* A sample application of the *improvement* to make sure all pertinent factors have been taken into account.

7. *Application.* The standardization and recording of the proposal; the training of the personnel who are to use the procedure; and the careful evaluation of the results of the proposal so as to provide more information for future use of Step 3 with other tasks.

It is interesting to note that these steps are really a summary of the procedure that should be used to solve any problem. The details of performing the various steps change with the type of problem, but the pattern remains the same. These steps should become a normal part of the designer's approach to procedure problems.

STANDARD PROCEDURE FOR PROCEDURE ANALYSIS

The following outline is the formal way of performing each of the seven steps of the scientific method. The designer should use it as the "standard procedure" for procedure analysis.

STEP 1—OBJECTIVE (Goal)

1. *Indefinite Goal*—If the problem is unspecified and its definition requires an examination of the problem area.
   1.1 To examine the problem from the viewpoint of organization.
      1.11 Study the Organization Manual and Charts for the activity. This will provide a simple, compact picture of the organization and its parts, and facilitate the definition and analysis of component procedures.
      1.12 If no manuals or charts are available, prepare them as out-

TABLE 23-1

OFFICE LAYOUT CHECKLIST

1. Does major work flow with a minimum of backtracking?
2. Has unused equipment been returned to stock?
3. Are machines, files, and reference materials placed at the desks or within easy reach of employees who use them?
4. Are related units near each other? Is transportation distance at a minimum among and within divisions? Branches? Sections? Are individuals having the most frequent contact located near each other? Does each worker receive his work from an adjacent desk?
5. Are desks (except those used by employees who work together) faced in the same direction?
6. Are desks which face in the same direction placed at least 2½ feet apart?
7. If employees work back to back, is there at least 4 feet between their chairs?
8. Are individuals who receive frequent visitors located near the office entrance?
9. Are employees placed in front of or within easy view of their supervisor?
10. Are main aisles at least 5 feet wide? Other aisles at least 3 feet wide?
11. Are desks faced away from the light?
12. Is heavy equipment placed against walls or columns?
13. Are private offices kept to a minimum? Are conference rooms available? Would railings or movable panel partitions serve in lieu of private offices, provided conference or committee rooms were available?
14. Are movable panel partitions, filing cabinets and other suitable pieces of furniture used to segregate and give semiprivacy to clerical activities?
15. Are employees who do the closest, most exacting work placed nearest the light?
16. Is lighting adequate?
17. Does the layout plan consider location of electrical outlets, etc? Does it allow space for opening doors, desk and file cabinet doors, etc?
18. Are valuable equipment and supplies placed in secure locations?

lined in detail in Chapter 27 or in outline form as required.

1.2 To examine the problem from the viewpoint of space arrangement.

    1.21 Prepare a layout chart of the activity.

        1.211 Utilize a blueprint of the floor space assigned, if available; locate equipment on the drawing.

        1.212 If a blueprint is not available, draw a diagram of the floor space assigned and locate the equipment on the floor plan.

        1.213 Trace the flow of the major documents between work places.

    1.22 Apply the Office Layout Check List of questions as shown in Table 23-1 on preceding page.

Fig. 23-1. Form used to show breakdown of the functions of a works center and the volume of work done in each function.

WORK ACTIVITY ANALYSIS SHEET		
ORGANIZATION *Inventory Control Section Production Control Dept.* SUPERVISOR *Dave Blair*		DATE *11 Jan. '60*
ACTIVITY NUMBER	ACTIVITIES	ACTIVITY UNITS AND/OR VOL. (OPTIONAL)
1	Maintains punched card tub files on each new material and finished product item.	12,000 stock keeping units
2	Maintains up-to-date reorder points on all items.	
3	Reports to Production Scheduling those items that are manufactured and have reached the reorder point.	100 per week
4	Reports to Purchasing those items that are purchased and have reached the reorder point.	600 per week
5	Reports Out-of-stock items to Back Order section.	
6	Pulls punched cards from tub files for all orders and requisitions and sends to Tabulating Dept.	200 per week
7	Misc.	

1.3 To examine the problem from the viewpoint of personnel and workload requirements.

    1.31 Prepare a Work Distribution Chart of the activity's personnel.

        1.311 Use Work Activity Analysis Sheet, Daily Clerical Utilization Chart, or Task List, depending on accuracy desired, as shown in Figures 23-1, 23-2, and 23-3.

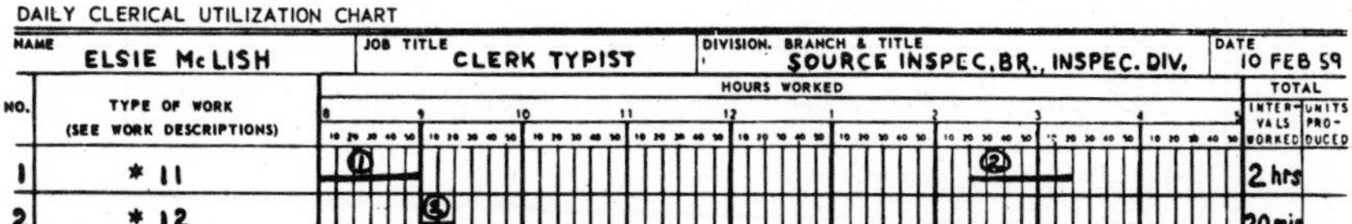

Fig. 23-2. Form used to show the time phasing of the work of a specific clerk in a clerical activity.

        1.312 Supply personnel concerned with ruled sheets or printed charts.

        1.313 Request personnel to maintain a log of the tasks they perform for an adequate period (generally one *normal* work week).

        1.314 Explain purpose in detail and assist in preparing standard terms for describing major tasks to facilitate later summarizing.

        1.315 Answer questions, and assist and supervise the collection of data.

        1.316 Summarize the tasks performed by each individual on a Work Distribution Chart such as shown in Figure 23-4.

TASK LIST					
NAME *Joan Lansfield*	WORKING TITLE *Card Puller*		GRADE *G-2*		
ORGANIZATION *Tub Files Inventory Control Section*	SUPERVISOR *Dave Blair*		DATE *11 Jan. '60*		
TASK NO.	DESCRIPTION		HOURS PER WK.	WORK COUNT	ACTIVITY NO.
1	Pulls card for each item on order or requisition.		20	4,000	6
2	Files cards in tub files.		10	1,000	1
3	Breaks.		2½		7
4	Personal time.		2½		7
5	Sorts orders for card pulling.		5	1,500	7
		Enter totals after last entry	40		

Fig. 23-3. Form used to show the breakdown of the work performed in a specific task in a works center with reference to the activities described on the work activity analysis sheet shown in Fig. 23-1.

1.32 Determine the most pressing needs from the above data by applying the following questions:

1.321 What activities take the most time?

1.322 Is there misdirected effort?

1.323 Are skills being used properly?

1.324 Are individuals doing too many unrelated tasks?

1.325 Are tasks spread too thinly?

1.326 Is work distributed evenly?

1.33 Designate the activities or areas needing detailed study.

1.4 To examine the problem from the viewpoint of equipment utilization or to confirm the results of the workload requirement survey (see 1.3) perform a work sampling study as outlined in detail in Chapter 24.

1.5 To examine the problem from the viewpoint of forms and reports.

1.51 Make a functional analysis of forms and reports on a Forms Analysis Chart such as shown in Figure 23-4.

1.511 List basic actions or functions across the top of the form.

1.512 List subjects or operational functions in left column.

1.513 Enter the number of forms that apply to each subject and function.

1.52 Subject all forms or reports in a particular block to detailed item analysis when of high frequency as shown in Figure 23-6.

FORMS ANALYSIS CHART

Page 1 of 1 Pages

ACTIVITY: Personnel Office
DATE OF ANALYSIS: 10 JULY 1959
ANALYZED BY: Management Staff

OPERATIONAL FUNCTIONS	ADJUST	APPLY FOR	ASSIGN	AUTHORIZE	BILL	CERTIFY	CLASSIFY or RATE	CONTRACT	DIRECT	FOLLOW UP	IDENTIFY	INSTRUCT or TEACH	ORDER or PROCURE	RECEIPT FOR	RECEIVE	RECORD	REPORT or NOTIFY	REQUISITION or REQUEST	ROUTE or TRANSMIT	TOTAL
1. Employee Relations	2		4						3								5		3	19
2. Salary and Wage	8	3	5	7		6	8		6	4						9	6	4	5	71
3. Employee Utilization	3	7	8	12		4			8	9	4		1	3	2	3	3	2	4	73
4. Personnel Administration				4		2		8	4	6	9		2	3	7	14		5	6	70
5. Training			6	4		8		5	6			15					3	3	8	58
6. Safety			4	6		3	4	2	3	6			3	4	1	2	8	2	6	54
7. Medical	6	2	2	5	3	8			4	5	2	3				7	9	1	9	66
8.																				
9.																				
10.																				
11.																				
12.																				
13.																				
14.																				
15.																				
TOTAL	19	12	25	38	7	31	12	15	21	37	12	15	21	9	6	28	40	22	41	411

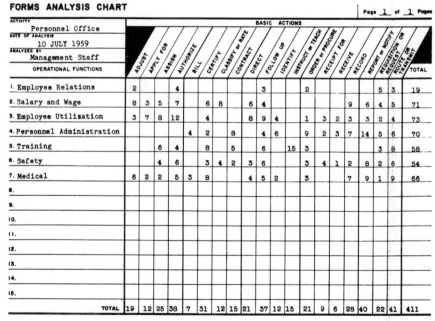

Fig. 23-4. Form used to make a functional analysis of forms and reports.

WORK DISTRIBUTION CHART

ORGANIZATIONAL UNIT CHARTED: **Cost Control**

☒ EXISTING ORGANIZATION ☐ RECOMMENDED ORGANIZATION

CHARTED BY: **Starr Brock** APPROVED BY: _____

Positions charted:

Name	Position	Grade
A. B. Seaward	Adm. Assistant	GS-9
Paul Smith	Fiscal Clerk	GS-7
A. B. Seaward, Jr.	Cost Clerk	GS-5
W. D. South	Cost Clerk	GS-4
S. N. North	Typist	GS-2

Activity totals (Work Count / Hours per Week):

No.	Activity	Work Count	Hours per Week
1	Prepare Regular Reports	15	30
2	Post Cost Records	85	83
3	Maintain Project Identification System	50	5
4	Prepare Special Reports	4	23
5	Miscellaneous		59
	TOTALS (man-hours)		**200**

Tasks by position:

Activity	Position (Name / Grade)	Task	Work Count	Hours per Week
1 Prepare Regular Reports	Paul Smith (GS-7)	Monthly Summary of Project Costs		5
1	Paul Smith (GS-7)	Monthly Report of underruns & overruns		1
1	A. B. Seaward, Jr. (GS-5)	Monthly Summary of Project Costs		5
1	A. B. Seaward, Jr. (GS-5)	(Monthly Summary of Project Costs, cont.)		1
1	W. D. South (GS-4)	Monthly Summary of Project Costs		4
1	W. D. South (GS-4)	Monthly Recap total Costs by Location		1
1	S. N. North (GS-2)	Type & Distribute Monthly Summary	1	10
1	S. N. North (GS-2)	Type & Distribute Monthly Cost Recap Report	1	2
1	S. N. North (GS-2)	Type & Distribute Weekly Report		4
2 Post Cost Records	Paul Smith (GS-7)	Post Labor and material costs		12
2	Paul Smith (GS-7)	Post committed Funds		3
2	A. B. Seaward, Jr. (GS-5)	Post Labor and material costs		12
2	A. B. Seaward, Jr. (GS-5)	Post committed Funds		3
2	A. B. Seaward, Jr. (GS-5)	Distribute Vouchers		
2	W. D. South (GS-4)	Post Labor and material costs	28	25
2	W. D. South (GS-4)	Post committed Funds	4	5
2	W. D. South (GS-4)	Funds		6
3 Maintain Project Identification System	S. N. North (GS-2)	Assign Project Nos.		1
3	S. N. North (GS-2)	Type Project Card when opened & file		2
3	S. N. North (GS-2)	Post Project Cards when closed & refile		2
4 Prepare Special Reports	A. B. Seaward (GS-9)	Prepare Special Reports		
4	Paul Smith (GS-7)	Prepare Special Reports		5
4	S. N. North (GS-2)	Type Special Reports		5
5 Miscellaneous	A. B. Seaward (GS-9)	Attend Meetings, conferences, etc.		
5	A. B. Seaward (GS-9)	Read mail and reports		
5	A. B. Seaward (GS-9)	Prepare Correspondence		
5	A. B. Seaward (GS-9)	Supervision		
5	A. B. Seaward (GS-9)	Miscellaneous		
5	Paul Smith (GS-7)	Review Mail & Distribute		5
5	Paul Smith (GS-7)	Miscellaneous		9
5	W. D. South (GS-4)	Attend Safety Meetings		4
5	W. D. South (GS-4)	Miscellaneous		2
5	S. N. North (GS-2)	Take Dictation		4
5	S. N. North (GS-2)	File Correspondence		3
5	S. N. North (GS-2)	Meetings and Personnel Duties		5

Column totals (each position = 40 hours): GS-9, GS-7, GS-5, GS-4, GS-2 — 40 each.

ANALYSIS: WHAT TAKES THE MOST TIME?...IS THERE MISDIRECTED EFFORT?...ARE SKILLS USED PROPERLY?...ARE THERE TOO MANY UNRELATED TASKS?...ARE TASKS SPREAD TOO THINLY?...IS WORK DISTRIBUTED EVENLY?

¹Work Count data is optional.

Fig. 23-5. Form used to summarize the information obtained from the task lists such as the one shown in Fig. 23-3.

FORMS ANALYSIS CHART OF RECURRING DATA

ACTIVITY: Supply of Class B Merchandise

DATE OF ANALYSIS: 11th January, 1960

ANALYZED BY: Dave Blair

TITLE OF FORM

ITEMIZED DATA	Requisition	Bal. of Store Card	Gen. Weekly Instr.	Shipping List	Shipping Ticket	Bill of Lading	Unit Price File	Store Ledger	Monthly Buy Sheet				TOTAL
FORM NO.	845			505	86								
1 Addressee	X			X	X	X							4
2 Name of Firm	X	X	X	X	X	X	X	X	X				9
3 Store Number	X			X	X	X			X				5
4 Order Number	X			X	X	X		X	X				6
5 Requisition Number	X			X	X	X		X	X				6
6 Date of Order	X			X	X	X							4
7 Required Date	X			X	X								3
8 Ship To	X			X	X	X							4
9 Item Number	X			X	X				X				4
10 Stock Number	X	X	X	X	X	X		X					7
11 Quantity Ordered	X			X	X				X				4
12 Nomenclature (Description)	X	X	X	X	X	X	X						7
13 Signature	X			X	X								3
14 Issues		X											1
15 Receipts		X											1
16 Balance		X											1
17 Posting Date		X						X	X				3
18 Items Added			X										1
19 Items Discontinued			X										1
20 Quantity Authorized			X										1
21 Unit Price				X			X						2
22 Quantity Shipped				X	X	X							3
23 Price Extension							X	X					2
24 Weight					X	X							2
25 Number of Packages					X	X							2
26 Date Shipment Prepared					X								1
27 Date Received					X								1
28 Date Shipped						X							1
29 Shortages/Overages									X				1
30													
31													
32													
33													
34													
35													
36													
37													
38													
39													
40													
41													
42													
TOTAL	13	7	6	15	18	13	4	6	8				90

Fig. 23-6. Form used to show detailed item analysis of a group or block of forms and reports used in a particular activity.

2. *Specific Goal*—If the problem has been specified or developed from one or more of the steps outlined above.

   2.1 Define the purpose of the study.

   2.2 Determine the affected organizational segments (see 1.1 above).

   2.3 Determine the affected personnel (see 1.3).

   2.4 Determine relationship to existing procedures and forms (see 1.5).

   2.5 Using experience, direction, or a balancing of needs to guide you, select one or more of the most desirable items below to "optimize." Note that it is difficult to optimize all factors at once.

      2.51 Fewer people.

      2.52 Fewer steps (in procedure or in using form).

      2.53 Less time on a step or steps.

      2.54 Less time in production (for over-all procedure to be completed).

      2.55 Less office space to be used.

      2.56 Less time for critical skills (or of critical personnel).

      2.57 Less time on critical equipment.

      2.58 Increased quality (accuracy).

      2.59 Less cost.

      2.510 Less skill on various steps.

      2.511 Better control.

      2.512 Other (intangible benefits, safety, etc.).

   2.6 Check the objective(s) selected with the supervisor or those affected or concerned.

   2.7 *Preliminary Analysis* (limiting factors)

      2.71 Make a general determination of the degree of change that is desirable or feasible by considering the following:

         2.711 How great is the actual or expected volume of the activity?

         2.712 How long will this activity last?

         2.713 How much time per unit is spent on the job?

         2.714 How much time is available to process a change?

         2.715 How much analysis time will be required?

         2.716 How much time would be lost in a change?

         2.717 How pressing is the need for change?

         2.718 How much retraining would be required?

         2.719 What is the possible saving?

         2.7110 What are the personalities involved?

         2.7111 Does policy allow a change?

         2.7112 Is any aspect beyond control? Who must be consulted?

      2.72 Evaluate each of the above at this initial stage.

      2.73 Based on the above evaluation, set limits for the study.

      2.74 Proceed to second main step, "Analysis."

METHODS	PROPOSAL SUMMARY TEAM #3	DATE 6/12/59

X PROCEDURE

TO: Mr. W. Smith	FROM: Stewart, Magagna, Shidler, Arscott, St. Germain	ESTIMATED ANNUAL SAVINGS: $ 5287.78	NET, X GROSS

SUBJECT: Processing of Purchase Orders In General Accounting. AREA: General Accounting Section Journal & Accounts Payable Unit

ADOPTION OF PROPOSAL WILL RESULT IN: (Check appropriate blocks)

X	New or Improved Procedure		Increased Production	X	Fewer Steps
X	Less Time to Complete Procedure		Fewer People	X	Less Time on Step(s)
	Less Time on Critical Skills		Better Control	X	Less Skill on Step(s)
	Less Time on Critical Equipment		Less Space	X	Increased Quality/Accuracy
	Intangible or Other (explain below)	X	Improved Forms	X	Reduced Cost

IF THIS PROPOSAL IS APPROVED, IT WILL BE NECESSARY TO: (SUMMARIZE)

1. Revise Form 261-1, Follow-up on Invoices.
2. Eliminate the Following:
   a. Form 14-1, Bills Register.
   b. Hold for Invoice File.
   c. Hold for Receiving Report File.
   d. All time-stamping of documents.
   e. Multiple handling of Purchase Orders and related documents by Voucher Examiner Supervisor.
3. Utilize the present Purchase Order Holding file as a Holding File for all documents.

FINANCIAL ASPECTS & EVALUATION OF CHANGE: (SUMMARIZE)

LABOR COSTS

		% PRESENT		% PROPOSED	
Supervisory Accountant	GS-11	6.0	$ 383.40	6.0	$ 383.40
Voucher Examiner Supr	GS- 6	48.0	1,958.40	24.0	979.20
Voucher Examiner	GS- 4	222.3	7,591.30	96.0	3,282.72
Mail Clerk	GS- 3	6.0	190.50	6.0	190.50
			$ 10,123.60		$ 4,835.82

7500	Present Annual Volume	$ 10,123.60	Present Labor Cost
$_____	Labor Rate (For Methods Use Only)	$ 4,835.82	Proposed Labor Cost
_____	Production Rate (For Methods Use Only)	5872	Present Manhours
_____	Proposed Production Rate (For Methods Use Only)	2746	Proposed Manhours
$ 1.35	Present Cost Per Purchase Order	3126	Net Savings Manhours
$ .64	Proposed Cost Per " "	$_____	Cost of Change
$ .71	Savings Per " "	_____	Time Required To Pay For Cost of Change

ATTACHMENTS (Insert Number of Sheets in Proper Boxes: Follow With Page Numbers on Line)

/ Page /	USE FOR METHODS ONLY	/ Page /	USE FOR PROCEDURES ONLY
_____	Present Methods Chart	X 3	Present Procedures Chart (SPAS Nr. 3)
_____	Preliminary Possibility Guide	X 4	Proposed Procedure Chart (SPAS Nr. 3)
_____	Detailed Possibility Guide	_____	Present Work Distribution Chart
_____	Proposed Methods Chart (Incl. Summary, Comparison)	_____	Proposed Work Distribution Chart
_____	Jig, Fixture, Work Place Sketches	_____	Forms Analysis of Recurring Data
_____	Cost of Change Estimate Sheet	_____	Typewriter Analysis of Forms
_____	Job Instruction Sheets	X 2	Summary Analysis of Proposal (SPAS Nr. 2)
_____	_____	_____	Proposed Equipment List & Details of Placement
_____	_____	X 5	Forms Design & Instructions For Use of Forms
_____	_____	_____	Space Layout (Before & After)
_____	_____	X 1	Narrative

Fig. 23-7. Form used to show the results of a procedures study in terms of monetary savings and intangible benefits.

STEP 2A—ANALYSIS (Procedures)

1. Use SPAS (Standard Procedure Analysis Set)
   1.1 SPAS No. 1 as shown in Figure 23-7 is the form used to summarize the results of the study on the basis of benefits and monetary savings.
   1.2 SPAS No. 2 is the symbol comparison sheet of the present and proposed procedure which summarizes the number of steps involved in these versions. It also shows the number of different steps involved, represented by the different symbols, in the handling of each type of document throughout the present and proposed procedure. (See Figure 23-8).

ANALYSIS SUMMARY — SPAS NR. 2

TITLE OF PROCEDURE: Preparation and issue of travel orders — DATE: 5 Jan 59 — ANALYST: W.A.S.

FORMS (PRESENT)	NUMBER CYS	PERS. HDLG	CYS FINALLY ▽	X	⊠	EXPRESS O	⅂	o	◇	▽	▼	FORMS (PROPOSED)	NUMBER CYS	PERS. HDLG	CYS FINALLY ▽	X	⊠	EXPRESS O	⅂	o	◇	▽	▼
DD 662	8	20	2	0	6	98	32	43	24	12	54	DD 662 Rev.	6	9	1	0	5	65	16	27	9	8	22
SF 1012	9	6	0	0	9	16	0	14	10	0	0	SF 1012	4	2	0	0	4	4	0	8	0	0	0
SF 1038	1	2	0	1	0	8	0	2	18	0	16	SF 1038	-	-	-	-	-	ELIMINATED	-	-	-	-	-
DA 455 Register	1	1	0	0	1	10	3	0	0	0	0	DA 455 Register	-	-	-	-	-	ELIMINATED	-	-	-	-	-
Add Tape	2	2	0	1	1	14	8	0	0	0	0	Add Tape	1	1	1	0	0	6	2	0	0	0	0
Local Travel log	1	2	0	0	1	9	6	0	0	0	0	Local Travlog	-	-	-	-	-	ELIMINATED	-	-	-	-	-

* Circle increases over present procedure

	FORMS	NUMBER COPIES	PERS. HDLG	COPIES FINALLY ▽	X	⊠	EXPRESS O	⅂	o	◇	▽	▼
PRESENT PROCEDURE TOTALS	6	22	33	2	2	18	155	49	66	52	12	70
PROPOSED PROCEDURE TOTALS	3	11	12	2	0	9	75	18	35	9	8	22
* DIFFERENCE	3	11	21	0	2	9	80	31	31	43	4	48

A. Place name or number of forms in "FORMS" columns. Also account for files, registers, etc. in the same column.
B. Place correct number or zero in proper boxes.
C. Use one line for each form, carrying across both "PRESENT" and "PROPOSED". Write the word "ELIMINATED" in the "PROPOSED" column if form is eliminated. Use the same line for a new form which replaces a present from.

LEGEND:
X DISPOSAL — ⅂ INFORMATION-OFF
▽ DELAY/STORAGE — X GAP/OR UN-CHARTED ACTIVITY — o MOVEMENT
▼ TEMP. STORAGE — O OPERATION — ◇ INSPECTION

Fig. 23-8. Form used to compare a present with a proposed procedure in terms of the number of steps and time.

1.3 SPAS No. 3 is used to make a flow chart in time sequence of the various tasks, operations, or steps in a procedure to facilitate a detailed analysis of that procedure. It is used for charting both the present and proposed procedure (see Figure 23-9).

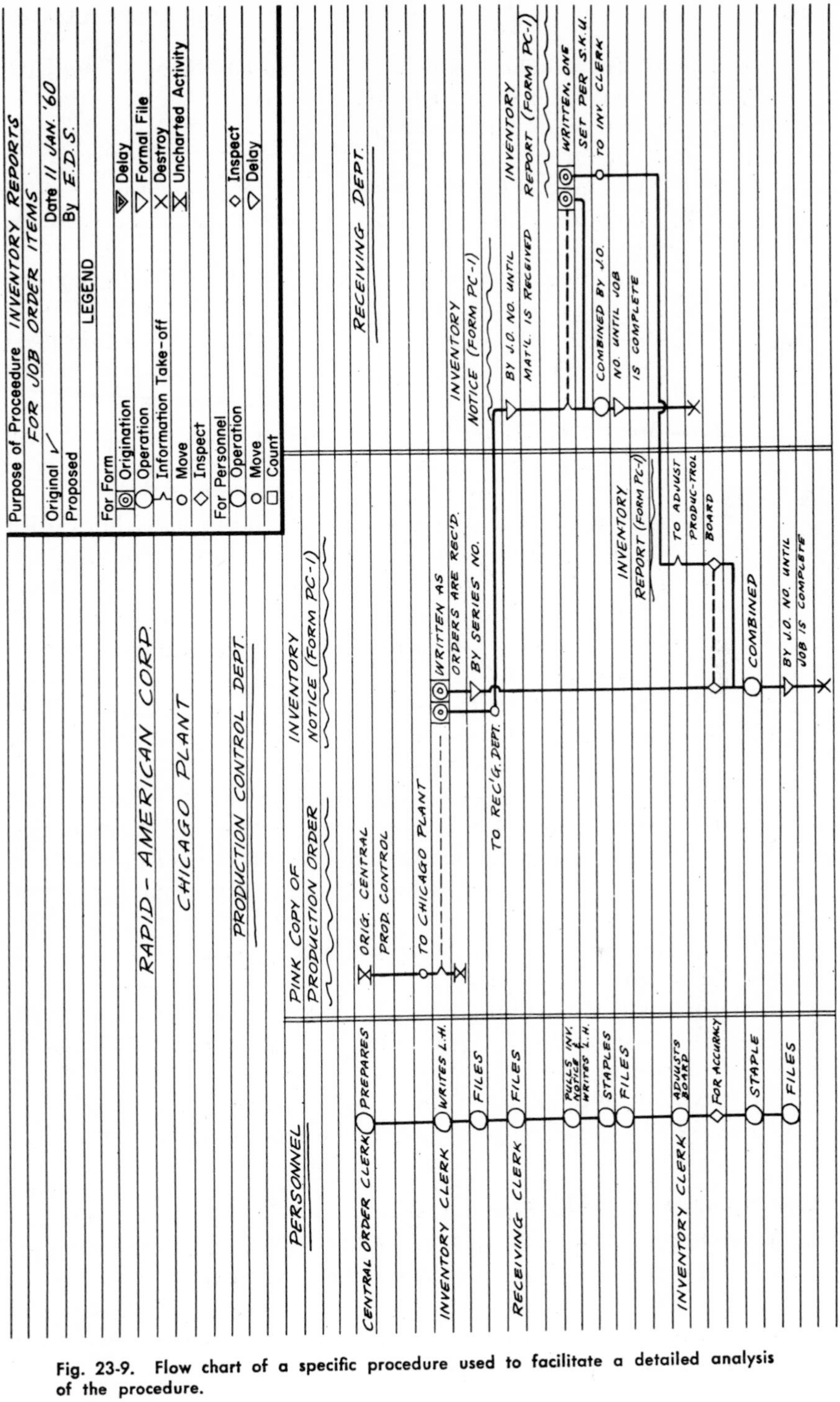

Purpose of Procedure **INVENTORY REPORTS**
FOR JOB ORDER ITEMS

Original ✓	Date 11 JAN. '60
Proposed	By E.D.S.

LEGEND

For Form
- ◉ Origination
- ○ Operation
- ↣ Information Take-off
- ○ Move
- ◇ Inspect

For Personnel
- ○ Operation
- ○ Move
- □ Count

- ▽ Delay
- ▽ Formal File
- ✕ Destroy
- ⊠ Uncharted Activity

- ◇ Inspect
- ▽ Delay

RAPID - AMERICAN CORP.

CHICAGO PLANT

PRODUCTION CONTROL DEPT.

RECEIVING DEPT.

INVENTORY NOTICE (FORM PC-1)

INVENTORY NOTICE (FORM PC-1)

INVENTORY REPORT (FORM PC-1)

PINK COPY OF PRODUCTION ORDER

INVENTORY NOTICE (FORM PC-1)

INVENTORY REPORT (FORM PC-1)

✕ ORIG. CENTRAL PROD. CONTROL

TO CHICAGO PLANT

TO REC'G. DEPT.

WRITTEN AS ORDERS ARE REC'D.

BY SERIES NO.

BY J.O. NO. UNTIL JOB IS COMPLETE

COMBINED

TO ADJUST PRODUC-TROL BOARD

BY J.O. NO. UNTIL MAT'L. IS RECEIVED

COMBINED BY J.O. NO. UNTIL JOB IS COMPLETE

WRITTEN, ONE SET PER S.K.U. TO INV. CLERK

PERSONNEL

CENTRAL ORDER CLERK ○ PREPARES

INVENTORY CLERK ○ WRITES L.H.
○ FILES

RECEIVING CLERK ○ FILES
○ PULLS INV. NOTICE L.H. WRITES L.H.
○ STAPLES
○ FILES

INVENTORY CLERK ◇ ADJUSTS BOARD
◇ FOR ACCURACY
○ STAPLE
○ FILES

Fig. 23-9. Flow chart of a specific procedure used to facilitate a detailed analysis of the procedure.

    1.4 The SPAS is numbered in the order of assembly for a completed study. The order of preparation is reversed.

2. For the analysis of an existing procedure:
    2.1 Gather the facts:
        2.11 Determine *What* happens to start the procedure.
        2.12 Determine *Where* in the organization the procedure actually begins.
        2.13 Determine *Who* first receives or initiates the documents in question.
        2.14 Determine *What* that person does, *Where* the action takes place, and *When* (how much delay time is involved before) the action occurs.
        2.15 Break down *What* takes place to determine *How* the action is accomplished. Each major operation or task has a series of steps which are required to accomplish it.
        2.16 Determine *Where* or to *Whom* the document is forwarded to *What* action occurs next. If it is forwarded, *How?* If it is filed, *How* is the file organized? If it is destroyed, *Why?*
        2.17 Discuss each document with the individuals concerned. Follow each copy of each document from beginning to final disposition.
    2.2 Using the symbols of Table 23-2 and SPAS No. 3, make a chart of the flow of the documents. Table 23-3 shows the symbols for charting personnel activities if desired.
    2.3 Indicate every separable activity of the form with a symbol and short explanation of action(s) taken. SPAS No. 3, Procedure Flow Chart, should show the sequence of actions and the personnel performing each action of an existing or proposed procedure.
    2.4 Evaluate the procedure by applying Step No. 3.

3. For the design of a new procedure:
    3.1 Determine the requirements of the new procedure for the affected organization(s), personnel, and equipment workload as indicated in Step 1.
    3.2 Lay out the procedure as indicated in paragraph 2.1 above.
    3.3 Treat as an existing procedure and attempt to improve by applying Step 3.

STEP 3A—EVALUATION (Procedures)

1. Three basic principles of work simplification should be kept in mind:
    1.1 All motions and activities should be productive.
    1.2 All motions and activities should be kept as simple as possible.
    1.3 All processes should be so arranged as to provide a smooth flow from one step or operation to the next.

TABLE 23-2

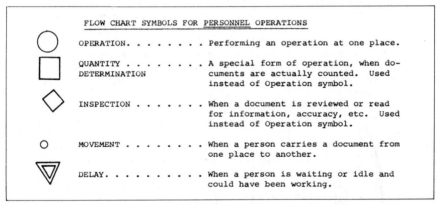

FLOW CHART SYMBOL AND PAPERWORK OPERATIONS

ORIGINATION. . . . . . . When a form is initiated.

OPERATION. . . . . Whenever a physical operation is performed on or to the document.

MOVE . . . . When document changes hands.

TEMPORARY DELAY. . . . . When document is sitting idle - not
being worked on.

INSPECTION . . . . Purely a <u>mental</u> process.

INFORMATION. . . . . . . Whenever data is taken off one docu-
TAKE OFF   ment posted on another.

FILE . . . . Whenever a document is actually filed.
Used instead of Operation symbol.

DESTROY. . . . . Whenever a document is actually destroyed.

GAP. . . . . Not pertinent to procedure being
charted.

CROSS OVER . . . . Whenever a vertical flow line crosses
a horizontal line.

ITEM CHANGE. . . . . . . Whenever a document takes on a different meaning than origianally intended.'

TABLE 23-3

FLOW CHART SYMBOLS FOR PERSONNEL OPERATIONS

OPERATION. . . . . Performing an operation at one place.

QUANTITY . . . . A special form of operation, when do-
DETERMINATION   cuments are actually counted. Used
instead of Operation symbol.

INSPECTION . . . . When a document is reviewed or read
for information, accuracy, etc. Used
instead of Operation symbol.

MOVEMENT . . . . When a person carries a document from
one place to another.

DELAY. . . . . When a person is waiting or idle and
could have been working.

Table 23-2 and 23-3. From M. E. Mundel, *Motion and Time Study, Principles and Practice*, (3rd ed., Englewood Cliffs, N. J.: Prentice-Hall, Inc., 1960).

2. Begin the evaluation with each form *originated* in the procedure. Apply appropriate questions from the checklist of Table 23-4.
3. Second major point for evaluation—those steps which add something to the document(s) in process should be examined in the light of the question in Table 23-4.

4. Next major point for evaluation—those steps where information is taken from the document in process and added to some other document should be examined in light of the checklist of Table 23-4.
5. Evaluate each remaining step of the procedure by applying the appropriate questions from the checklist of Table 23-4.
6. After applying the checklists, compare the procedure as presently designed with the appropriate policies. *Do existing policies prescribe a procedure for this activity?*

   6.1 *If so:*

      6.11 Does the present procedure comply with the policies?

      6.12 Do the policies create a bottleneck in handling the workload?

      6.13 Has a waiver to the policies been requested to expedite the handling or authorize the simplification of the procedure?

   6.2 *If not:*

      6.21 Do the policies permit (liberal) interpretations?

      6.22 Have you an interpretation from proper authorities?

      6.23 Are your suggestions within this realm of interpretation?

TABLE 23-4
CHECKLIST FOR IMPROVEMENT OF PROCEDURES

1. Does each form originated serve a real purpose? Verify it.
2. Is the form necessary? Can it be eliminated, combined with another form, or replaced by a copy of another form?
3. Is each step or operation actually necessary? If not, eliminate it.
4. Does someone sign all copies? Is it necessary? How can this be avoided?
5. Are information, fact, and data supplied in suitable condition for use?
6. Information going from one form to another suggests more copies. Are all information take-offs necessary? If so, which are to be given priority for redesign?
7. Is this operation duplicated at any point in the procedure? Is it necessary? Why?
8. Is there excess checking? Can calculated risks be taken to eliminate "paper massaging" and substitute spot checks of individual units?
9. Are all copies getting equal use? Can forms be kept in action, out of "alibi" file baskets?
10. Are all files and filing actions necessary? Are there duplicate files?
11. Can any filing operations be eliminated, shortened or expedited, or changed in sequence?
12. Are any forms destroyed? This suggests the possibility that they were unnecessary in the first place or that the reproduction of the information should not have occurred.
13. What would happen if a form were lost?
14. Can any movements of papers be eliminated, shortened, or expedited? How was the movement accomplished? Who carries the form? Could another more economical, faster means of transportation be utilized?
15. What interruptions or delays can be reduced or eliminated entirely?
16. Each step should have a reason for being performed by itself. Can certain steps be combined to form a single step?
17. Can the originator perform additional work or supply more information that would make subsequent operations easier?

18. Can the vendor or customer be consulted to help make joint operations easier and more economical?
19. Is the sequence of steps the best possible? Each step should have an ideal place in the sequence. Where should it be performed?
20. Can the operation be done more economically in another organizational segment or by another person?
21. Are functions such as transcription, stenography, typing, duplicating, recording, sorting, assembling, and so on brought together and dispensed as a central service to the entire organization?
22. Is centralization of functions feasible? Weigh the advantages and disadvantages.
23. If the operation is changed, what effect will it have on other operations in the system?
24. Can forms be sorted into order simultaneously with this operation for use in the next operation?
25. Is the operation a bottleneck in the procedure? If so, can work be scheduled differently?
26. Does one person have too much of the procedure?
27. Are the steps spread too thinly among personnel?
28. Can any part of the work be performed during waiting time or idle time?
29. Can any operation(s) be subdivided and the various parts added to another operation?
30. Will specialization of skills increase productivity and retain enough variety in the job to maintain job interest?
31. Are as many steps as possible assigned to the lowest classified person capable of performing the job?
32. Are standardized methods or practices employed throughout the organization?
33. Each step should be as easy as possible. How can it be simplified? Study the monotonous manual jobs, such as sorting, assembling, folding, and distributing papers in the light of motion economy:
    a. Develop simultaneous actions for both hands.
    b. The paths followed by the arms should be smooth and continuous.
    c. The work should be so arranged that the simplest possible motions can be used.
    d. Whenever possible, use both hand and foot motions.
    e. Transport materials and tools with a sliding or gravity motion.
    f. Locate materials and tools in the normal work area (this is the area within reach when elbows are placed at the edge of the desk and are pivoted in an arc).
    g. Locate materials and tools so that the best sequence of motions may be used.
    h. Pre-position materials and tools whenever possible.
    i. Release hands for productive work by providing fixtures whenever possible.
    j. Make sure that the work level is high enough to permit the employee to use good posture in sitting or standing.
34. Have the employees who carry out the procedure been consulted for their views on changes which might be indicated?
35. Is there effective mechanization of office functions? Is the work load processed expeditiously?
36. Is the equipment appropriate for the job? Can job performance be expedited by such special devices as ten-key decimal tabulators for statistical typing?
37. Is the equipment placed at the proper height for efficient operation? Studies show, for example, that typewriters can be operated most efficiently at a height of about 30 inches rather than the usual 26 inches.
38. Are all personnel doing the same kind of job provided with comparable equipment?

39. Can gathering and sorting aids be used for paper assembly tasks?
40. Will electrically-driven equipment prove more economical than manually-operated equipment?
41. Are machines being properly maintained by qualified maintenance men?
42. Is there a turnover plan in effect to insure the replacement of machines when they reach a certain age?
43. Is the most efficient duplicating system being used? For example, the spirit process is less expensive for short-run requirements than the stencil process.
44. Can machines be operated at maximum capacity? If not, can machines be used jointly by two or more organizational segments?
45. Is proper use being made of dictating machines as a means of saving clerical time?
46. Would the work flow more easily and quickly if the office equipment and desks were arranged differently?
47. Are desk tops and drawers kept orderly to improve office appearance and to reduce the time spent in seeking papers?
48. Can pneumatic tubes or other form-carrying systems be used?
49. Can automatic typewriters be used where quantities of letters must be typed?
50. Has an effort been made to standardize the supplies and machines being used?

STEP 2B—ANALYSIS (Forms)

1. *For the analysis of an actual form:*
   1.1 Obtain a copy of the form blank.
   1.2 Obtain a copy of the form completely filled out. This should be a representative sample of each entry on the form to show the average amount of space required for each entry.
   1.3 Determine the sequence of writing of all items on the form.
   1.4 Discuss the form with individuals concerned to determine:
      1.41 Type of equipment used in preparation and subsequent operations.
      1.42 Type of information, sequence, and manner of receiving information from which the form is prepared.
      1.43 Information taken from the form, the sequence and manner of furnishing information, and the requesting office or activity.
      1.44 Operations performed on form during its use.
      1.45 Distribution of form and number of copies to each activity.
      1.46 Purpose of each copy distributed.
      1.47 Manner and method of filing, length of time to be filed, and type of equipment used for filing.
      1.48 Manner and method of binding, if form is bound.
   1.5 Obtain a sketch of the work places at which form is used.
   1.6 Determine effectiveness of form.
   1.7 Evaluate the form by applying Step 3, "Evaluation."
2. *For the design of a new form:*
   2.1 Determine the necessity of form and number of copies required. Explore the possibility of an existing form as indicated under Step 1, "Objective," paragraph 1.51. Verify the number of copies required by usage as indicated in paragraph 1.14 preceding.

2.2 Arrange items in proper sequence by related groups according to source documents.

2.3 Arrange items to fit equipment.

2.4 Provide for ease of sorting.

2.5 Provide for ease of routing.

2.6 Provide for ease of filing.

2.7 Determine final disposition of form.

2.8 Develop a form to meet the above requirements. Follow principle of straight line flow for motion economy.

2.9 Determine material requirements, printing instructions, and controls.

2.10 Treat as an existing form and attempt to improve by applying the Evaluation Checklist, Table 23-5.

2.11 Evaluate the cost of the form.

2.12 Prepare instructions for use of form.

STEP 3B—EVALUATION (Forms); use Table 23-5.

TABLE 23-5

CHECKLIST FOR FORMS IMPROVEMENT AND CONTROL

1. Is the form actually necessary? If so, is all the information requested actually required?

2. Can any of the present blanks on the form be preprinted with fixed data and numbers so as to reduce the number of items which must be filled in?

3. Can all blanks on the form be arranged in the same sequence as the source document, thus reducing the amount of time required to enter data?

4. Are all blanks aligned so that in filling it out typewriter tabs can be used, thus reducing the need for irregular moves from one portion of the form to another in entering data?

5. Is the amount of space allotted for the entry of data sufficient?

6. Can pre-assembled sets of forms be used which will permit invoices, accounting copies, loading copies, and billings to be prepared simultaneously?

7. Will use of continuous forms and one-time carbons expedite preparation?

8. Has the spacing on the forms been adapted to the machines that will be used in filling them out?

9. Is the size of the form adequate for the purposes required and is it conducive to easy filing?

10. Can different colors of paper be used for different forms to facilitate sorting, etc.?

11. Can selective boxes be used instead of blanks, thus permitting check marks to be used in completing the form?

12. Can the form be completed satisfactorily and expeditiously in pencil?

13. If the form is to be filled out by typewriter, is it necessary to have ruled lines?

14. Should there. be a notation at the top of the form as to how many copies must be prepared and where each copy must go?

15. Must the form itself be printed or would it be just as satisfactory if it were prepared on the office duplicator?

16. Is there central control over the authorization, design, and supply of new forms?

17. Are forms ordered in reasonable yet economical quantities?

18. Are suggestions from users concerning the efficiency of forms encouraged?

STEP 4—IMPROVING

1. If the analyst has followed Step 3 carefully and applied the improvement formula (eliminate, combine, change sequence, or simplify), alternate ways of performing the job will have resulted.
   1.1 Evaluate each alternate.
   1.2 Select the "best way."
   1.3 Draw up the proposal.
      1.31 If procedure, use procedure analysis chart SPAS No. 3 (Step 2A, paragraph 1.3).
      1.32 If form, use Forms similar to Figures 23-5 and 23-6.
   1.4 If procedure is modified, make comparison of present and proposed procedure on SPAS No. 2 (Step 2A, paragraph 1.2).
   1.5 If form is modified, make comparison of present and proposed form on typewriter analysis sheet (applicable to handwritten forms).
   1.6 Combine all changes in a report showing the relative value of both the present and proposed procedures or forms. Use SPAS No. 1.

STEP 5—TEST

1. Treat your proposal with the items from Step 3 as if it were an original.
2. Discuss final result with all concerned as a "proposal." Work up through affected segment(s), adjusting proposal if necessary. Adequately "consult" supervisor and workers, obtaining participation rather than acquiescence.
3. Submit through channels.
4. Await authorization to proceed.

STEP 6—TRIAL

1. Distribute instruction sheets for new forms or new steps of procedure as necessary, after obtaining authorization to do so.
2. Discuss with each person concerned, after obtaining supervisor's permission to do so. Obtain cooperation.
3. Insofar as possible:
   3.1 Instruct.
   3.2 Demonstrate.
   3.3 Observe, praise success; correct as necessary.
   3.4 Check performance of those concerned shortly after instructing so as to provide assistance if necessary before improper method becomes habitual.
4. Report to management on results obtained.

NOTE: Whenever possible, it is advisable to put in a new procedure on a "dry run" or trial basis. Sometimes, when the operation is large enough, part of a group can be working on the new procedure while the balance continues on the old until the "bugs" are out of the trial run. Then it is a simpler matter to put the new procedure into the balance of the group. This will greatly decrease the loss in production due to change-over on a new procedure.

STEP 7—APPLICATION

1. Make procedure chart of new method a matter of record.
2. Enter new material into standard practice files as necessary.
3. Make sure of routine supply of forms and materials.
4. Check after short installation period and rectify any defects or deviations.
5. Evaluate results and report these to management.

Summary   The basic "tool" of any systems and procedures analyst and designer is the technique of procedure analysis. This technique provides the designer with a systematic approach to the solution of problems in procedures design and forms design. It is essential that the designer of Production Control systems have a working knowledge of the systematic method and its relationship to procedure analysis in order to assure greater degrees of success the first time in improving or designing Production Control systems.

24

Work Sampling

In Chapter 14, work measurement and its relationship to Production Control was discussed; in Chapter 23, procedure analysis and its relationship to Production Control was discussed. Both are "tools" of management having an important role in effective and efficient Production Control. This chapter will be devoted to another such important management "tool"—work sampling.

Work sampling is a method of taking random observations of any activity to determine the actual amount of time being taken to perform any or all parts of the activity. This management tool was first reported by the British statistician L. C. Tippett in 1935. Robert Morrow described it in this country in 1940 under the name "Ratio Delay Study." In 1952, C. L. Brisley described this technique as "Work Sampling"; this term has become widely accepted since then.

Work sampling has a very broad field of application and usefulness, including many phases of Production Control. It has opened up a field for fact finding which would otherwise be impracticable. Probably its greatest usefulness is in measuring indirect labor, which cannot, in many cases, be measured by means of the stop-watch method. Other examples of its use to illustrate its broad field of application include:

1. Office and Clerical Operations
2. Engineering Activities
3. Dental Operations
4. Machine Utilization
5. Materials-Handling Equipment Utilization
6. Grocery Store Activities
7. Service Station Personnel Utilization
8. Hospital Activities

These are only a few of the vast areas of application of this practical management tool.

The mathematical theory behind work sampling will not be detailed in this chapter. Only the general theory will be outlined along with the step-by-step approach for using work sampling in practical application. Those readers who are interested in the mathematics may consult the references listed at the end of this chapter.

Briefly, work sampling consists of taking instantaneous observations of an activity at random intervals of time, tallying the observations on an appropriate observation sheet, and calculating the per cent of the tallies under each descriptive category to the total number of observations made. This percentage approximates the percentage of the time of each part of the activity defined to the total time of all parts of the activity. The steps in this procedure will be described in more detail later.

The general theory of work sampling is based upon the law of probability. It can be explained in the same manner as the chance of getting "heads" in the toss of a coin or the chances of getting any particular number in throwing dice. If a coin is tossed enough times, a head will come up half the time. Similarly, in the throwing of a single die, each number will come up one sixth of the time if it is thrown often enough. Each number on the die is equally likely to occur. Accordingly in work sampling there must be some method of selecting the time periods at which observations are made so that each is equally likely to be selected —in other words, that each is selected at random. There must also be assurance that enough observations or samples are taken so that the percentages of time that are obtained are within the desired limits, or tolerances. As will be seen, these things are very easily accomplished without any complicated mathematics or routine.

Following is a list of steps used in making a work sampling study, after which each step is described in more detail:

1. Define the objectives of the study.
2. Break down the activity into categories to be studied.
3. Design the observation form.

4. Determine the elapsed time period of the study and the frequency of making observations.
5. Determine the total number of observations to be taken.
6. Set up the random time selection method.
7. Train the observers.
8. Inform people who are to be observed.
9. Collect the data.
10. Evaluate the data's validity and reliability.
11. Take action on conclusions drawn from the data.

Step 1. Define the objectives of the study. Much time and effort can be wasted by not thoroughly defining why the study is being made. Usually, too little information is gathered while the study is being made, and after finishing the study it is found that additional information would have been desirable and could have been obtained without any additional effort.

In other cases, the whole study may be worthless from the standpoint of using the information obtained. A good question to ask is, "When I have the information from the study, would I possibly change any ways of doing things?" To illustrate this point, the president of a small plant was sure that a forklift truck was needed to handle material in certain areas of the plant. He requested that a work sampling study be made to determine how much manual handling was done of the material that could be done with a forklift truck. After the study was completed, in about six weeks, it was established that the expenditure for a forklift truck would not be justified. However, the president was so determined that one was needed that he bought it anyway. One would suspect that the president was just kind-hearted and did not like to see his employees doing heavy lifting. Obviously, the work sampling study was a waste of time.

Step 2. Break down the activity into categories to be studied. A good rule of thumb on this step is, "It is better to have just a little too much information than not enough." It is usually easy to combine information after the study is completed, but once it is completed, it is impossible to break it down into any finer categories. For instance, a study of machine downtime would require a list of all the causes for downtime. It may even be necessary to spend a day or so observing the machine to determine the various reasons for downtime so that a list of the reasons can be made. Extra spaces can always be left on the observation form so that additional categories can be added.

Step 3. Design the observation form. It would be difficult and certainly impracticable to design a *standard* observation form for work sampling. In most instances it is found that it is more practical to set up the form for the specific study that is to be made. Figures 24-1 and 24-2 show sample forms that were used in two different studies.

SUBJECT: DEPT. A DISPATCHING OFFICE	OBSERVER: L Holland								
DATE:	8:00	9:00	10:00	11:00	1:00	2:00	3:00	4:00	TOTAL
1. CLERICAL:									
a. Sorting	/		//		/		/		5
b. Filing-Dispatch	//	/	///	/		/		/	9
c. Filing-Office		/			/		/	/	4
d. Posting	/			//		/	/	/	6
e.									
2. COMMUNICATIONS:									
a. Telephone	/				/			/	3
b. Talking-Foremen	ЖЖ	/		//		/		///	12
c. Talking-Shop Pers.	/		//		/		/		5
d. Talking-Other		//		///	/	/	/	/	9
e.									
3. IDLE:									
a. Waiting									0
b. Rest Period									0
c. Out of Room	/				/		/		3
4. MISCELLANEOUS:									
a. Absent									0
b. Personal	/	ЖЖ	//		/	////	/	//	16
c.									
TOTAL									72

Fig. 24-1. Work sampling observation form designed for the specific job studied showing sampling tallies.

Fig. 24-2. Work sampling observation forms designed for multi-position observations.

WORK SAMPLING STUDY
DAILY OBSERVATION RECORD
OBSERVER _____
DATE _____

ELEMENT DESCRIPTION / MACHINE	MACH. A	MACH. B	MACH. C	MACH. D	MACH. E	MACH. F	MACH. G	MACH. H	MACH. I	TOTALS
OPERATING: HAND FEED										
GAUGING										
WAITING										
PREPARING STOCK										
MACHINE IDLE: STOCK HANDLING										
GAUGING										
TOOL SHARPENING										
CLEAN-UP										
MACHINE SET-UP										
PERSONAL										
ABSENT										
DELAY: TALK WITH SUPR.										
TALK WITH INSP.										

SUBJECT: *DEPT. A DISPATCHING OFFICE*

DATE: *WEEK ENDING November 6, 1959*	MON	TUE	WED	THUR	FRI	TOTAL OBSERVATIONS		TOTAL PER CENT	
I. CLERICAL									
a. SORTING	1	4	6	2	5	18		5.0	
b. FILING-DISPATCH	3	9	6	6	9	33		9.2	
c. FILING-OFFICE	5	3	5	3	4	20		5.6	
d. POSTING	12	8	12	5	6	43		11.9	
TOTAL	21	24	29	16	24		114		31.7
2. COMMUNICATIONS									
a. TELEPHONE	2	2	1	2	3	10		2.8	
b. TALKING-FOREMEN	10	9	13	13	12	57		15.8	
c. TALKING-SHOP PERS.	14	10	9	3	5	41		11.4	
d. TALKING-OTHER	2	10	6	7	9	34		9.4	
TOTAL	28	31	29	25	29		142		39.4
3. IDLE									
a. WAITING	1	4	1	3	0	9		2.5	
b. REST PERIOD	5	0	0	8	0	13		3.6	
c. OUT OF ROOM	8	4	5	4	3	24		6.7	
TOTAL	14	8	6	15	3		46		12.8
4. MISCELLANEOUS									
a. ABSENT	—	—	—	—	—	—			
b. PERSONAL	9	9	8	16	16	58		16.1	
TOTAL							58		16.1
GRAND TOTAL	72	72	72	72	72		360		100.0

Fig. 24-3. Work sampling observation summary form used to summarize daily and weekly the data collected from the individual observation sheets.

In addition to the form for recording original observations, it is necessary to have a summary sheet for the original data. An example of the summary sheet is shown in Figure 24-3. If only one observer is making the study, then the daily summaries can be put on the original data observation sheet. However, if more than one observer is taking observations, it will be necessary to have a daily summary sheet.

Step 4. Determine the elapsed time period of the study and the frequency of making observations. There are a number of factors which have to be considered in determining the elapsed time period of the study and the number of observations that an observer should make each day, or each week. These factors are such things as:

1. How quickly it is desired to obtain the information.
2. How long the job will exist in its present manner.
3. Who makes the observation.
4. How long it takes to make a round of observations.

Obviously, the frequency of observations is governed by the elapsed time period available to make the study and the accuracy desired from the data. The desired accuracy is explained in Step 5. As long a period of time as possible should be used to collect data. The frequency then is primarily dependent on factors 3 and 4. For instance, if the foreman of a department is to be trained to take the observations, it is not desirable to demand too much of his time. If, however, making observations is to be the full-time job of one or more persons, then many more observations can be made of each type of work. Generally speaking, it is desirable to make an observation at least once for each hour of the day, or still better, at least ten observations should be made each day. Sufficient elapsed time is also necessary to obtain the desired accuracy.

Step 5. Determine the total number of observations to be taken. This step must be combined with Step 4 because the number of observations must be known in order to determine the frequency of observation. The total number of observations depends on the accuracy desired in the final answers. The larger the number of observations, the greater the accuracy. A quick way to determine the total number of observations is as follows:

1. Make a guess of the answer expected from the main categories of the study.
2. Referring to Table 24-1, read the number of observations required for that answer and for the desired degree of accuracy.

Example: In a Production Control dispatching study, it was estimated that approximately 20 per cent of the time would be used for answering the foremen's questions. By referring to the table opposite the number 20 under "p in per cent" and for an error of 5 per cent of p, we read that 6,400 readings are required. In this case, a ten-week observation period was decided upon so that 640 observations had to be made each week. For a six-day week, 108 observations had to be made each day to start with. After a week of observation, it was found that the original estimate was high and that the value was closer to 15 per cent. Again referring to Table 24-1, a p of 15 shows an N of 9,067 for an error of 5 per cent of p. This meant that the number of observations had to be increased to 301 per day for a 10-week period of 6 days per week.

Step 6. Set up the random time selection method. The importance of selecting observation times in a random manner has already been emphasized. This can be done in several ways but there are two ways which are particularly easy to understand and apply:

1. The use of random sampling numbers. All of the references at the for their use. Their use can be taught to observers in a very short time. end of this chapter have tables of random numbers with explanations

TABLE 24-1

VALUES OF N (NUMBER OF OBSERVATIONS) AT 95/100 PROBABILITY OF NOT EXCEEDING ERROR INDICATED, FOR VALUES OF p (PER CENT OF ACTIVITY)

p in per cent	Values of N — Error			p in per cent	Values of N — Error		
	1 per cent	1 per cent of p	5 per cent of p		1 per cent	1 per cent of p	5 per cent of p
1	396	3,960,000	158,400	51	9,996	38,431	1,537
2	784	1,960,000	78,400	52	9,984	36,923	1,477
3	1,164	1,293,000	51,720	53	9,964	35,472	1,419
4	1,536	960,000	38,400	54	9,936	34,074	1,363
5	1,900	760,000	30,400	55	9,900	32,727	1,309
6	2,256	626,667	25,067	56	9,856	31,429	1,257
7	2,604	531,429	21,257	57	9,804	30,175	1,207
8	2,944	460,000	18,400	58	9,744	28,966	1,159
9	3,276	404,444	16,178	59	9,676	27,797	1,112
10	3,600	360,000	14,400	60	9,600	26,667	1,067
11	3,916	323,636	12,945	61	9,516	25,574	1,023
12	4,224	293,333	11,733	62	9,424	24,516	981
13	4,524	267,692	10,708	63	9,324	23,492	940
14	4,816	245,714	9,829	64	9,216	22,500	900
15	5,100	226,667	9,067	65	9,100	21,538	862
16	5,376	210,000	8,400	66	8,976	20,606	824
17	5,644	195,294	7,812	67	8,844	19,701	788
18	5,904	182,222	7,289	68	8,704	18,824	753
19	6,156	170,526	6,821	69	8,556	17,971	719
20	6,400	160,000	6,400	70	8,400	17,143	686
21	6,636	150,476	6,019	71	8,236	16,338	654
22	6,864	141,818	5,675	72	8,064	15,556	622
23	7,084	133,913	5,357	73	7,884	14,795	592
24	7,296	126,667	5,067	74	7,696	14,054	562
25	7,500	120,000	4,800	75	7,500	13,333	533
26	7,696	113,846	4,554	76	7,296	12,632	505
27	7,884	108,148	4,326	77	7,084	11,948	478
28	8,064	102,857	4,114	78	6,864	11,282	451
29	8,236	97,931	3,917	79	6,636	10,633	425
30	8,400	93,333	3,733	80	6,400	10,000	400
31	8,556	89,032	3,561	81	6,156	9,383	375
32	8,704	85,000	3,400	82	5,904	8,780	351
33	8,844	81,212	3,249	83	5,644	8,193	328
34	8,976	77,647	3,106	84	5,376	7,619	305
35	9,100	74,286	2,971	85	5,100	7,059	282
36	9,216	71,111	2,844	86	4,816	6,512	261
37	9,324	68,108	2,724	87	4,524	5,977	239
38	9,424	65,263	2,611	88	4,224	5,455	218
39	9,516	62,564	2,503	89	3,916	4,944	198
40	9,600	60,000	2,400	90	3,600	4,444	178
41	9,676	57,561	2,302	91	3,276	3,956	158
42	9,744	55,238	2,210	92	2,944	3,478	139
43	9,804	53,023	2,121	93	2,604	3,011	120
44	9,856	50,909	2,036	94	2,256	2,553	102
45	9,900	48,889	1,956	95	1,900	2,105	84
46	9,936	46,957	1,878	96	1,536	1,667	67
47	9,964	45,106	1,804	97	1,164	1,237	50
48	9,984	43,333	1,733	98	784	816	33
49	9,996	41,633	1,665	99	396	404	16
50	10,000	40,000	1,600				

From M.E. Mundel, Motion and Time Study: Principles and Practice, Third Edition (Englewood Cliffs, N. J.: Prentice-Hall, Inc., 1960), p. 149.

2. The "chip draw." Depending on who is to take the observations, it is sometimes advantageous to simplify greatly the selection of the random times. This can be done by making a very simple chip draw. If it is decided, for instance, that ten observations will be made each day and that the minimum time that it takes to make an observation round is five minutes, then the working day can be divided up into five-minute intervals. These five-minute time intervals can be written on poker chips or beads, or similar objects, placed into a tumbler or bowl and ten chips drawn each day. The times written on the chips are recorded in chronological order and the observer immediately has his schedule for the day. This is particularly useful if the foreman or supervisor is to be used as the observer. There is little chance that the time periods selected can be misinterpreted.

Step 7. Train the observers. This is an extremely important step, particularly when more than one observer is to be used on a given study. It is essential that each observer thoroughly understand the importance of taking instantaneous observations and that all of the categories in the study are properly described in writing. It is also important that he understand the general theory of work sampling, the purpose of the study, and the use of the final results. Whenever possible, it is desirable that the observer is the person in charge of the activity studied. Such a person taking an "honest" study cannot question the results. The summaries and data presentations, however, should be made by the statistician, engineer, or other staff person.

Step 8. Inform the people who are to be observed. This step is a requirement in many union shops. However, even if the shop is nonunion, it is important from the psychological point of view. People usually are more willing to go along with something if they understand it. If they get the idea that the study is being made to criticize them, it is very unlikely that cooperation will ever be obtained. Keeping everything open and above-board lends greatly to creating a cooperative attitude.

Step 9. Collect the data. It is worthwhile to again point out the importance of instantaneous observations in collecting the data. The observer should also be cautioned on "waiting for something to happen." It is also important for the observer never to guess what is being done. If he is in doubt, he should inquire either of the person being observed or of the supervisor about the nature of the activity at the time of the observation. The data collected should be summarized at the end of each day. This can be done on the original form or on a separate form if more than one observer is making the study.

Step 10. Evaluate the data's validity and reliability. The validity of the data depends primarily upon the understanding of the principles and purpose of the study by the observer and those being observed. If the observations are properly made and those observed are fully cooperative, then the data should be valid. For instance, if the observer is in charge of the activity on which a study is made, he might have a tendency to "color" his observations to favor himself. A check of the data with the individual being observed might identify this situation. Also, if the observer is taking his observations at particular times throughout the day, those observed might be waiting for him and, as a result, the data would not be valid.

The reliability of the data can be tested by the use of Table 24-1. Another way of testing reliability is by use of a control chart. This will not be described in detail in this text and the reader is again referred to the publications listed at the end of this chapter for a description. An example of a control chart is shown in Figure 24-4.

Step 11. Take action on conclusions drawn from the data. All of the work sampling studies in the world are of little value unless something is done with the data. For instance, if the study is made for the purpose of determining the idle time of an operator on a machine, definite steps should be immediately taken to reduce the amount of idle time by taking corrective action on its causes. These causes, of course, would be identified in the work sampling study.

Another very important aspect is a follow-up at a later date on the corrective action taken to determine whether or not some definite improvement was obtained. This can be done by taking another work sampling study and comparing it to the first.

If it is necessary to sell management on the idea of taking corrective action, then it might be worthwhile to present the data in some graphical form to highlight the results of the study, particularly in those areas on which most attention should be concentrated.

The preceding steps should give the reader an idea of what is involved in work sampling. As can be seen, there is nothing mysterious about it and even inexperienced personnel can use it with good results. Following is the result of a study made by inexperienced people on their first use of work sampling.

CASE STUDY: TIME DISTRIBUTION OF PRODUCTION CONTROL DISPATCHING OFFICE

Figures 24-1 and 24-3 are the actual samples of the data obtained. The final per cent time distribution was as follows:

1.	Clerical	31.7
2.	Communications	41.2
3.	Idle	13.0
4.	Miscellaneous (personal)	14.1

The study was under the direction of an industrial engineer but the observations were made by the Production Control manager and later by

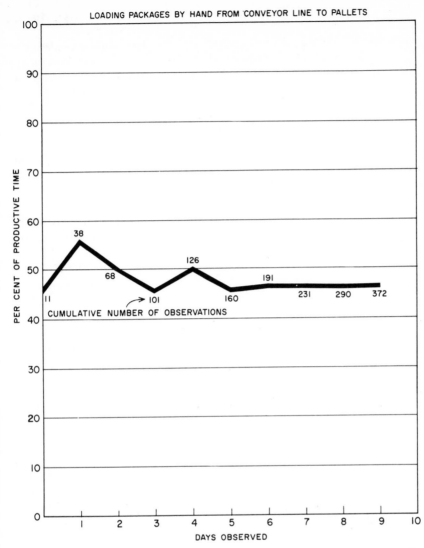

Fig. 24-4. Work sampling study control chart—per cent of productive time. The levelling out of the chart line indicates that further sampling would not significantly change the results.

the dispatching staff itself. The results pointed out the excessive personal and idle time. Closer supervision and a sincere effort by the staff reduced the personal and idle time considerably. The results of the study were indisputable by the staff because the staff participated in taking the observations.

Summary   Work sampling is a very simple and useful "tool" for many uses in Production Control. The Production Control manager and the designer of the Production Control system should have a working knowledge of work sampling and make full use of it.

REFERENCES

1. Robert E. Heiland and Wallace J. Richardson, *Work Sampling* (New York: McGraw-Hill Book Co., Inc., 1957).
2. Ralph Barnes, *Work Sampling,* Second Edition (New York: John Wiley & Sons, Inc., 1957).
3. C. L. Brisley, "Work Sampling," *Factory Management and Maintenance* (July 1952), p. 83.
4. Bertrand Hansen, *Work Sampling,* (Englewood Cliffs, N. J.: Prentice-Hall, Inc., 1959).

25

Evaluating A Production Control System

Many organizations find themselves in the position of realizing that their Production Control system is not as effective as it should be, yet they cannot determine the exact cause of the trouble. In such a situation there is usually a host of apparently unrelated symptoms such as missed delivery dates, "lost" jobs, and equipment idle even while the back orders pile up. To make matters worse, everyone in the organization "knows" how to correct the problems but everyone's solution is different. The only way that a situation like this can be remedied is to make a systematic evaluation of Production Control. This chapter will consider the methods of evaluation available. Once the system has been adequately appraised, then the causes for ineffective operation can be isolated and corrected.

There are basically only two different methods of evaluation: qualitative and quantitative. The definitions of these two methods are:

Qualitative Evaluation is a method of appraisal based entirely upon the subjective opinion of the person making the evaluation.

Quantitative Evaluation is a method of appraisal based upon the numerical measurements of the item being evaluated.

In order to illustrate the difference between the two evaluating techniques, consider the situation that most of us have faced at one time or another: the appraisal of the relative merits of different automobiles. Most of us would first make a qualitative evaluation of the general appearance and styling of the various automobiles. The results of this first appraisal would be expressed in a series of adjectives such as "beautiful," "nice," and even "ugly." Some of the other features of the automobile that would be judged subjectively would be:

1. Comfort
2. Ease of handling
3. Appearance of the interior
4. Prestige factor,
   and so forth

The adjectives used to describe our various opinions cannot be assigned numerical values, but this does not decrease the importance of the qualitative evaluation. In the case of the automobile the intangible factors, which cannot be expressed numerically, are probably as important as the mechanical features expressed with numbers.

Normally, we do not select an automobile on the basis of qualitative evaluation alone. One of the most important comparisons we make is the cost of the various cars. In this case we have numbers by which to make our appraisal. It is possible to determine with absolute accuracy the purchase price of each vehicle and the actual differences among the vehicles. This is not a matter of judgment, but rather a numerical fact. There are other characteristics which can be measured and expressed in numbers.

1. Gasoline mileage
2. Horsepower rating
3. Top speed
4. Turning radius,
   and so forth

All of these factors have numbers associated with them and these numbers aid in the evaluation, but too many people believe that since the factors are expressed in numbers the final evaluation is completely objective. This is not true. Once the numbers are available, we then must subjectively appraise the numbers. Is an automobile with a 300 horsepower engine better than one with a 250 horsepower engine? Certainly the first one has more apparent power but is this desirable? Is the performance of the more powerful car significantly higher than the second car? Will we ever have use for the extra power? In this simple example, it is obvious that a measurable factor cannot be judged on the basis of numbers alone. It must be kept in mind that sooner or later a subjective appraisal will have to be made regardless of the type of evaluation used. The advantage of a quantitative evaluation is the fact that the numbers aid in making the final appraisal.

The ultimate goal of evaluating a Production Control system is to determine if the system is operating effectively or not, and if not, to locate the cause of the ineffective operation. In order to achieve this goal it is necessary to look for the symptoms of substandard performance, evaluate the importance of the symptoms and finally determine the deficiency in the Production Control system. As an example, if an organization consistently shipped 10 per cent of all of its customers' orders late, is it good or bad? To answer this question it is necessary to know the answers to such questions as:

1. What type of item does the organization produce?
2. How late were the orders? A day? A month?
3. What kind of performance record do the competitors have?

The actual evaluation of any particular Production Control system must be done on an individual basis. It is impossible to establish a universal set of norms to be applied to all systems, but it is possible to illustrate the methods of performing a qualitative and a quantitative evaluation. The remaining portion of this chapter will be devoted to considering the applications of these techniques.

QUALITATIVE EVALUATION

When evaluating a Production Control system qualitatively, it is necessary to compare the system against a set of standards or norms. In Chapter 3, a list of basic principles was developed and discussed. These principles are applicable to all Production Control systems, and therefore can be used as the basis for evaluation. The six principles are:

1. The system must furnish timely, adequate, and accurate information.
2. The system must be flexible to accommodate necessary changes.
3. The system must be simple and understandable to operate.
4. The system must be economical to operate.
5. The system must force prior-planning and corrective action by the user of the system.
6. The system must permit "management by exception."

In order to use these principles as the basis for evaluating a Production Control system, it is necessary to discuss the system with all of the people involved. By asking questions and securing opinions it will soon become evident whether the system under consideration complies with the principles or not. The six principles should provide a framework for formulating pertinent questions and categorizing the answers and opinions.

Before considering the type of questions to be asked, an important point must be brought out. As mentioned above, it is necessary to discuss the system with *all* of the people involved with the system, when making a qualitative evaluation. The questions should not be directed exclusively to the head of the Production Control system, as is frequently done. The information secured from this source is bound to be misleading. The head of the Production Control department cannot know all of the operational problems encountered by the people working with the system. In order to find out about these operational problems, the scheduling clerks, inventory control clerks, dispatchers, foremen, and so forth should be consulted. Only in this way can a true picture be secured.

1. *The system must furnish timely, adequate, and accurate information.*

    1.1 How much time elapses before the Production Control section receives and acts upon the progress reports being made out by the shop?

    1.2 How long does it take the shop to receive the "firm" schedule once it has been prepared?

    1.3 Are orders ever completed without material or labor having been charged to them?

    1.4 How often are progress reports received with the wrong order number or production count, or omissions on them?

    1.5 Could the information on the progress reports be consolidated or summarized?

2. *The system must be flexible to accommodate necessary changes.*

    2.1 If a rush order should arrive in the Production Control section now, what actions would have to be taken? How much time would this require?

    2.2 If an order were cancelled now, what action would have to be taken? How much time would this require?

    2.3 If the activity of the facility were to increase or decrease 25 per cent, what changes would have to be made in the Production Control system?

    2.4 What is the effect on the schedule when a machine breaks down? How long does it take to establish a new schedule?

    2.5 How much idle equipment time is caused by a schedule change?

3. *The system must be simple and understandable to operate.*

    3.1 How long does it require to train people for the various duties in the Production Control section?

    3.2 Do the people understand how their function fits into the over-all system?

    3.3 How many different paper forms are there in the system?

    3.4 How many different reports go out of the Production Control system?

    3.5 If any of the Production Control staff should become temporarily unavailable, can someone else take over his job?

4. *The system should be economical to operate.*

4.1 What is the ratio of Production Control staff to direct labor workers?
4.2 How much time is being spent on making out reports as compared to planning the activity of the facility?
4.3 How much time do the foremen spend in dispatching and progress reporting?
4.4 Are the commercial data processing systems *really* paying for themselves?
4.5 What would happen if the Production Control staff should be reduced 10 per cent?

5. *The system must force prior-planning and corrective action by the users of the system.*

5.1 Is there sufficient time between the start of the planning phase and the action phase to allow real planning or is most of the work done on a "crash" basis?
5.2 When corrective action is required, can the staff of the Production Control section make the necessary adjustments or do higher-level management decisions have to be made?
5.3 Is it possible to accumulate a larger group of jobs so that an overall effective plan can be established?
5.4 Are most exceptional situations covered by formal or informal company policies?
5.5 Is there a current activity forecast? Is it used?

6. *The system must permit "management by exception."*

6.1 Where is "management by exception" used?
6.2 What percentage of the progress reports simply confirm the accomplishment of the planned activity?
6.3 How many reports to management are being made by the Production Control section? Could "management by exception" be used?
6.4 Could progress reporting by the shop be on the exception principle?

After completing an extensive qualitative analysis, a complete picture of the Production Control system will be obtained. In addition to this, clear indications of the causes of the difficulties should be available. The people actually operating the system are in the best position to know what the defects are. Although they may not be fully aware of the overall problems, they frequently have good suggestions for alleviating a portion of the difficulties.

The disadvantage of using the qualitative approach in analyzing and evaluating a Production Control system is that the results are based primarily upon subjective judgments. As an example, if in the analyst's opinion the system is not simple and understandable, it would be very difficult for him to "prove" that he is right to someone with the opposite point of view. Also, if a change is made in the Production Control system to correct a problem, it would be hard to measure the effect of the change at some future date.

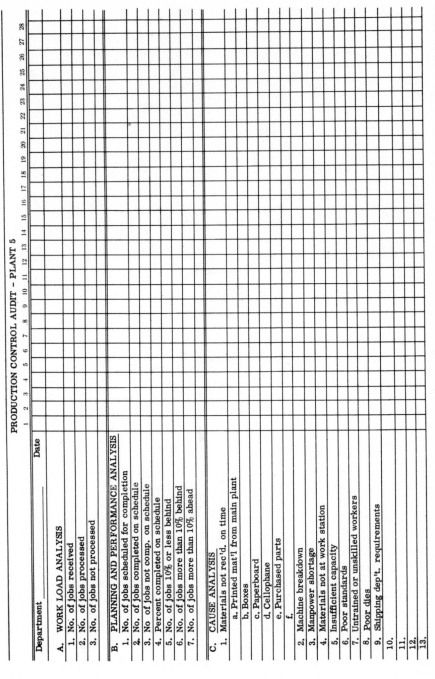

Fig. 25-1. Typical audit sheet factors.

282

QUANTITATIVE EVALUATION

In order to evaluate a Production Control system quantitatively, it is necessary to limit the appraisal to only the measurable factors. This generally restricts the technique to the point where it must be considered an adjunct to a qualitative analysis. It is, however, a very important adjunct which should be used whenever the time and money can be justified.

As was the case with qualitative evaluation, it is impossible to establish a universal list of measurable factors which are equally applicable to all Production Control systems, but the general method of performing a quantitative evaluation can be considered. In Figure 25-1, a list of typical audit sheet factors for quantitatively analyzing a Production Control system is shown. The factors in Figure 25-1 are general in nature and can be used as a pattern for developing an audit sheet for a particular situation. The audit list is broken down into three major sections.

1. Work load analysis
2. Planning and performance analysis
3. Cause analysis

Each of these sections will be considered separately.

Work load analysis   In order to interpret the results of a quantitative analysis of a Production Control system, it is first necessary to establish what the conditions were when the analysis was made. The first portion of the audit sheet is devoted to this. In the future, when a comparison of the effectiveness of any change in the system is made, this information will place the comparison on an equivalent basis.

Planning and performance analysis   The second portion of the audit sheet is devoted to the analysis of the actual performance of the facility as compared to the planned and the required performance. A word of explanation should be made at this point about the difference between the planned and the required performance. The former is the anticipated accomplishments based upon the adjusted standard time (see Chapter 15). The latter is the performance needed to meet the customers' delivery dates. It is entirely possible that plans established with little regard to the customers' requirements might be successfully fulfilled by the facility yet would be completely unacceptable from the management's point of view. By securing information in the manner shown on the data sheet, the over-all effectiveness of the system can be ascertained.

Cause analysis   The last portion of the audit sheet is concerned with establishing the cause of performance variance: The established schedule was not realistic (item 3.11 on the audit list) or the facility did not perform in accordance with the established schedule. If the schedule is not realistic, it is clearly the fault of the Production Control system. On the other hand, if the facility does not perform in accordance with a realistic schedule, it may or may not be attributable to a defect in the Production Control system. The audit sheet separates the causes so that a faster analysis can be made.

As was pointed out previously, the advantage of using the quantitative methods of evaluation is the fact that the numbers aid considerably in appraising the situation. The major disadvantage of this technique is the fact that it requires more time and money than the qualitative methods.

Whenever an evaluation of a Production Control system is to be made, it is generally advisable to use both the qualitative and the quantitative methods of evaluation. The two methods augment each other, and a better understanding of the problems can be secured. The evaluation should not be the end of the analysis, however, because it only uncovers the symptoms of a defective system. Once the symptoms are known then the Production Control function which must be redesigned will be apparent. The actual redesign should follow the principle laid down in this text.

Evaluating the Production Control system should never be considered completed. It is a continuing job. As time goes along, subtle changes occur in an organization which will cause any Production Control system to become obsolete. It is not wise to wait until the system causes major problems before it is evaluated and corrected. Evaluation and modifications should be done periodically in order to maintain the most effective control over the facility.

Summary   The two basic methods of evaluating a Production Control sytsem, qualitative and quantitative, have been considered in this chapter. It has been pointed out that the evaluation should be a continuous program rather than "once in a great while" as is the practice in so many companies. The passage of time brings about subtle changes both in the Production Control system and in the organization's operation. These changes necessitate continuing evaluation and modification of the Production Control system in order to maintain the most efficiently run plant.

26

The Place of Production Control

Up to now, we have devoted all of our attention to:

1. Developing the fundamental functions and principles of Production Control.
2. Describing the equipment and techniques used in Production Control.
3. Describing the methods of analyzing and evaluating Production Control systems.
4. Describing how to use and improve Production Control systems.

Obviously, we can do none of these unless the organization is designed and set up to perform all of the functions and carry out the responsibilities assigned. We have stressed a number of times that all of the functions of Production Control have to be performed by someone at sometime. This applies to all types of work activities as described in Chapter 5. It is also true whether the system is formal or informal. It should be kept in mind that in evaluating an activity for the purpose of determining the degree of Production Control required, the choice might be that an informal system is the most feasible under the existing conditions. As was previously pointed out, our objective is to develop the "best" system of Production Control, everything considered. However, the problem of "organizing" is only present where there is a need for a formal Production Control function. Formal Production Control was defined as "centralizing and coordinating the various functions of Production Control for better supervision at all levels, so as to seek and sustain the best known methods of Production Control at any given time."

This chapter and the next are devoted to defining and describing principles and factors in "organizing" for Production Control. These chapters might have been put at the beginning of the text rather than at the end, reasoning on the one hand that the over-all role of Production Control in the complete organization must first be known before its specific function and responsibilities are studied. On the other hand, it can be reasoned, as the authors have done, that the reader is in a much better position to understand Production Control and the organizing of functions within the Production Control department if a knowledge of all of the specific parts of the whole are first known.

There are two main steps in organizing for Production Control:

> Step 1. Determining the place of Production Control in the total organization. This involves the following action:
> 1.1 Defining the over-all goals required in Production Control.
> 1.2 Determining the range of activities to be included in the Production Control organization.
> 1.3 Determining the reporting level of Production Control.
> 1.4 Defining the specific authority delegated to the Production Control function.
> 1.5 Describing the relationships of the Production Control organization to other functions in the total organization.
> Step 2. Designing the Production Control organization within the framework defined in Step 1. This involves the following action:
> 2.1 Defining the *specific* functions within the Production Control organization.
> 2.2 Defining the type of personnel to perform each function and determining where they are to be obtained.
> 2.3 Determining the size of the Production Control staff.
> 2.4 Fitting men to the functions.
> 2.5 Training personnel to perform the functions.
> 2.6 Planning and controlling the Production Control activities.
> 2.7 Establishing the methods of evaluating the performance in each function.

In this chapter, we will develop Step 1; in the following Step 2.

OVER-ALL GOALS

Most companies have three main objectives or goals which they are trying to achieve:

1. Satisfactory customer service with maximum quality and minimum cost
2. Maximum profit on production
3. Minimum capital investment

It can be seen that, in order to achieve these goals, there will be conflicting departmental viewpoints in the organization. In order to satisfy the customer, the sales department might want excessive inventories on hand at all times in order to provide quick delivery. In addition, it might ask for better quality than the customer requires thus increasing costs, and it might place pressure to reduce the selling price in order to obtain sales more easily. On the other hand, the factory may want to produce at no more than acceptable quality, and place pressure to get longer runs and uninterrupted schedules in order to reduce unit costs. Still a third point of view would be that of the controller or treasurer, who is interested in maintaining the lowest possible investment in inventories and equipment, which would reduce the effectiveness of both sales and production. It is not at all unusual that each group has concern only for the goals for which it is responsible.

Obviously, in companies where these widely different interests exist, there must be some means of coordination among the various departments. In companies of sufficient size and complexity, the establishment of a separate production planning and control function may be the solution to the problem. The Production Control department acts as the coordinating body.

RANGE OF ACTIVITIES

In Table 3-1, page 18, the range of activities of Production Control organizations is listed; many companies include other functions which are not ordinarily considered to be Production Control functions. The following is a list of the functions of Table 3-1 plus additional functions sometimes included:

1. Sales forecasting
2. Order writing
3. Product design
4. Process planning
5. Routing
6. Material control
7. Tool control
8. Machine loading
9. Scheduling
10. Dispatching
11. Data processing
12. Expediting
13. Receiving

14. Shipping
15. Materials handling
16. Purchasing
17. Timekeeping
18. Traffic
19. Stockroom operations

The individual company must review its organization and determine which of the foregoing functions are to be placed in the Production Control department. At the same time, the division of the other responsibilities must be made to other departments. Surveys have been made by research groups to determine what functions are normally placed in the Production Control department but the only thing that was proven is that there is no set pattern. In the final analysis it is relatively unimportant which functions are placed in the Production Control department as long as the functions are related or similar and are within the capabilities of the people running them. The most important point is that the organization has been carefully reviewed and assignments made on the basis of particular company circumstances, philosophies, and individual capabilities.

REPORTING LEVEL

In the vast majority of cases, it will be found that the manager of a Production Control department will argue the point that his department should report at as high a company level as possible in order to assure maximum operating efficiency and to prevent undue influence by any particular group. However, the final determination of the reporting level should depend upon which planning and control functions are to be given to the Production Control group, and its authority. The greater the number of functions performed in Production Control, the greater the degree of authority required, and hence the higher the reporting level. For example, the complexity of Production Control in the job order and special project types of activities demands a high degree of authority in the function.

A point to remember is that in many cases, particularly in the smaller companies, the reporting level of the separate Production Control department is many times a matter of "growing up." This is particularly true if the Production Control department is relatively new in the company. In cases where the Production Control manager feels that he is reporting to too low a level in order to do an effective job, he should strive to prove that he can do a top notch job at the level in which he operates. When management discovers that the present area of authority and responsibility is handled with a high degree of effectiveness, the manager is in an excellent position to take on added responsibilities and

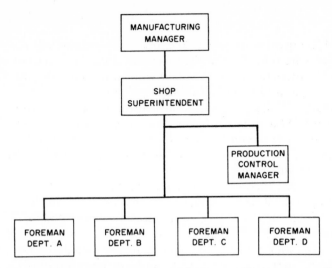

Fig. 26-1. Organization chart, showing production control manager reporting to shop superintendent—typical in small plants.

be elevated to a position of greater authority. Figures 26-1, 26-2, and 26-3 show examples of the Production Control manager reporting to three different levels in different organizations. Figure 26-1 shows a Production Control manager reporting to the shop superintendent or general foreman. This level of reporting is usually found in relatively

Fig. 26-2. Organization chart showing production control manager reporting to works manager—typical in medium-sized plants.

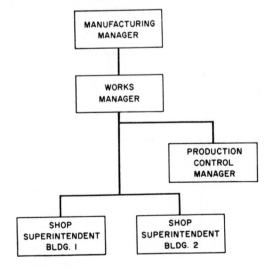

small companies in which the operations are fairly simple and the products are few. Most of the functions of Production Control are placed in other departments in the plant while a minimum of functions remain in the separate Production Control department.

Figure 26-2 shows a Production Control manager reporting to the works manager. This situation is found often in medium-sized plants that have a fair degree of complexity of operations and a fairly large number of products. The Production Control manager is on the same level as the shop superintendent.

Figure 26-3 shows an example of the Production Control manager reporting to the general manager of manufacturing. In this case, he would have a status equal to that of a works manager. This type of organization is often found where there is more than one plant producing the same type of product. The operations are usually of the job order or special project type with a wide range of items produced. Most of the range of activities shown previously in this chapter would be included under the Production Control manager. He has a very strong coordinating responsibility and a high level of authority.

In some very large plants there may be a "super Production Control" function. The central Production Control staff may be set up to provide assistance in designing and implementing systems of Production Control within the various plants of the company. Their main function is to assure top management that there is effective Production Control within

Fig. 26-3. Organization chart showing production control manager reporting to the general manager—typical in large plants.

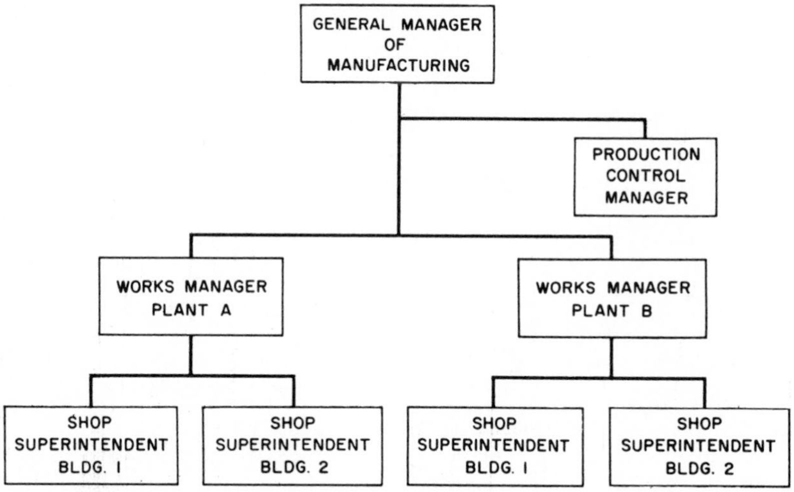

each one of the divisions. The Production Control managers in the plant would report to the central Production Control manager who in turn reports to the Vice President in charge of manufacturing or to the President of the company.

SPECIFIC AUTHORITY

The specific authority of the Production Control department would obviously depend upon the factors of goals, range of activities, and reporting level. Once these have been determined the specific authority then can be defined with little difficulty. Following is an example of a position description for the Production Control manager of a large plant.

TOOL DIVISION
Production Control Manager

DESCRIPTION OF POSITION

It is the responsibility of the Production Control manager of the Tool Division to plan, organize, and direct the following activities within that division:

1. Stock control records:
   1.1 Raw materials
   1.2 Purchases parts
   1.3 Manufactured parts
   1.4 Castings
   1.5 Finished stock
2. Authorization for, or requisitioning of:
   2.1 Production orders for tools or parts to be manufactured
   2.2 Castings to be produced in foundry division
   2.3 Raw material, parts, and some supplies to be purchased
3. Stock-keeping and stores of:
   3.1 Raw material and purchased parts
   3.2 Manufactured parts
   3.3 Castings
4. Production Control and scheduling:
   4.1 Long-range scheduling
   4.2 Detailed schedules
   4.3 Providing sales department with delivery promises
   4.4 Dispatching and expediting of orders or work in process
5. Preparation of reports:
   5.1 Monthly report of sales and production on each item
   5.2 Work-ahead reports
   5.3 Special reports
6. Posting of operations reported by workers on time sheets to office copies in order to:
   6.1 Provide accurate record of progress of orders
   6.2 Eliminate duplications in reporting of operations by workers
7. Preparation for, and taking of the annual physical inventory in accordance with instructions issued by the auditor

He analyzes the relative urgency of all orders or work ahead through consultation with the Tool Division sales manager; and plans schedules accordingly, insofar as economical plant operations and capacity will permit.

He counsels and cooperates with his superior on operations, problems, studies, or surveys, and in the formulation of plans or procedures in the Tool Division.

RESPONSIBILITIES:

Planning and Organizing

The Production Control Manager of the Tool Division:

1. Plans and makes provision for the final selection, placement, and training of personnel for his department.
2. Plans and organizes, in consultation with his superior and other executives, department heads, or subordinates affected, efficient and economical handling of all his departmental activities, consistent with established company policies, rules, or procedures.
3. Plans, in consultation with his superior, the financial division, and subordinates, simplified procedures, systems, and forms consistent with sound accounting principles.
4. Recommends to his superior plans that will contribute to the efficient and effective operation of his department or the Tool Division in general:
   4.1 Broad plans and procedures for the effective scheduling and flow of production.
   4.2 Plans for increased or decreased production capacity or production quotas.
   4.3 Broad plans for production output and flow of work to meet specific production schedules.
   4.4 Other plans which will contribute to the increase in output or expedite delivery of orders.
5. Plans and organizes for his department:
   5.1 The formulation and preparation of written "standard practice" procedures and instructions.
   5.2 The delegation or assignment of duties or functions to the persons most qualified to handle them.
6. Establishes, in consultation with his superior and the Tool Division sales manager, the production quotas to be used as a basis for scheduling, and the sales objectives and policies to be observed.
7. Plans, in consultation with his superior and Tool Division foremen:
   7.1 Breakdown of production schedules by departments or sections of departments as necessary.
   7.2 Standard scheduling intervals by departments or sections on all standard items.
   7.3 Production capacities to be observed.
   7.4 Scheduling procedures in accordance with sales objectives insofar as economical plant operations will permit.
8. Plans, in consultation with his superior, policies and procedures to be observed in controlling size of the inventory. Plans purchase requisitions and production authorization accordingly.

Coordinating

The Production Control manager of the Tool Division is expected to:

1. Coordinate and integrate the activities under his direction with those

of all departments served, affected by, or affecting his activities.

2. Counsel with his superior and the office coordinator on:

    2.1 Clerical or stenographic help—

        21.1 Rate ranges and proposed change in rates

        21.2 Increases or decreases in size of force

        21.3 Interviewing, screening, and hiring, in coordination with the personnel department

        21.4 Extraordinary working hours

        21.5 Grievances, quits, discharges, or transfers

        21.6 Vacation schedules

    2.2 Office policies, rules, or procedures to be observed

    2.3 Compliance with Federal or State laws governing employment

    2.4 Requirements of office supplies or equipment

    2.5 Maintenance and care of office equipment

    2.6 Office layouts and working conditions

3. Coordinate with the auditor in planning procedures and forms consistent with sound accounting principles to maintain accounting or financial controls. Cooperate in surveys and studies of procedures or systems initiated by his superior or the auditor for the purpose of simplifying and coordinating same.

4. Keep superintendent, division or department heads, and foremen fully informed as to the functions, objectives, and procedures of his department. Counsel with other executives and supervisors on activities and procedures which involve or affect them.

5. Counsel with his superior and Tool Division foremen on the establishment or revision of:

    5.1 Intervals of time to be allowed in scheduling.

    5.2 Plans for production output and flow of work to meet production schedules.

    5.3 Records or reports essential to the efficient operation of the Production Control system.

6. Counsel and coordinate closely with the sales manager of the Tool Division on sales objectives and policies to be kept in mind in planning Production Control activities.

7. Counsel and coordinate closely with the routing and standards department on mutual problems.

Controlling and Directing

The Production Control manager of the Tool Division:

1. Establishes for his department, in consultation with his superior and the office coordinator:

    1.1 The standards of performance and conduct expected from his personnel, conforming to company policy.

    1.2 The methods of evaluating and reporting deviations from such standards.

2. Directs and supervises the personnel and activities of his department. Observes, checks, and controls his departmental operations and conduct to conform with company policy, keeping his superior informed on performance and results.

3. Makes recommendations to his superior on the following for his department:

3.1 Increased or decreased manpower requirements. He requisitions help within approved limits.

3.2 Changes in wages or rates conforming with company policy and established rate ranges.

3.3 Discharge or transfer of unsatisfactory workers.

3.4 Upgrading or promotion of worthy individuals.

4. Approves final selection and placement of personnel within his department; and delegates to his staff the extent of their duties, responsibilities, and authority. Directs the instruction and training of his personnel, except group training programs sponsored by the company.

5. Utilizes all means at his disposal to promote a spirit of unity and team play within and between all departments and takes steps to correct any maladjustments which may arise within or involve his department.

6. Directs within his department the maintenance and filing of all records on an orderly and systematic basis.

7. Directs the operation of the following:

7.1 Stock control system and records.

7.2 Stock-keeping of materials and parts.

7.3 Preparation and maintenance of the scheduling system.

7.4 The posting to office copies of orders operations reported by workers.

8. Controls the size of the Tool Division inventory by exercising judgment on:

8.1 Approving requisitions or orders for materials, parts, and finished tools.

8.2 Directing the release of shop orders for production in accordance with schedules based on sales requirements.

It must be remembered that one of the most important things about the authority given to the Production Control department, is that it must be backed up by top management. The department must, in turn, accept its responsibilities and provide the leadership which is required.

RELATIONSHIPS TO OTHER MANAGEMENT FUNCTIONS

Regardless of the level at which the production planning and control department may report in an organization, there are many important relationships which exist between Production Control and other elements of the organization. The most important of these are:

1. The sales organization
2. The purchasing department
3. The traffic department
4. Materials handling function
5. Plant engineering and maintenance
6. New product development
7. Industrial engineering
8. Production management

In defining the position description of the production manager and the description of the Production Control department, it is very important to define the relationships that exist between the Production Control department and the functions listed above. Without the coordination of Production Control with these groups, it would be impossible for Production Control to operate effectively. Because of these relationships, it is vital that the designer of the Production Control system understands what the relationship is and assures that there is a proper coordination and communication between Production Control and the various co-ordinating functions. It would be impossible to describe the specific routines in detail that exist in the relationship between Production Control and these various functions. These will vary greatly with the type of operation and from company to company. However, the basic principles as described in this text are applicable in all cases. Only the niceties of practical applications must be determined. When the nature of the specific information and requirements of the Production Control are known, the basic relationship of the other functions becomes relatively simple. For example, if Producton Control is required to set up operation schedules it requires standard times for the various steps in the operation. If an industrial engineering department exists which sets the standards, then it is obvious that Production Control must set up a relationship with industrial engineering in order to obtain these standards for scheduling purposes.

Summary   Production Control techniques and "tools" are of little value unless an organization is designed to use these tools in the proper manner. There are two steps in organizing for Production Control:

    Step 1. Determining the place of Production Control in the total organization.

    Step 2. Designing the Production Control organization within the framework of Step 1.

This chapter fully develops the requirements of Step 1. Chapter 27 is devoted to Step 2.

    Step 1 requires the determination of:

1. Over-all goals
2. Range of activities
3. Reporting level
4. Specific authority and responsibilities
5. Relationships to other plant and management functions

All of these requirements are discussed fully in this chapter. It is imperative that the designer of the Production Control system have a thorough understanding of them and apply them conscientiously.

27

Designing The Production Control Organization

In Chapter 26, we described the various principles and factors which affect the place of the Production Control function in the over-all organization. In this chapter we will assume that the level or place of Production Control in the organization has been determined and that the problem is now one of designing the Production Control organization itself. We will assume that we will have a formal Production Control function with all of the basic activities described in Table 3-1.

Before proceeding with the specific factors involved in the design of the Production Control organization, the reader should first be familiar with some of the basic principles of organization. These basic principles are applicable to any *kind* of organization and to any *level* of organization.

1. *Over-organization should be avoided.* No organization unit should be set up unless it has a distinct job to

perform. A sufficient volume of work to occupy enough people to justify a supervisor should also be required before a separate organizational element is created.

2. *Objectives, goals, and functions should be clearly stated.* The objectives and goals of the organization determine the types of functions to be performed. It is essential that they be clearly stated, preferably in writing.

3. *All necessary functions should be assigned to specific units of the organization.* An organization exists for one purpose—to make sure that all functions essential in accomplishing the objectives and goals are assigned and carried out. An organization which excludes necessary functions, major or minor, is seriously handicapped in achieving success.

4. *Functions assigned to a unit should be related or similar.* This principle is based on the fact that most people have limited ranges of interest and capabilities and everyone does certain things better than he does others. The tendency, especially in a rapidly growing organization, is to assign new functions to whoever is most available. Also, when a man is transferred to another job in the organization, he frequently takes some of his old duties with him, even if they are not related to his new assignment. This, of course, results in a hodge-podge of unrelated tasks for many individuals.

5. *Each function should be assigned to only one organizational unit.* Violation of this principle causes duplications in operations. Very often, these conditions are created through lack of knowledge of the fact that the same work is being performed by another part of the organization. There are, of course, individual instances when duplication of operation may be justified. The physical location of various buildings of a plant may necessitate decentralizing reproduction operations or files and records operations. In each case, however, the functions which have been decentralized should be under the technical supervision of a central organization.

6. *Responsibilities assigned should be specific, clear-cut, and understood.* Functions should be assigned to the units concerned in clear, simple language. If two different units perform related functions, special care must be taken to draw a firm line of separation where the authorities and responsibilities of one unit end, and those of other units begin. This will prevent jurisdictional complaints and duplication of effort.

7. *Responsibility for a function should be matched by the authority necessary to perform that function.* Frequently, delegation of responsibility fails to carry with it commensurate authority or it is accompanied by such checks as to make impossible the satisfactory performance of the assigned functions. In this situation, one will find that before an action can be taken, the supervisor at any level must clear with higher authority. This type of organization makes frustrated "glorified errand-boys" of su-

pervisors and generally results in a reduction of initiative and resource-fulness.

8. *Authority to act should be as close as possible to a level at which the work is performed.* By following this principle the occasion for review by higher authority and duplication of staff at higher levels is reduced. There are certain practical limitations, of course, that must be placed upon decentralization of authority and responsibility. These limitations are necessary to assure that the basic policies of the entire organization are coordinated and carried out, and that in certain instances standard procedures are followed throughout.

9. *The nature of the activity at each organizational level should be the determining factor in deciding upon the type of organization.* Specialized skill or knowledge required, and a degree of coordination and control necessary in the handling of the work, must be considered. If the knowledge required is primarily of a functional nature (e.g., accounting, storage, and the like), specialized knowledge of the function is needed. In other activities—purchasing, for example—an over-all knowledge of the commodity itself is necessary.

If the emphasis is placed on a uniform handling of functions with respect to all commodities, the functional type of organization is suitable. Where the emphasis is placed on the product or commodity, the commodity type of organization may be more desirable.

Careful analysis and accurate determination should be made to determine whether the function should be subordinated to the commodity being handled, or whether the commodity itself is of such a nature that it can be subordinated to the functions necessary in its handling.

Such factors as size, performance of the organization, availability of space and personnel, processes to be utilized, "mechanized or manual," and philosophy in top management will influence the type of organization.

In arriving at any decision, the problem is to determine which organization will most efficiently obtain the objectives and goals using the simplest procedures.

10. *The number of people reporting to a chief should not exceed the number which can effectively be coordinated and directed.* No fixed formula exists by which one can determine the number of people who, under varying circumstances, and in all types of organization, can work effectively under the supervision of one man. A great deal depends upon the character of work being done. A foreman may supervise 20 or 30 people performing the same type of manual operation, whereas the head of a technical organization unit may be unable to supervise more than 3 or 4 people. For the average type of administrative unit, it is generally agreed that the number of people that can be directly supervised is somewhere between 4 and 8. When a supervisor has more than 8 per-

sons reporting to him, there is a possibility that the supervisor's ability is wasted, owing to the dilution of his efforts over too wide a span.

11. *All supervisors should know who reports to them; all subordinates should know to whom they report.* Violation of this principle is a clear indication of confusion in assignment of responsibilities and definition of the functions to be performed. Conflicting orders, dual responsibility, and lowered morale inevitably result from such a situation and seriously affect the performance of the entire organizational unit.

12. *Heads of organizational elements should control the activity through attention to important policy problems rather than through review of routine actions.* The necessity for speedy action is such that it is better to delegate discretionary authority to subordinates at the risk of a few mistakes then to retard performance by a system of reviews. Subordinates should be selected and trained so that they can be trusted to report unusual policy problems to their supervisors, and to act promptly and decisively on routine functions without reference to higher authority. Failure to follow this principle not only lowers morale and delays action, but also overburdens the head of the activity with numerous petty details that prevent him from exercising effective supervisory control.

These general principles must be understood and followed by the designer of the Production Control organization. In addition, he should fulfill certain specific requirements in order to complete the design of an organization which will maximize the possibility for success. These are as follows:

1. Define the specific functions within the framework of objectives, goals and responsibilities laid down by top management.
2. Define the type of personnel and determine where they are to be obtained.
3. Determine the size of the Production Control staff.
4. Fit the men to the functions.
5. Set up the necessary training program for personnel to perform the functions.
6. Set up plans and controls for the Production Control activity.
7. Establish the methods of evaluating the performance in each function within the Production Control activity.

DEFINE THE SPECIFIC PRODUCTION CONTROL FUNCTIONS

There are two descriptions which are required in order to define a specific function:

1. Position description
2. System or procedural description

The position description is defined in a manner similar to a position description of the Production Control manager in the previous chapter. These descriptions are usually used whenever the activity involves a supervisor. As described in a primary principle of organization, there should be a sufficient volume of work to occupy enough people to justify a supervisor before a separate organizational element is created. If this principle has been followed, then a position description for the supervisor with his responsibilities and authority is written. The specific functions which are performed within this phase of the activity are described in a procedures manual.

Of course, the procedure cannot be written until the system has been designed. In any formal Production Control organization, the procedures for all systems should be in writing and placed in a procedures manual.

DEFINE THE TYPE OF PERSONNEL AND DETERMINE WHERE THEY ARE TO BE OBTAINED

The various functions of Production Control require personnel of different types. Some of the functions require very little contact with other departments while others primarily constitute contact with other departments in the plant. For example, the dispatcher, the expeditor, and the scheduler require a great amount of contact with other departments and other people in the plant. The characteristics or personal qualities which are required in these jobs are:

1. *Tact.* For example, the scheduler or dispatcher must assign work in the proper sequence to the department to which he is assigned, but he is not the supervisor of that department. Obviously, he must get along with the people or he will not be successful.

2. *Aggressiveness.* The dispatcher or the expeditor must be aggressive if he is to be assured that his department does what the plan calls for instead of following its inclinations. Of course, he must combine tact with aggressiveness.

3. *Ability to work under pressure.* The expeditor, dispatcher and many others in Production Control have constant pressure on them. They must all recognize the importance of meeting schedule dates and deadlines. Everything must be done in a hurry but it must be done correctly in order to have a smooth running function.

There are, of course, the other personal qualities such as intelligence, conscientiousness, and initiative that every supervisor is looking for in everyone. There must certainly be a high degree of all of these qualities in the Production Control employees. Most of these characteristics can be identified through personnel testing programs and interviews by qualified people. Many of the larger companies today have trained psychologists who devote their full time to this type of testing and interviewing. Even without trained interviewers, most plants can identify needed characteristics in applicants if the characteristics have been properly defined in terms of the functions of Production Control.

The personnel to fill the various positions can be obtained either from within the plant or from the outside. Many plants have the policy of selecting Production Control personnel from administrative, industrial engineering, accounting, and production employees. The testing and selecting of employees from within the plant is the same as for those obtained outside of the plant. For many of the functions in Production Control, college training, particularly in business administration or industrial engineering, is desirable for the key jobs. Many plants that have well established Production Control departments use this department as a training ground for college graduates to learn the business. Production Control proves to be an excellent training ground because it provides the employee with the opportunity of becoming involved in almost every activity in the plant. Many of the positions in Production Control require a very thorough knowledge of some particular part of the product or business. In these cases it might be most desirable to select people from the product engineering or mechanical engineering departments or even exceptional production workers to fill the positions.

DETERMINING THE SIZE OF THE PRODUCTION CONTROL STAFF

The sizes of the Production Control staffs of different companies will vary as much as the difference in the types of activities. It becomes apparent that the size of the Production Control staff must be determined on an individual plant basis. The size is completely dependent upon the functions that are to be performed and the quality of the personnel employed.

It is certainly axiomatic that nothing should be "bottlenecked" in the Production Control function because of the lack of manpower alone. However, it must still be determined whether the manpower that is available is producing at an acceptable rate of efficiency.

If the Production Control staff is of sufficient size, then work standards on the various functions might be feasible. The standards might be set through the use of the work sampling techniques described in Chapter 24. Once the standards have been set, it is a simple matter to determine the efficiency of the personnel. The number of people in each position can be determined on the basis of standards. The other, but less sensitive, method of determining the size of the staff would be by means of the methods of evaluation described in Chapter 25. This chapter describes one of the methods of evaluating the effectiveness of the system rather than the people performing the functions of the system.

FITTING MEN TO FUNCTIONS

We have previously discussed the importance of defining the type of personnel required in each one of the Production Control functions and where they may be obtained. One of the most important jobs of the Production Control manager is to evaluate the strengths and weaknesses of the individuals who are responsible for each activity in the department. It is a continuous job to maintain the highest degree of competence in each position consistent with the salary structure of the department. Much has been said about the "square peg in the round hole." There is literally no end to the unpleasant things that happen to the square peg in the round hole. When a job doesn't fit the man there is a loss not only to the man but to his organization as well. In fitting different kinds of people to organizational roles it must be done so that the organization is helped instead of hurt. One general rule that should always be followed is that *in assigning members to their respective roles in an organization, the members' preferences, as far as possible, should be considered.* Each member has a pattern of needs he strives to gratify. Each organizational role has different kinds of activities connected with it, such as work, decision-making, and communications activities. Each role has different opportunities for making friendships and for winning recognition. Each role has different privileges and prerogatives attached to it. The problem in any organization is to place members with various personality needs in roles that provide opportunities for meeting these needs.

TRAINING PERSONNEL TO PERFORM PRODUCTION CONTROL FUNCTIONS

Clearly the first requirement in the training program is that the trainees are capable of absorbing the training offered. This, of course, is dependent upon how well the job of fitting the men to the functions was performed as described in the previous paragraph. We will assume here, that the proper personnel have been selected for training in the various functions of Production Control. There cannot be too much argument that some sort of orderly training program is desirable for Production Control employees. The results of greater skill and decreasing turnover among higher-grade personnel are obvious. Because few plants can afford a formal off-the-job training program, the formal training program will primarily consist of on-the-job training. For supervisory personnel this might be supplemented with some off-the-job training.

With the exception of the clerical personnel, very few of the skilled Production Control employees should be expected to stay on any particular job for a long period of time. *The key to a good training program is the opportunity for steady advancement of those trained.* The resulting high turnover in Production Control is not necessarily harmful.

Because the functions of the department are procedurized, training is not too difficult or lengthy. If the people are properly selected for intelligence and other characteristics, the trainees take hold quickly enough to minimize inefficiency in the department and the over-all benefits to the company clearly outweigh the disadvantage of the turnover. With each job properly defined, an employee can be placed in the lowest classification of job for training and systematically advance along every job in the Production Control department until he reaches the highest position available.

The starting salary of many jobs in Production Control is usually not high enough to attract high caliber personnel unless they are assured of opportunities for advancement. The time that an employee would stay on the job of entry might vary from six months to two years. During the on-the-job training program, it is possible to weed out those employees who do not demonstrate a potential for advancement to better positions or to supervisory positions within Production Control or elsewhere in the company. An advancement program as outlined makes it possible for a company to acquire good personnel and to upgrade employees with a minimum of training. Each job that is set up in a progression is the best possible training for the next job.

During this training program, it is important to have some means of merit-rating the employee. This will point out potential problems; the supervisor can work with the individual employee on those points in which he needs help. The earlier this can be discovered, the better for all concerned in the long run.

To install such a training program for employees requires a great deal of effort, but once it is set up and running, it will save time and trouble.

PLANNING AND CONTROLLING THE PRODUCTION CONTROL ACTIVITIES

The Production Control department is no different from any other "production" department with respect to the need for the planning and controlling of its activities. The principles which apply to all other activities also apply to the Production Control department, particularly when it is a large one. It should not be necessary to go into detail on how to plan and control the activities of the Production Control department. Certainly, there should be no one better qualified to set up such plans and controls than the Production Control manager. However, it is not uncommon to find the "doctor getting sick."

ESTABLISHING THE METHODS OF EVALUATING
THE PERFORMANCE OF PRODUCTION CONTROL

This particular point was fully described in Chapter 25. To sum up the previous points described, it can be emphasized that to get the various units of the Production Control department in harmony with each other and with other units of the company becomes a major function of the leadership of the Production Control department.

Summary   The design of the Production Control *organization* is an extremely important part of the design of the Production Control *system*. The designer can be assured of greater success by following basic principles of organization and meeting specific requirements for implementing the organizational structure. None of these rules and requirements can be overlooked if a close-knit, smooth-running organization is desired.

The effect of good techniques and procedures in a system can be quickly nullified by lack of attention to the organization—the personnel—that will carry out the responsibilities of Production Control. The design of the organization must be accomplished through the fulfillment of certain specific requirements in order to assure success.

Principles of organizational design and specific requirements for Production Control organizations are detailed in this chapter.

Statistical Control Charting

In order to have an understanding of the basic rationale of the activity control techniques discussed in Chapters 7 and 16, a brief description of the fundamental statistics will be given. All justification will be made on an intuitive basis rather than on a mathematical one. An absolute minimum of statistical mathematics will be discussed or described. The reader would be well advised to refer to one or more of the excellent books in statistical quality control before applying the charts to an actual situation. Basic assumptions have to be made for different situations, and a discussion of these assumptions is beyond the scope of this book.

The measured properties of all systems are subject to a certain amount of variation due to chance alone. With sufficient observation and analysis it is possible to discover and define the stable "system of chance causes" inherent in any particular system. Variations within this stable pattern are inevitable. Significant deviations *from* this inherent pattern may be detected by statistical means and the cause of the deviations assigned to some exogenous factor or factors.

As an example of a system, consider the manufacturing of heavy metal pins one inch in diameter using a grinding machine. It is impossible to produce pins exactly one inch in diameter. With sufficient care and precision of measurement small variations from pin to pin can be detected regardless of the process itself. If a large number of pins were measured and the distribution of sizes made, a curve similar to Figure I-1 would probably be found. Statisticians call this bell-shaped curve a "normal curve." This is the stable system of chance causes inherent in the manufacture of one-inch-diameter ground pins.

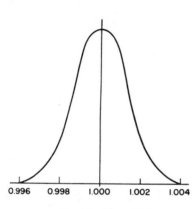

Fig. I-1.   Normal curve.

Any normal curve can be completely described by two factors: a measure of central tendency ($\overline{X}$, the arithmetic mean) and a measure of dispersion or spread (σ, Greek *sigma*, the standard deviation). The arithmetic mean is what is commonly known as the average. Expressed algebraically it is $\overline{X} = \dfrac{\Sigma X_i}{n}$. The standard deviation is the less well known root-mean-square deviation of the observed numbers from their arithmetic mean. The algebraic expression of standard deviation is:

$$\text{Standard Deviation} = \sigma = \sqrt{\frac{\Sigma (X_i - \overline{X})^2}{n}}$$

$$= \sqrt{\frac{(X_1 - \overline{X})^2 + (X_2 - \overline{X})^2 + \cdots + (X_n - \overline{X})^2}{n}}$$

The mathematical properties of σ allow us to make predictions or statements about the system described by the normal curve. Certain specific proportions of the production will, in the long run, fall within various sigma limits. These proportions are:

X $\pm$ σ —approximately 68   per cent of the production
X $\pm$ 2σ —approximately 95   per cent of the production
X $\pm$ 3σ —approximately 99.7 per cent of the production

Returning to the pin example, assume that the measurements of ten representative pins were found to be: 1.000, 1.001, 1.000, 1.002, 0.999, 1.000, 0.998, 1.001, 1.000, 0.999.

The arithmetic mean can be determined in the conventional manner as follows:

$$
\begin{aligned}
X_1 &= 1.000 & X_6 &= 1.000 \\
X_2 &= 1.001 & X_7 &= 0.998 \\
X_3 &= 1.000 & X_8 &= 1.001 \\
X_4 &= 1.002 & X_9 &= 1.000 \\
X_5 &= 0.999 & X_{10} &= 0.999 \\
\hline
5.002 & & 4.998
\end{aligned}
$$

$$
\begin{aligned}
& 5.002 \\
& 4.998 \\
\Sigma X_i\ & \overline{10.000} \\
X &= \frac{\Sigma\ X_i}{n} = \frac{10.000}{10} = 1.000
\end{aligned}
$$

The standard deviation would be calculated in the following manner:

$$
\begin{aligned}
(X_1 - \overline{X})^2 &= (1.000 - 1.000)^2 &= (0)^2 &= 0.000000 \\
(X_2 - \overline{X})^{2-} &= (1.001 - 1.000)^2 &= (0.001)^2 &= 0.000001 \\
(X_3 - \overline{X})^2 &= (1.000 - 1.000)^2 &= (0)^2 &= 0.000000 \\
(X_4 - \overline{X})^2 &= (1.002 - 1.000)^2 &= (0.002)^2 &= 0.000004 \\
(X_5 - \overline{X})^2 &= (0.999 - 1.000)^2 &= (-0.001)^2 &= 0.000001 \\
(X_6 - \overline{X})^2 &= (1.000 - 1.000)^2 &= (0)^2 &= 0.000000 \\
(X_7 - \overline{X})^2 &= (0.998 - 1.000)^2 &= (-0.002)^2 &= 0.000004 \\
(X_8 - \overline{X})^2 &= (1.001 - 1.000)^2 &= (0.001)^2 &= 0.000001 \\
(X_9 - \overline{X})^2 &= (1.000 - 1.000)^2 &= (0)^2 &= 0.000000 \\
(X_{10} - \overline{X})^2 &= (0.999 - 1.000)^2 &= (-0.001)^2 &= 0.000001 \\
\hline
\Sigma\ (X_i - \overline{X})^2 & & &= 0.000012
\end{aligned}
$$

$$
\sigma = \sqrt{\frac{\Sigma(X_i - \overline{X})^2}{n}} = \sqrt{\frac{0.000012}{10}} = \sqrt{0.0000012} = 0.001
$$

The proportion of pin sizes would be as follows:

$$
\overline{X} \pm 1\sigma = 1.000 \pm 1(0.001) = 0.999 \text{ to } 1.001\text{—}68 \quad \text{per cent of the pins}
$$
$$
\overline{X} \pm 2\sigma = 1.000 \pm 2(0.001) = 0.998 \text{ to } 1.002\text{—}95 \quad \text{per cent of the pins}
$$
$$
\overline{X} \pm 3\sigma = 1.000 \pm 3(0.001) = 0.997 \text{ to } 1.003\text{—}99.7 \text{ per cent of the pins}
$$

The acceptance of this property of the standard deviation is the fundamental premise upon which many statistical techniques, including control charting, is based.

Assume you wish to check the operation of the manufacturing process by selecting and measuring one pin. The pin happens to measure slightly more than 1.003 inches in diameter. Could it be possible that the process is still producing pins with an X of 1.000 and a σ of 0.001? The answer to this question is yes, it is possible. We know that by chance alone approximately three pins out of every two thousand will have a diameter at least this large. But, the likelihood of selecting one of these three pins is so small that we arbitrarily choose to assume that the grinding process has changed. The odds are very good (approximately 650 to 1) that this assumption is correct. Basically, all that statistics is interested in is the determination of the *likelihood* of various events. If the pin had measured 1.000 inches in diameter, we would have assumed the process is unchanged because this is a highly likely event. Similarly, if the pin turned out to be 0.999 inches in diameter, we would assume that nothing had changed.

The Least-Squares Method Of Linear Regression

In many types of analysis, such as forecasting the future activity of a facility, it is desirable to establish the average relationship between two variables. One way of doing this is to plot the individual observed relationships between the variables on a piece of graph paper and then draw in the line of best fit by eye. The trouble with this procedure is that it is based entirely on judgment and no two people will place the line in the same place.

The least-squares method of linear regression is simply a method of mathematically determining the single line which best describes a group of data.[1] To describe the linear relationship, the equation for a straight line is used.

$$Y_r = a + bX$$

---

[1] The name of this technique is derived from the fact that the regression line will reduce the sum of the squared deviations of each observed point from the line to a minimum. The least-squares technique can also be used for curvilinear relationships but only the simple linear case will be considered here. The reader can consult any of the standard statistical texts for further discussion.

Where

Y_r = The numerical value of a point on the line for any given value of X
a = The numerical value of the Y intercept for the line
b = The numerical value of the slope of the line

The regression formulae used to determine these values are:

$$a = \frac{\Sigma X^2\, \Sigma Y - \Sigma X\, \Sigma XY}{n\Sigma X^2 - (\Sigma X)^2}$$

$$b = \frac{n\Sigma XY - \Sigma X \Sigma Y}{n\Sigma X^2 - (\Sigma X)^2}$$

These formulae appear to be much more formidable than they actually are. Their use can be illustrated in an example.

A stationery company manufactures a line of grammar-school note-books and paper supplies for distribution in the Midwest. It is felt that a relationship exists between the kindergarten enrollment in any given year and the sales of notebooks two years later. The following data are known for the last ten years.

Year	Kindergarten Enrollment	Dollar Sales
1	4,000,000	$800,000
2	4,500,000	790,000
3	4,200,000	810,000
4	5,500,000	840,000
5	5,800,000	810,000
6	5,500,000	860,000
7	6,200,000	910,000
8	7,200,000	890,000
9	6,700,000	910,000
10	7,900,000	860,000

In order to establish a linear regression the data are placed in a special table as follows:

Sales Year	Kindergarten Enrollment Two Years Ago (10^6) X	Dollar Sales (10^5) Y	X^2	Y^2	XY
3	4.0	8.1	16.00	65.61	32.40
4	4.5	8.4	20.25	70.56	37.80
5	4.2	8.1	17.64	65.61	34.02
6	5.5	8.6	30.25	73.96	47.30
7	5.8	9.1	33.64	82.81	52.78
8	5.5	8.9	30.25	79.21	48.95
9	6.2	9.1	38.44	82.81	56.42
10	7.2	9.6	51.84	92.16	69.12
	$\Sigma X = 42.9$	$\Sigma Y = 69.9$	$\Sigma X^2 = 238.31$	$\Sigma Y^2 = 612.73$	$\Sigma XY = 378.79$

The calculations for the Y intercept and the slope of the line using the regression formulae are:

$$a = \frac{\Sigma X^2 \Sigma Y - \Sigma X\ \Sigma XY}{n\Sigma X^2 - (\Sigma X)^2} = \frac{(238.31)\ (69.9) - (42.9)\ (378.79)}{8\ (238.31)\ -\ (42.9)^2} = 6.17$$

$$b = \frac{n\Sigma XY - \Sigma X\Sigma Y}{n\Sigma X^2 - (\Sigma X)^2} = \frac{8\ (378.79)\ -\ (42.9)\ (69.9)}{8\ (238.31)\ -\ (42.9)^2} = 0.48$$

Thus the equation of the least-squares linear regression line becomes:

$$Y_r = a + bX = 6.17 + 0.48X$$

The regression line and the sales forecast for next year for the stationary company is shown in Figure II-1. Since last year's kindergarten enrollment was 6,700,000 children, the graph is entered at this point on the X axis. A line is drawn vertically to the regression line and then horizontally to the Y axis. The anticipated sales is $939,000. It would, of course, be possible to substitute the X value (6,700,000) into the equation to determine the same Y value.

$$Y_r = 6.17 + 0.48\ (6,700,000) = \$939,000$$

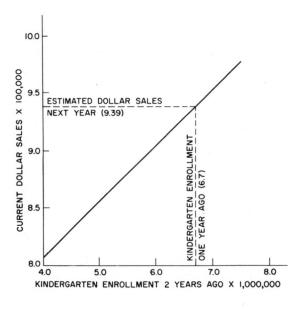

Fig. 11-1. Least-square linear regression line used for forecasting next year's dollar sales.

The Modi Method of Linear Programming

The classical example of distributing goods from a number of warehouses to numerous markets at the lowest total transportation cost is a convenient example to explain the mechanics of the Modi method of linear programming. It is important to realize that this movement of goods is *not* the *only* type of problem that can be solved by this technique. In Chapter 16 the method of scheduling by linear programming is shown. This appendix is provided to give the reader a basic understanding of the simple mechanics. There are many excellent books devoted exclusively to the techniques for those persons interested in pursuing the topic in more detail.

A distributor is faced with the problem of supplying five major market areas: M_a, M_b, M_c, M_d, and M_e from three existing warehouses: W_1, W_2, and W_3.[1] He is

---

[1] For brevity, the market and warehouse locations are denoted by letters rather than names. M_a might have been called Chicago, M_b New York, W_1 Detroit, etc.

naturally interested in supplying the needs of all the markets, but at the same time he would like to be sure that the total freight bill is no larger than absolutely necessary. To analyze the problem objectively, the distributor has compiled the data in Figures III-1, III-2, and III-3.

To set up and solve a distribution problem by linear programming, a special table, like the one shown in Figure III-4, must be used. One row in the table is provided for each of the warehouses. The number of units of the product in the warehouses is shown at the right of each row. Similarly, one column is provided for each market with the market demand shown at the bottom of the column. The intersections of the warehouse rows and the market columns represent the various available freight routes. The unit freight cost associated with each route, taken from Figure III-3, is written in the upper left-hand corner of the intersections. The extra blank row and column will be required for the mechanics of solving the problem and will be considered later.

WAREHOUSE	SUPPLY AVAILABLE
W_1	40 UNITS
W_2	80 UNITS
W_3	120 UNITS
TOTAL	240 UNITS

Fig. III-1. Available supply of goods.
Fig. III-2. Demand requirements for goods.
Fig. III-3. Unit distribution costs in dollars per unit. Note: Warehouse 2 is located in the same community as Market C.
Fig. III-4. Initial modi table.

MARKET	DEMAND REQUIREMENT
M_A	10 UNITS
M_B	20 UNITS
M_C	30 UNITS
M_D	80 UNITS
M_E	100 UNITS
TOTAL	240 UNITS

	M_A	M_B	M_C	M_D	M_E
W_1	7	10	5	4	12
W_2	3	2	0	9	1
W_3	8	13	11	6	14

$\downarrow R_i \ \nearrow K_j$	M_A	M_B	M_C	M_D	M_E	
W_1	7 (10)	10 (20)	5 (10)	4	12	40
W_2	3	2	0 (20)	9 (60)	1	80
W_3	8	13	11	6 (20)	14 (100)	120
	10	20	30.	80	100	

In solving a distribution problem by linear programming it is necessary to start with a feasible solution and then make improvements on the solution. There are many ways of developing the initial feasible solution. One way is to begin in the upper left-hand corner of the table in Figure III-4, and systematically satisfy the markets' demands with the stock available in the warehouses. Caution must be exercised to avoid shipping too many units to any market and that only the actual number of units available in the warehouse are used. Starting in the upper left-hand corner, M_a requires 10 units and W_1 has 40 units available. By shipping 10 units from W_1 to M_a, the market requirements are filled and there are still 30 units left in W_1. Moving to the right, the requirements of M_b, 20 units, are met from the remaining 30 units in W_1, leaving 10 units in the warehouse. Moving to the right again, M_c requires 30 units. The 10 units left in W_1 are shipped to M_c and the remaining 20 units required by M_c are supplied by W_2. This process of developing a feasible solution is continued until all of the market requirements are met and the stock in the warehouses is exhausted.

At this point the mechanics of solving a linear programming problem must be introduced. There will appear to be many details to remember, but with a little practice the steps will become automatic.

As an aid to learning, imagine that each of the "route" squares in Figure III-4 is a puddle of water. Imagine further, that each of the circle numbers of the first feasible solution—i.e., 10, 20, 10, 60, 20, and 100—represents a stone placed in the square. It is possible to move around the table stepping on the stones but *not* stepping in the squares with no stones in them. There is one important restriction as to how the moves along the stones can be made: It is possible to move along the rows and columns (horizontally and vertically), but *not* diagonally. This concept of moving around the table might appear irrelevant, but it will be an essential phase in determining the optimum solution to the problem.

If the distribution table is set up correctly and the linear programming rules outlined below are followed without error, the positions of the stones will always have a very important property. It will always be possible to stand on a stone and look at one side of any non-stepping-stone square and then by moving vertically and horizontally along the stones, end up on another stone looking at a second side of the same non-stepping-stone square. When making the moves along the stones a minimum number of jumps should be made, i.e., jump over any intermediate stones and always make a right angle turn at the end of each jump.

To illustrate the last paragraph, a few examples will be considered. The route square for Warehouse "ONE" and Market d (the square with the four dollar unit cost) is a non-stepping-stone square. By standing on the "60 stone" we can look at the bottom of the W_1-M_e square. Moving horizontally to the "20 stone" and then vertically to the "10 stone" it is possible to see the W_1-M_d square from the side. The W_1-M_e square is also a non-stepping-stone square. By standing on the "100 stone" we can look at the bottom of the W_1-M_e square. Taking the path, 100 stone horizontally to 20, vertically to 60, horizontally to 20, and vertically to 10, we end up looking at the left-hand side of the W_1-M_e square. One more example would be the W_3-M_a square. Standing on the "10 stone" in the upper left-hand corner, we can look at the top of the W_3-M_a square. Moving horizontally to the 10 (remember the minimum number of moves), vertically to 20, horizontally to 60, and vertically to 20, it is possible to look at the right-hand side of the square. The reader should practice the stepping-stone moves by checking the remaining non-stepping-stone squares. The rules are:

1. Move along the stones only horizontally and vertically.
2. At the end of each move, make a right angle turn.

Once the stepping-stone moves have been mastered, the next thing to learn is how to fill in the R and K squares. This is simple. The following rules will determine the values.

1. Assign a zero to any one R or K square.
2. For each stepping-stone square, the sum of corresponding R and K values must be equal to the unit distribution cost of the stepping-stone square.

In Figure III-5, a zero was arbitrarily placed in the R square of the first row. Selecting the stepping-stone square in the upper left-hand corner, the "10 stone," and following the second rule we have: zero plus the K value of the M_a column must be equal to seven: $O + K = 7$; this K value must be 7. Moving to the right to the "2 stone," $O + K = 10$; the K value of the M_b column must be equal to 10. For the next "10 stone," $O + K = 5$. K for M_c equals 5. Moving vertically downward to the "20 stone," $R + 5 = 0$; the R value of the W_2 row must equal -5. Continuing this process for all the stepping-stones, a unique set of R and K values would result. The initial placement of the zero was arbitrary.

R_i \ K_j	M_A	M_B	M_C	M_D	M_E	
	7	10	5	14	22	
W_1 0	7 (10)	10 (20)	5 (10)	4 10 12 10		40
W_2 −5	3	2	3 0 (20)	9 #3 (60)	1 16 ➙(60)	80
W_3 −8	8	13	11	6 (80) #2 (20)	14 (40) #1 (100)	120
	10	20	30	80	100	

Fig. III-5. First program change.

If it was placed in any other location, a different set of unique numbers would result, but the following operations would still be exactly the same.

The reason for introducing the R and K numbers is to have a medium for systematically improving the initial distribution program. If the initial distribution program is not the best, it is logical to assume that one of the routes currently not being used, should be incorporated into the program. This means that the non-stepping-stone squares should be checked for possible use. The following is the check to use:

1. For each non-stepping-stone square, sum the corresponding R and K values and subtract the unit distribution cost of the non-stepping-stone square.
2. If the result of Step 1 is a negative number, the introduction of the route *would not* decrease the program's distribution cost.
3. If the result of Step 1 is a positive number, the introduction of the route *would* decrease the program's distribution cost.
4. Whenever a positive number results from Step 1, write the number in the upper right-hand corner of the non-stepping-stone square. After checking all of the non-stepping-stone squares, select the square with the largest positive number in the upper right-hand corner as the route to be introduced into the distribution program.

A few illustrative examples of the above rule should clear up any questions about the operation of the rule. Consider the non-stepping-stone square $W_1\text{-}M_d$. The R value is 0 and the K value is 14. The sum of the two is $0 + 14 = 14$. Subtracting the cost value of this square, 4, $(14 - 4 = 10)$ gives a positive value of 10. This positive value is written in the upper right-hand corner. The calculations for the square to the right would be $0 + 22 = 22, 22 - 12 = 10$. This positive value is written in the upper right-hand corner. For the square $W_2\text{-}M_a$ the calculations would be: $-5 + 7 = 2, 2 - 3 = -1$. Since this is a negative value, the initial solution cannot be improved by using this route, therefore, this square is disregarded. After completing the same calculations for the remaining non-stepping-stone squares, the largest positive number $(+16)$ will be found to occur in the $W_2\text{-}M_e$ square. We should use the route in preference to some route that we are already using. The next question is what route already being used, should *not* be used?

To answer this question, the stepping-stone moves, discussed earlier, will be applied. By standing on the "100 stone," it is possible to see the bottom of the $W_2\text{-}M_e$ square, (the route which we should use in preference to some route that we are already using). Moving to the "20 stone" and up to the "60 stone," in accordance with the stepping-stone rules set up before, it is possible to see the side of the $W_2\text{-}M_e$ square. These moves, "100 stone" to the "20 stone" to the "60 stone," make up the stepping-stone path.

The last set of rules for the distribution solution is:

1. Number the stones along the stepping-stone path being used, starting with the number one for the first stone. (If the moves have been correctly made there will *always* be an odd number of stones.)
2. Find the odd-numbered stone along the stepping-stone path that has the smallest numerical value.
3. Move this stone from its present square to the square coming into the solution.
4. Add the numerical value of the stone found in Step 2 to the even-numbered stones along the stepping-stone path.
5. Subtract the numerical value of the stone found in Step 2 from the odd-numbered stones along the stepping-stone path.

Returning to the example in Figure III-5, the "100 stone" would be the number one stone, the "20 stone" would be the number two stone, and the "60 stone" would be the number three stone. The odd-numbered stone with the smallest numerical value is the number three stone, the "60 stone." This stone is moved into the W_2-M_e square. The 60 value is then added to the "20 stone" and subtracted from the "100 stone." The results of these moves are shown in Figure III-6. Notice that in Figure III-6, by following the rules *to the letter,* the warehouse supply and market demands are not violated. This will always be the case.

The distribution program given in Figure III-6 is an improvement over the initial program developed in Figure III-4, but it is not necessarily the "best" program available to the distributor. It might be improved. To determine if the program can be further improved, the steps used on the initial program are repeated, i.e.:

1. Determine the R and K values.
2. Find the non-stepping-stone square with the largest positive number for $(R + K)$ — unit cost number.
3. Establish the stepping-stone path and number the stones.
4. Move the smallest odd-numbered stone into the new route path. Add the stone's numerical value to the even-numbered stones, subtract it from the odd-numbered stones.
5. Set up a new table and repeat the operations.

	K_j	M_A	M_B	M_C	M_D	M_E	
R_i		7	10	5	-2	6	
W_1	0	7 #1 (18)	10 (20)	5 (20) #2 (18)	4	12	40
W_2	-5	3	2	3 0 (10) #3 (20)	9	1 (70) #4 (60)	80
W_3	8	8 7 (10)	13	5 11	3 6 (80)	14 (30) #5 (40)	120
		10	20	30	80	100	

Fig. III-6. Second program change.

	$R_i \backslash K_j$	M_A 8	M_B 18	M_C 13	M_D 6	M_E 14	
W_1	−8	7	10 (10) #5 ⊗20	5 (30) #4 ⊗20	4	12	40
W_2	−13	3	2	3 0 #3 ⊗8	9	1 (80) #2 ⊗70	80
W_3	0	8 (10)	13 5|11 (10)	2 6 ⊗80		14 (20) #1 ⊗30	120
		10	20	30	80	100	

Fig. III-7. Third program change.

It is readily apparent that the solution of the distribution problem by linear programming involves a series of cyclical steps. The final table showing the best answer to the problem can be recognized by the fact that no improvement in the program can be realized by introducing any of the non-stepping-stone routes. The exact number of cycles or tables necessary to solve this type of problem is impossible to predict, but the number is finite and never exceeds a reasonable amount. In the illustrative example, four tables were required as indicated in Figures III-5 through III-8.

A very interesting fact should be noted in the final table, Figure III-8. Although Warehouse 2 and Market C are located in the same community, the best over-all freight minimizing solution clearly shows that it would not be economically feasible to use this zero cost route. At the same time, the most expensive route, W_3 to M_e, should be used in order to minimize the freight bill. Why is this so? Linear programming solves the entire problem at one time, balancing all of the alternative plans and coming up with the best over-all solution.

	$R_i \backslash K_j$	M_A 8	M_B 13	M_C 8	M_D 6	M_E 14	
W_1	−3	7	10 (10)	5 (30)	4	12	40
W_2	−13	3	2	0	9	1 (80)	80
W_3	0	8 (10)	13 (10)	11	6 (80)	14 (20)	120
		10	20	30	80	100	

Fig. III-8. Optimal program.

Data Processing Equipment

CARD PUNCHING is the basic method of converting source data into punched cards. The operator reads the source document and, by depressing keys, converts the information into punched holes. The machine-feeds, positions, and ejects the card automatically. The operator's primary concern is to depress the proper keys in the correct sequence. This is basically the same kind of function as typing or other key-driven operations. Card punches equipped with printing mechanisms automatically interpret the punched information at the top of the card directly above the hole being punched.

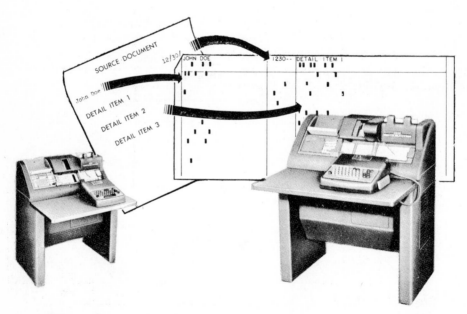

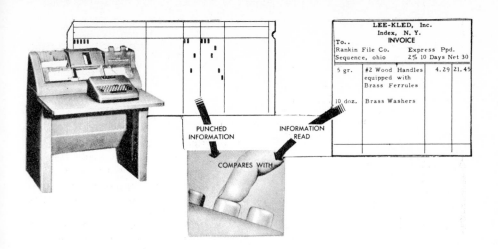

LEE-KLED, Inc.		
Index, N. Y.		
INVOICE		
To..		
Rankin File Co.	Express Ppd.	
Sequence, ohio	2% 10 Days Net 30	
5 gr.	#2 Wood Handles equipped with Brass Ferrules	4.29	21.45
10 doz.	Brass Washers		

PUNCHED INFORMATION

INFORMATION READ

COMPARES WITH

CARD VERIFYING is simply a means of checking the accuracy of the original key punching. A second operator verifies the original punching by depressing the keys of a verifier while reading from the same source data. The machine compares the key depressed with the hole already punched in the card. A difference causes the machine to stop, indicating a discrepancy between the two operations.

A notch in the upper right edge of the card indicates that it has been key punched and verified correctly. A notch directly above a column signifies that the punching of that column is in error.

This is basically the same type of function as typing or other key-driven operations.

DUPLICATING is automatic punching of repetitive information from a master card into a group of succeeding detail cards. This is normally performed as part of the card-punching function. Instead of depressing keys repetitively for common information (such as entry date, which is to be punched in every card), the operator punches the common information only once in the first card of each group, and it is automatically punched into all remaining cards for the group. This reduces the work per card, insures consistency of common data, and increases productivity of the operator.

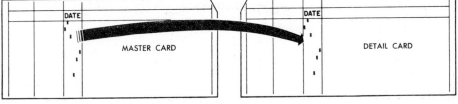

DATE

DATE

MASTER CARD

DETAIL CARD

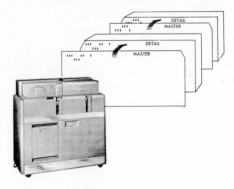

GANG PUNCHING is the automatic copying of punched information from a master card into one or more detail cards that follow it.

In single master-card gang punching, one master card precedes all detail cards to be punched with the same information.

Where information changes from one group of cards to the next, interspersed gang-punching methods may be used. A master card precedes each group of detail cards. Information in the master card is automatically selected for punching into all following detail cards until a new master is read. The punching pattern then changes to conform with the new master. Gang punching can be performed separately or in combination with reproducing and summary punching for both alphabetical and numerical information.

REPRODUCING from one card to another is like copying from one record to another. Information from one set of punched source cards is automatically punched into another set of cards. The two sets of cards are fed through the machine synchronously.

The comparing feature proves agreement between originals and reproductions. Differences are automatically indicated.

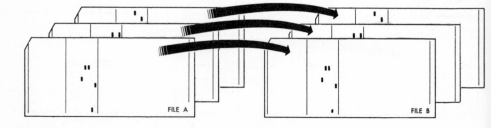

FILE A   FILE B

MARK-SENSED PUNCHING is the automatic punching of a card by means of electrically-conductive marks made on the card with a special pencil.

Thus, original facts may be recorded anywhere—in the office, plant or field, by workmen, timekeepers or field workers—and these facts are translated directly into punched-hole form.

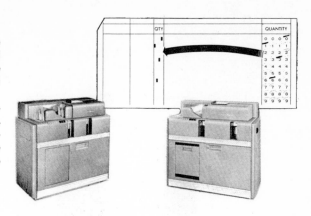

INTERPRETING is the translation of punched holes into printed information on the face of the card.

Alphabetical or numerical information can be printed in many different positions on the same card from which it is read. Common data can be repetitively printed on a group of detail cards from punched information on a master card.

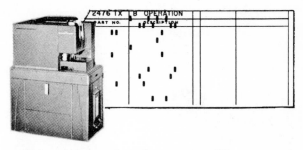

Interpreting is advantageous when punched cards are used as documents on which additional information is written or marked, or wherever reference to filing operations is involved.

END PRINTING converts punched information into bold printing across the end of the card simultaneously with gang punching, summary punching, reproducing, and mark-sensed punching. This is similar to interpreting, and makes possible quick reference to the card.

Cards are printed in this manner for use in prepunched files where cards are stored on end, or in attendance-card racks for convenient reference and selection.

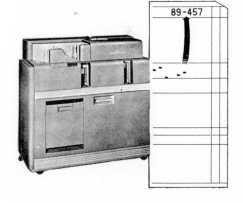

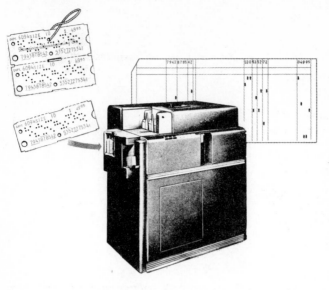

TICKET CONVERTING is the process of changing prepunched ticket stubs into cards. The ticket is made up of a basic section and one or more stubs that are numerically prepunched and printed with identical information.

When a transaction occurs, a stub is detached from the ticket and put into a receiver; the receiver is then placed directly in the ticket converter. The ticket stubs are fed from the receiver, and cards are punched with the corresponding information. A typical application of the ticket converter is in merchandising where price tickets often represent the greatest volume of transactions.

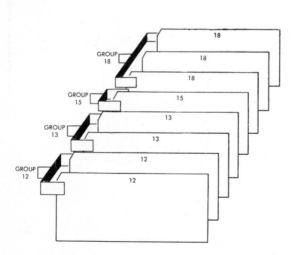

SORTING is the process of grouping cards in numerical or alphabetical sequence according to any classification punched in them. To group cards by account, for instance, they are sorted into account sequence. This makes possible summarizing the cards by account.

A fast, automatic machine process thus is provided for arranging cards for the preparation of various reports—all originating from the same cards, but each requiring a different sequence or grouping of information.

SELECTING is the function of pulling from a mass of data, certain items that require special attention. Selection of individual cards is accomplished automatically by either the sorter or collator, according to the type of selection. Typical selections are:

Cards punched with specific digits   Cards between two specific numbers
Certain type of cards for a specific date   First card of each group
All cards containing a specific number   Last card of each group
All cards higher than a specific number   Unmatched cards
All cards lower than a specific number   Cards out of sequence

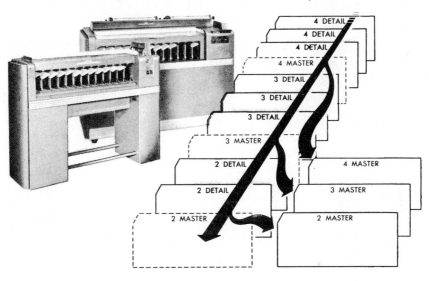

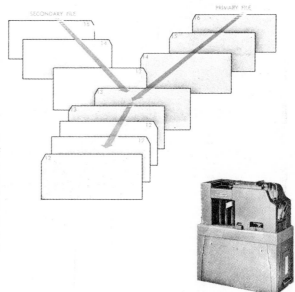

MERGING is the combining of two sets of punched cards into one set of given sequence. Both files of cards must be in the same sequence before they are merged. This function makes possible automatic filing of new cards into an existing file of cards. It is a faster method than sorting to use in placing related cards together.

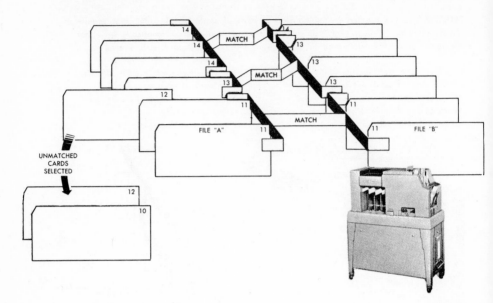

MATCHING is a checking function used to check the agreement between two sets of cards. Groups of cards in one file are compared with similar groups in a second file. Unmatched cards or groups of cards in either file may be selected or separated from the files.

This function is frequently performed in conjunction with merging.

DETAIL PRINTING is the printing of information from each card as the card passes through the machine. The function is used to prepare reports that show complete detail about each transaction.

During this listing operation the machine adds, subtracts, cross-adds or cross-subtracts and prints many combination of totals.

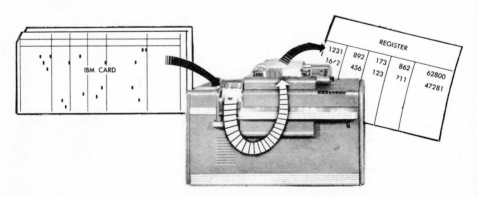

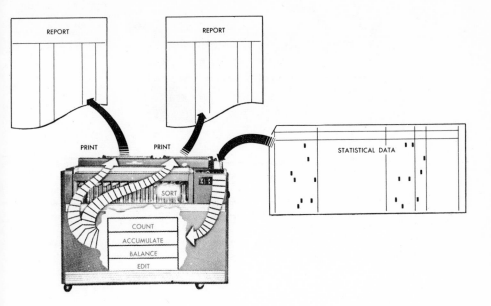

STATISTICAL WORK is essentially a problem of counting units in many different classifications. At the same time it is frequently desirable to accumulate certain quantities or amounts, check or edit for consistency or reasonableness, and balance counts to the control totals to check the accuracy of the summaries.

All of these functions are performed by the Electronic Statistical Machine to produce printed summaries. This machine also performs sorting and card-arranging operations.

GROUP PRINTING is the accounting-machine function that summarizes groups of cards and prints the totals on a report. Totals may involve adding, subtracting or crossfooting. Information read from punched cards is entered into counter units; at the end of each group of cards, the totals are read out of the counters and printed on the report. This function is used in preparing all types of reports requiring summarized totals. Complete descriptive information identifies all totals.

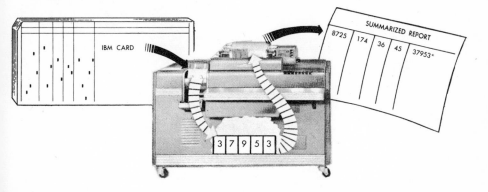

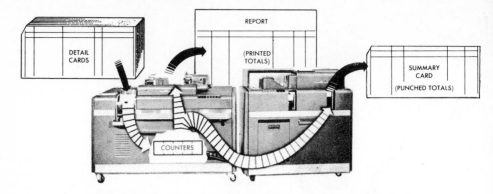

DETAIL CARDS

REPORT

(PRINTED TOTALS)

SUMMARY CARD

(PUNCHED TOTALS)

COUNTERS

SUMMARY PUNCHING is the automatic conversion into punched-hole form of information developed by the accounting machine. Summary punching is used for two purposes:

1. To carry balance figures forward. To do this, it is only necessary to include the previous total-to-date card with the current card or cards, and, while a current report is being run, summary punch new balance-to-date cards. These are saved for the next balance-to-date operation when the process is repeated.

2. To reduce card volume and carry summary data. Summary cards reduce peak-load periods due to accumulated card volume, and can be used as entries to general ledger accounting.

CALCULATING is the computing of a result by multiplication, division, addition, or subtraction. Any combination of these calculations can be performed—often in one run. Factors to be calculated may be read from each card, or series of cards, emitted by a device within the machine, or be developed by the accumulation of a series of calculations. One or several results are punched in each card or in a trailer card which

follows a group of cards carrying the factors. Many routines allow automatic checking to prove accuracy of calculations. For example, to check the punched result, an A X B calculation can be cross-proofed against a B X A calculation during the same run.

A × B ÷ C + D − E = R

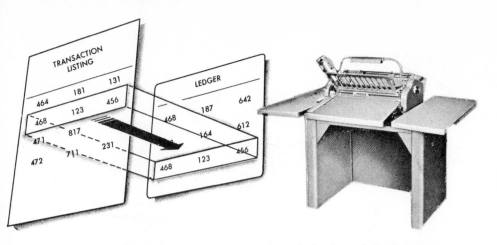

FACSIMILE POSTING is the process of transferring by a duplicating process a printed line on a report to a ledger or other record sheet. These may be posted from a transaction listing previously prepared on the accounting machine.

Typical uses of this function are the posting of customer ledgers, employees' earning records, and stock ledger cards.

CARD-TO-CARD TRANSCEIVING makes possible instantaneous and accurate duplications of punched cards over telephone and telegraph networks between locations separated by either just a few miles or thousands of miles. A switch on the machine halts card transmission at any time to permit direct voice communication over the same telephone circuits connecting the sending and receiving units.

The machines at either end of a circuit are identical and can be used interchangeably for transmitting or receiving.

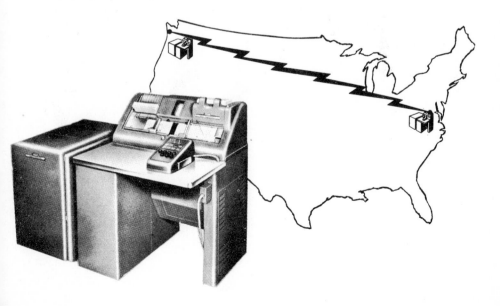

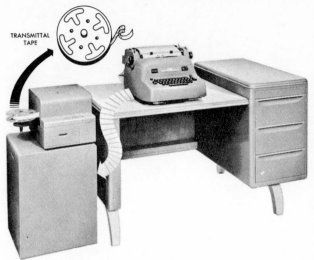

TRANSMITTAL TAPE

TYPEWRITER TAPE PUNCHING

is a means of recording information in code onto a tape by use of a special typewriter. As a document is being created on the typewriter, any or all of the typed information can be recorded on the 8-channel tape. The tape can be easily transported to other locations and processed through a tape-to-card punch to transfer the information into punched cards.

In general, the typewriter tape punch can be used to prepare any document now created on a typewriter and later used as a source document for key punching and key verifying cards; this eliminates the need for the latter two functions.

Typical applications of this machine are for billing, order writing, personnel changes, address changes, insurance policies, railroad accounting, journal vouchers, purchase orders, receiving reports, directories, inventory control, check reconciliation, and many more.

TAPE READING is a process of feeding coded tapes through a tape-to-card punch to convert the coded information into punched cards. Tapes can be prepared on the typewriter tape punch or on the card-controlled tape punch; the latter is capable of punching tape that can be transmitted by telegraph.

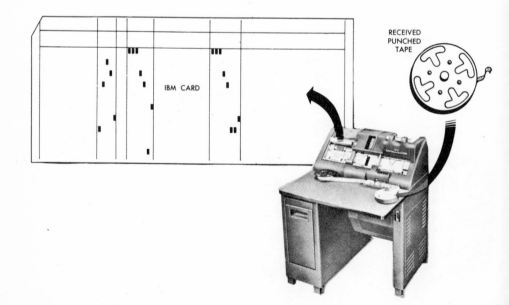

IBM CARD

RECEIVED PUNCHED TAPE

Questions And Problems

Chapter 1   QUESTIONS

1-1. What is Production Control? What are the elements of control? Describe each element in detail.

1-2. Describe a specific example of an activity in which the elements of control are present.

1-3. Name at least 5 examples of activities, other than those listed in Chapter 1, to which planning and control can feasibly be applied.

1-4. Describe why "production" control is needed in each of the examples in Question 3.

1-5. Describe the logical benefits that would be obtained in each example in Question 3.

1-6. Describe a specific example of an activity having:

    a. A formal Production Control system.

    b. An informal Production Control system.

    c. A combination formal and informal Production Control system.

1-7. Describe the step-by-step approach to designing the "best" system of Production Control for a specific activity.

1-8. What is the probable result of using the "case study" approach in designing a Production Control system? Explain the reasons for this result.

Chapter 2   QUESTIONS

2-1. List as many "rules" of human relations as possible that should be followed in designing and implementing a Production Control System.

2-2. Describe fully the role of top management in the "design" of Production Control systems.

2-3. Describe fully the role of foremen, supervisors, and workers in the "design" of Production Control systems.

2-4. How can the value of Production Control be substantiated? Describe quantitative and qualitative methods.

2-5. Describe a specific case of a human relations problem in designing a Production Control system for an activity and how it could be or was overcome.

Chapter 3   QUESTIONS

3-1. List the requirements of a qualified Production Control systems designer.

3-2. Describe how a Production Control systems designer would usually obtain the qualifications in Question 1.

3-3. Describe each of the functions in Table 3-1 in a specific activity with which you are familiar.

3-4. Visit a local manufacturing plant and describe each of the functions of Table 3-1 as it exists in the company. Determine the responsibility and authority for each of the functions.

3-5. Do the same as in Question 3-4 in a non-manufacturing company.

3-6. Describe how to determine whether or not each of the 6 principles of sound Production Control are being followed in any given system.

3-7. Describe a case example of the application of the "management by exception" principle.

Chapter 4   QUESTIONS

4-1. For each of the four basic methods of operation name at least two examples other than those listed in Chapter 4.

4-2. Why is the type of work activity and method of operation important in the design of a Production Control system?

4-3. Why are the physical plant and facilities important in the design of the Production Control system? Give specific examples.

Chapter 5   PROBLEMS

5-1. Visit a local plant or business which has either a continuous-process or a similar-process type of production and describe the characteristics of the Production Control system consistent with those described in Chapter 5.

5-2. Visit a local plant or business which has either a job-shop or special-project type of production process and follow same procedure as in Problem 5-1.

Chapter 6 QUESTIONS

6- 1. Should all elements within an organization base their plans on the same forecast? In some organizations different departments do develop their own forecasts independently. What would cause this? What would be the result?

6- 2. Discuss how both the short- and long-range forecasts would be used in developing planning data? List some of the possible effects of inaccurate or inadequate forecasting.

6- 3. Explain the difference between short- and long-range forecasting. Which of the forecasting techniques are most applicable for long-range forecasting? Which for short-range?

6- 4. In what situations is the historical estimate of forecasting most useful? Describe how the historical estimate technique would be used in conjunction with other techniques for developing planning data?

6- 5. In what situations is the sales-force estimate most useful? How can it be modified for staff or service activities?

6- 6. Indicate one principal limitation on the trend line and other mathematical techniques. In what types of industries would this be a significant limitation? In what types of industries would this limitation not be significant?

6- 7. How may sales figures be adjusted for price changes if one wishes to draw a trend line?

6- 8. When is the market research technique used to develop forecasting data? What are the most important problems confronted when this technique is used?

6- 9. Differentiate between current and leading correlators. Which is the most desirable for forecasting purposes? List four examples, other than those given in the text, of industries and correlators which can be used to forecast sales for them. Indicate whether the correlators are current or leading.

6-10. If an organization has been supplied with prior knowledge for the major elements of production has the need for forecasting at the local level been eliminated? If not, what requirements must be forecast?

6-11. Differentiate between periodic and continuous forecasting and list the advantages and disadvantages of each type.

6-12. What is the first function to be considered when designing a Production Control system? Why may the designer happen to overlook this function?

Chapter 6 PROBLEMS

6-1. Design a form to be used by salesmen to record their estimate of sales for a future period.

6-2. Listed below are the sales for the past ten years of cast iron plumbing castings for a medium sized foundry.

Year	Sales
1	$ 100,000
2	105,000
3	95,000
4	101,000
5	110,000

Year	Sales
6	115,000
7	120,000
8	125,000
9	130,000
10	132,000

A five per cent price increase went into effect at the beginning of the fifth year. Another five per cent price increase went into effect at the beginning of the eighth year.

(1) Using the tenth year as a base, convert the data to a consistent selling price.

(2) How could the need for converting the data to a base year be avoided?

(3) Plot the converted sales data and draw a trend line for the next three years (assume no price increase).

(4) What is the sales forecast for the 11th year? For the 13th year?

(5) Using the converted sales data and the least-squares method, derive a formula for the trend line.

(6) Using the formula obtained in the preceding step, determine the sales forecast for the 11th year and for the 13th year. Compare this answer with that obtained in (4). What was the magnitude of error when the trend line was drawn by "eye"?

6-3. For the same product line in the foundry in the preceding problem, the sales by month are listed for years 8, 9, and 10:

Month	Year 8	Year 9	Year 10
January	12,500	14,300	13,200
February	13,750	11,700	13,200
March	11,250	11,700	13,200
April	11,250	10,400	11,880
May	10,000	9,100	10,560
June	8,750	7,800	9,240
July	7,500	6,500	7,920
August	7,500	9,100	6,600
September	8,750	10,400	7,920
October	10,000	11,700	10,560
November	11,250	13,000	11,880
December	12,500	14,300	15,840
	125,000	130,000	132,000

(1) Giving double weighting to the last year's data, establish a forecast by month for the 11th year. Total sales for the year should be that determined in Part 6 of Question 2.

(2) What two forecasting techniques are you applying?

6-4. The management of the same foundry feels that there is a high degree of correlation between their sales in *tons* and the number of residential housing starts. Sales in tons and number of units started for the last ten years are listed below:

Year	Sales (Tons)	Million Residential Unit Starts
1	100.0	5.0
2	105.9	5.3
3	95.0	4.9
4	101.0	5.0

Year	Sales (Tons)	Million Residential Unit Starts
5	104.5	5.2
6	109.8	5.5
7	114.0	5.7
8	112.5	5.6
9	117.0	5.8
10	118.8	5.9

(1) Plot the data and draw a line of best fit to all the points.

(2) Assuming that sales data were only available in dollars, what would have been necessary?

(3) The National Association of Home Builders has forecast 6.0 million residential unit starts for the next year (Year 11). Based on the correlation chart, what should the forecast tons of castings be for the company?

(4) Using the least-squares technique, develop the correlation formula for the two variables.

(5) Using the formula developed in the preceding step and a forecast of 6.0 million residential unit starts for the next year, estimate the number of tons of castings which should be sold by the company.

(6) Compare the answers obtained in steps (3) and (5). What was the magnitude of error as a result of using the "draw by eye" technique? What is one disadvantage to using the mathematical formula other than the effort involved?

6-5. Prepare a report for the management of this company. Recommend a forecast in terms of dollar sales and tons of production. This forecast should be based on the results of Problems 2 and 4. (In year 1 revenue was $1,000 per ton. With this information the results of problem 2 can be compared.)

Chapter 7 QUESTIONS

7- 1. What is meant by forecast control?

7- 2. When should comparisons between actual and forecast orders be made?

7- 3. Even if actual total dollar sales and forecast total dollar sales compare favorably, what other factors must be considered?

7- 4. What are the two basic forms of corrective action which may be taken when actual sales are less than forecast sales?

7- 5. Why is it important to minimize adjustments in the forecast?

7- 6. What will be the result of insufficient adjustments in the forecast?

7- 7. Why is it desirable to compare cumulative orders received with cumulative orders forecast rather than to consider each month separately?

7- 8. What factors would management take into consideration when establishing confidence coefficients for forecast control limit analysis?

7- 9. Other than assuming that the forecast is incorrect when an "out of control" situation occurs, what other possibilities should be investigated?

7-10. Discuss the most important limitations on the forecast control limit technique.

Chapter 7 PROBLEMS

7-1. Use the monthly sales data from Problem 3 of Chapter 6 to develop the cumulative monthly percent distribution for the receipt of sales for the cast-iron product group. (Give Year 10 double weighting in computing the average cumulative per cent.)

7-2. A forecast of $135,000 sales has been made for the product group in problem 1. In the past, the average order has been for $216.16. Management has selected the following confidence coefficients for use in analyzing forecast versus actual sales for this product group:

	January-June	July-December
For upper control limit:	.80	.70
For lower control limit:	.90	.80

Determine the upper and lower confidence intervals in sales dollars and as a percentage for both the first half and last half of the year.

7-3. Draw the control chart for Problem 2.

7-4. Actual dollars ordered each month in the following year (Year 11) are as follows:

January	14,856	July	82,180
February	22,119	August	93,413
March	38,638	September	99,312
April	56,971	October	107,242
May	60,013	November	127,763
June	77,193	December	138,653

Using the data from Problems 1 and 2, prepare a table similar to Table 7-3 and analyze actual sales activity versus the forecast. Plot cumulative sales on the control chart prepared for problem. In what month does the forecast go "out of control"?

7-5. Referring to Problem 4 and the data of preceding problems, prepare a new sales forecast beginning with the month following the out-of-control situation. Base your new forecast on the cumulative per cent of orders normally received by the end of the month which created the out-of-control situation. Determine the new upper and lower confidence limits. Analyze the remaining months of the year.

7-6. What would be the effect of using a confidence coefficient of .50 for both the upper and lower limits during the entire year? How many times would the forecast have been revised? How much more accurate would the final forecast have been?

Chapter 8 QUESTIONS

8- 1. Define "work authorization."

8- 2. How may the authority which is mentioned in the definition occur?

8- 3. List five "objectives" of work other than those mentioned in the text.

8- 4. Differentiate between work authorization and dispatching.

8- 5. What is the basic requirement for successful performance of the work authorization function? Why?

8- 6. In what way are work authorization, accounting, and performance evaluation related?

8- 7. What are the two basic ways in which work is generated?

8- 8. What are the advantages of an externally-generated work situation?

8- 9. In what types of activities are internally-generated work situations to be found?

8-10. What are some of the problems that occur as a result of internally-generated work situations?

8-11. Should all work be "forced" through a formal Production Control system?

8-12. Who should decide what work should bypass the formal system? What important control must be established?

8-13. List several methods of bypassing the formal production control system.

8-14. How is budgetary control related to the work authorization function?

Chapter 9   QUESTIONS

9- 1. Define "product planning."

9- 2. What are the three basic steps in product planning? Illustrate for a nonmanufacturing activity other than those discussed in the text.

9- 3. Why is a schematic bill of materials prepared?

9- 4. What are the factors to be considered when making a decision on whether to make or buy a component part? How should the problem be resolved?

9- 5. Define "value analysis."

9- 6. What are the five basic questions asked in the value analysis approach?

9- 7. What is the most important single factor in the value analysis approach?

9- 8. If a separate organization is established for value analysis, what must be avoided? How should this be done?

9- 9. What is the disadvantage of the committee approach to value analysis?

9-10. Why can't the product planning function always be carried out to the maximum extent?

9-11. Why should the product planning function always be carried out to the maximum possible extent? List some of the disadvantages of incomplete product planning.

Chapter 10   QUESTIONS

10- 1. Define "process planning." Define "routing."

10- 2. When will process planning automatically include the function of routing? When will scheduling include the function of routing? When will dispatching include the function of routing? Give examples other than those mentioned in the text.

10- 3. What is the objective of process planning?

10- 4. What is the prerequisite information which must be available to the process planner if he is to do a competent job?

10- 5. Why is knowledge of quality requirements important to the process planner?

10- 6. Why is the degree of skill of the personnel important to the process planner?

10- 7. What are the three basic steps in process planning? Illustrate for a nonmanufacturing type of activity.

10- 8. What factors will control the detail in which the formal process plan is developed?

10- 9. What information should be developed in process planning in regard to the operations to be performed? What information in regard to materials-handling requirements?

10-10. Why is it important to specify the personnel required when developing a process plan?

10-11. How can the estimates of the time requirements for an operation be improved?

10-12. What is the basic concept involved in breakeven analysis?

10-13. What is one major limitation on the breakeven analysis technique?

10-14. What is the advantage of absorbing much of the methods improvement activity in the process planning function?

10-15. Where will the process planning function be performed if this responsibility is not assigned to a staff activity?

10-16. What are the advantages of having the process planning function performed by a staff?

10-17. If the process planning function is performed within the line, what should be investigated?

10-18. What should determine the amount of process planning that will be performed by the line rather than by the staff?

10-19. What problem is likely to be encountered when analyzing the process planning function in a nonmanufacturing type of activity?

Chapter 10   PROBLEMS

10-1. Design a form for recording all the information that would be developed in the "Production Control manager's dream" for the process planning function.

10-2. A manufacturing organization has received an order to produce 40,000 steel boxes and has the alternative of producing the order in their present "job-shop" type layout or of setting up a production line. It has been established that there is no likelihood of repeat orders. Costs are as follows:

> Producing with present "job shop" conditions:
> Set-up cost $15,000
> Variable cost per unit: $4.80
> Producing with production line:
> Set-up cost $100,000
> Variable cost per unit: $4.00

2.a What is the breakeven point for this situation?
2.b Illustrate graphically the breakeven-point analysis.
2.c What is the total cost for the order quantity under each condition?

10-3. An organization has received an order to produce 100,000 bracket assemblies and can produce the order with their present layout or can set up a production line. Costs are as follows:

> With present production layout:
> Engineering $2,000
> Tools   4,000
> Set-up   3,000

                    Variable unit cost $7.00 per assembly
          With production line:
                    Engineering $30,000
                    Special Equipment   20,000
                    Training   5,000
                    Relayout   25,000
                    Variable unit cost $5.00 per assembly

   3.1 What is the breakeven point in this situation?
   3.2 Illustrate graphically the analysis.
   3.3 What is the total cost for the order quantity under consideration?
       What would have been the loss if the wrong decision had been
       made in selecting the alternatives?
10-4. Formulate an expression for applying the breakeven-point analysis
   technique to the determination of confidence intervals for forecast
   control limit analysis. Define the variables.

Chapter 11   QUESTIONS

11- 1. What are the two main sections of the material control function?
11- 2. What related activities can help to minimize material control
       problems?
11- 3. What is meant by "exploding" a bill of materials?
11- 4. How should the volume to be produced affect the scrap allowance?
       Illustrate.
11- 5. Describe how the "tickle file" procedure discussed in this chapter
       would be employed throughout the entire procurement cycle.
11- 6. What is the first requirement for a sound material control program?
       What are the three important factors within this requirement?
11- 7. What constant information should appear on the material control
       card?
11- 8. What are the three sections of a proper material record card?
11- 9. What is the purpose of including an "Allocated" column on the
       material record card?
11-10. Under what conditions should columns be added to the material
       record card for "Amount Purchased" and "Total Amount Due"?
11-11. What is the purpose of the recap portion of the material record?
11-12. Indicate three activities (other than those given in this chapter) in
       which the two-bin material control system would be applicable.
11-13. Why can't the material record card be completely eliminated?
11-14. Indicate three activities (other than those given in this chapter) in
       which the "supermarket" material control system would be appli-
       cable.

Chapter 11   PROBLEMS

11-1. Design all the necessary forms for an effective requisitioning and
      associated material control record keeping procedure.
11-2. Design a simple Production Control system for controlling a material
      record keeping operation.
11-3. On a sheet of paper containing all the columns indicated in Figure
      11-6, record the following transactions and the proper adjustments
      to all balance columns:

Opening Balances: Due In: 5,000   (Purchase Order 1.)
July   1:   Balance: 3,000
                       Available: 3,000
July   2: Received 2,000 units on Purchase Order 1.
July   3: Allocated 1,000 units on Factory Order 1.
July   8: Allocated 4,000 units on Factory Order 2.
July 12: Placed Purchase Order 2 for 5,000 units.
July 13: Issued 1,000 units on Factory Order 1.
July 19: Received 3,000 units on Purchase Order 1.
July 22: Issued 2,000 units on Factory Order 2.
July 23: Received 2,000 units on Purchase Order 2.
July 29: Issued 2,000 units on Factory Order 2.
July 31: Placed Purchase Order 3 for 2,000 units.
July 31: Issued 2,000 units on Factory Order 3.

11-4. Write a complete procedure including the design of necessary forms and record cards for a two-bin material control system.

11-5. Write a complete procedure for a supermarket material supply system. Include provisions for restocking of the "market."

Chapter 12   QUESTIONS

12- 1. Define material management. In what type of material control situations are the techniques of material management applied?

12- 2. What is wrong with the concept of always producing in very large lots to minimize set-up and procurement costs?

12- 3. What is meant by the variable costs of inventory? What are the two principal components of this cost?

12- 4. How is average inventory normally determined? How does this factor relate to carrying costs?

12- 5. What is the objective of the economic order quantity calculations?

12- 6. What is the common practice when application of the economic order quantity formula indicates that a quantity in excess of one year's supply should be procured? Why?

12- 7. What is the first step in any systematic inventory analysis?

12- 8. What is the purpose of classifying inventory items according to dollar usage?

12- 9. Can the economic order quantity formula always be applied? If not, what alternatives are available?

12-10. What two factors must be compared when determining an economic order quantity in conjunction with quantity discounts?

12-11. What two factors are usually considered in establishment of the reorder point?

12-12. What activities should be included when determining procurement lead time?

12-13. Theoretically, how should safety stock requirements be determined? What is wrong with this approach? What alternative can be employed?

12-14. How does the order quantity affect the safety stock requirements?

12-15. What would be required to insure that a stock-out would never occur?

12-16. In most organizations, what would be the relationship between the classification of inventory items by dollar usage and by effects of a stock-out?

Chapter 12   PROBLEMS

12-1. Determine the economic order quantity for the following situation:
Procurement cost: $50.00
Carrying charge: 12 per cent
Monthly usage: $100

12-2. Determine the economic order quantity for this situation:
Procurement cost: $50.00
Carrying charge: 16 per cent
Monthly usage: 75 pieces
Cost per unit: $2.00
Express the answer in terms of months' supply, units, and dollars.

12-3. How much is the annual storage cost as determined by the solution to Problem 12-2? (Disregard any reserve quantity.)

12-4. How much is the annual procurement cost as determined by the solution to Problem 12-2?

12-5. Determine the procurement cost of a purchased item if the following quantity discount is given by the vender:
$8.50 per unit for order of more than 200 units
  8.20 per unit for orders of more than 500 units

12-6. Determine the economic order quantity for the following purchased item:
$5.10 per unit for orders of more than 500 units
  5.05 per unit for orders of more than 1000 units
Cost of paper work: $11.20 per order
Monthly usage: 100 units
Carrying charge: 16 per cent

12-7. Determine the reorder point for the following situation:
Average usage during lead time: 81 units
Order quantity: 3 months' supply
Acceptable probability of stock-out: Once every 5 years

12-8. Determine the reorder point for the following situation:
Economic order quantity: 2700 units
Procurement lead time: 2 months
Average monthly usage: 450 units
Acceptable probability of stock-out: Once every 10 years

12-9. The management of a midwest concern has established the policy of turning all inventory over 4 times a year. The reason for this policy is to release working capital for an expansion program to be undertaken in the near future. Although it will be impossible for the reader to analyze all of the products for this problem, 3 typical products will be considered in order to evaluate this inventory policy:
Item A (purchased)
    Price: $15.00 per unit for 500 or more
            14.50 per unit for 1000 units or more
    Procurement cost: $10.00 per order
    Usage: 300 units per month
    Lead time: 1/4 month
Item B (manufactured)
    Price: $8.00 per unit
    Machine Set-Up cost: $350.00 per order

Paper Processing cost: $10.00 per order
Usage: 1,000 units per month
Lead time: 1 month
Item C (manufactured)
Price: $25.00 per unit
Machine Set-Up·Cost: $140.00 per order
Paper Processing cost: $10.00 per order
Usage: 900 units per month
Lead time: ½ month

The storage charge for all items has been estimated to be 22% of the cost of the items per year. This figure includes the rent for the warehouse, taxes on the stored items, loss of interest on inventory money, and so on. To assure immediate delivery for all parts in storage, management has also established the policy of having a reserve (cushion) of one month's supply.

(1) Determine the order quantity, order point, and the total annual cost (set-up plus storage) for this sample of the company's inventory based on present policy.

(2) Using the economic order quantity formula, determine the best order size for each item in the sample.

(3) If management would accept one stock-out every 5 years for each item, what is the correct reorder point for each item?

(4) Determine the total annual cost of the inventory policy given in 9.2 and 9.3. Compare this cost with the current inventory policy.

(5) Determine the appropriate order quantity to minimize costs if the total inventory value could not be changed from the present amount. Compare the cost with the present policy and with the cost when the basic formula is used.

(6) Determine the appropriate order quantity for each item to minimize costs if the total number of orders processed annually must remain the same as at present. Compare the cost with the present policy and with the cost when the basic formula is used.

Chapter 13   QUESTIONS

13-1. When should the tool requirements be determined?
13-2. What is included in the function of tool control?
13-3. Why is it normally difficult to control the production of "custom-built" tools?
13-4. What approach should be used in scheduling and loading tool rooms?
13-5. What should be the basis of the decision in regard to the decentralization of tool cribs?
13-6. What is the disadvantage of the brass ring tool control system?
13-7. Describe the purpose of each copy of the McCaskey tool control form.
13-8. What is the result of improper tool control?

Chapter 13   PROBLEMS

13-1. Design an inventory control system based upon the McCaskey tool control type of forms and system.

Chapter 14   QUESTIONS

14- 1. Define "work measurement."
14- 2. List and explain the 4 major uses of work measurement.
14- 3. In what way is the scheduling function dependent upon work measurement?
14- 4. In what way is the measurement of effectiveness related to the measurement of utilization?
14- 5. In what way are methods improvement and work measurement related?
14- 6. What are the 6 basic steps in stop-watch time study?
14- 7. How may "standard" be defined?
14- 8. What criteria should be used in selecting a work unit?
14- 9. When should the operation be rated?
14-10. What is the purpose of adding allowances to the base time?
14-11. What is meant by predetermined time systems?
14-12. Why is the predetermined time system technique of particular importance to Production Control?
14-13. Of what value is the predetermined time system technique in a custom-built type of activity?
14-14. Should work sampling be used to establish standards for the purpose of incentive payment? Discuss.
14-15. How can the "quality" of technical time estimates be improved?
14-16. What are the principal disadvantages of the historical data technique?
14-17. What technique should be used when there is not a work unit, or the work units are too numerous or varied to measure?
14-18. What is meant by the standard data approach to work measurement? Give examples other than those discussed in the text.
14-19. How may standard data be developed?
14-20. Of what value is standard data to the Production Control manager? Discuss.
14-21. When is rating performed if a predetermined time system is used.

Chapter 14   PROBLEMS

14-1. Listed below are the 5 elements of an operation which was time studied:

Element	Average Observed Time	Rating
1.	.24 minutes	105%
2.	.13	100
3.	.41	95
4.	.36	110
5.	.08	120

The allowance factor for the department in which the operation is performed is 22 per cent.
(1) What is the base time for the operation?
(2) What is the standard time for the operation?
14-2. What technique discussed in this chapter could be used to establish the allowance factor used in problem 1?

14-3. How could the standard established by time study in Problem 14-1 be checked with another technique discussed in this chapter?

14-4. Listed below are 7 elements of an operation which was time studied:

Element	Aver. Observed Time	Occurrence per work unit	Rating factor
1	1.80 minutes	0.1	80 %
2	.14	1.0	90
3	.34	0.5	85
4	.27	1.0	95
5	.21	1.0	90
6	.09	1.0	105
7	.12	0.05	95

The allowance factor for the department in which the operation is performed is 17 per cent.

(1) What is the base time for the operation?

(2) What is the standard time for the operation?

Chapter 15   QUESTIONS

15- 1. Give the type of situations in which loading could be used while scheduling would be impracticable.

15- 2. In what type of situation would you suggest using informal loading and scheduling? Formal loading and scheduling?

15- 3. What is meant by "unit of measure work"?

15- 4. What are some of the factors which must be considered when determining what facilities should be loaded or scheduled?

15- 5. Indicate the principal limitations of master scheduling. Under what type of situation would it be used?

15- 6. What is the purpose of maintaining an open order file?

15- 7. Under what circumstances should a graphical load chart be developed?

15- 8. Explain why order scheduling is the most difficult to maintain. When should this technique be used?

15- 9. What is the purpose of only partially loading a facility when loading by schedule period?

15-10. Explain the role of the foreman when loading by schedule periods.

15-11. Why is it important to convert the actual clock hours into standard hours?

15-12. Discuss the reasons why, when a facility is to be unloaded, standard hours must be used and not the actual clock hours.

Chapter 15   PROBLEMS

15-1. The Dade Manufacturing Company has as one of its lines of products a patented fixture which prevents doors from slamming. It is produced in various styles and sizes.

The Lambert Construction Corporation has a job underway which calls for a fixture which is very similar to that manufactured by the Dade Company; however, none of the stock fixtures have the proper dimensions.

Today the Dade Company received a request from the Lambert Corporation for a price and the earliest possible shipping date which can be guaranteed on an order of 3000 fixtures of the type in question. A complete set of drawings showing the required dimensions is also submitted. You, as production engineer of the Dade Company, are asked to forecast the earliest possible shipping date under the following conditions:

a. Only one machine shall be assigned to each operation.

b. No overtime. Working schedule is five 8-hour days per week.

c. There shall be a four-hour move factor injected between the end of one operation and the beginning of the next.

d. This order shall be given priority over any other work now in process.

e. No work shall be performed on a part until it is necessary to do so in order to complete the order at the earliest time.

The fixture is to be packed individually and consists of five parts. Parts No. 1, No. 2, and No. 3 form one sub-assembly which is designated as sub-assembly No. 1. Parts No. 4 and No. 5 form another sub-assembly which is designated as sub-assembly No. 2. Sub-assemblies No. 1 and 2 are assembled into the final assembly or completed unit. After a careful study of the limiting dimensions and a conference with the tool room, you determine that special tools will be needed for: Part No. 2—Operation 2; Part No. 3—Operation 3; Part No. 5—Operation 2. The tool room estimates that the following times will be required for the securing of these tools:

Tools for Part No. 2 ....32 hours
Tools for Part No. 3 ....48 hours
Tools for Part No. 5 ....64 hours

Further analysis of the specifications furnished by the Lambert Corporation reveals that a special steel stock is required for Parts No. 2 and No. 4. The Purchasing Department advises you that it will take a period of twelve calendar days to secure the desired stock.

The first available date that you can begin the schedule for this job is one week from today.

The Standards Department furnishes you with the following operation standard times:

Part No. 1
    Operation 1—Standard Time—1.64 min/piece
    Operation 2—Standard Time—2.50 min/piece
Part No. 2
    Operation 1—Standard Time—1.22 min/piece
    Operation 2—Standard Time—1.20 min/piece
    Operation 3—Standard Time—0.65 min/piece
Part No. 3
    Operation 1—Standard Time—1.10 min/piece
    Operation 2—Standard Time—2.24 min/piece
    Operation 3—Standard Time—1.97 min/piece
Part No. 4
    Operation 1—Standard Time—1.66 min/piece
    Operation 2—Standard Time—1.28 min/piece
    Operation 3—Standard Time—2.81 min/piece
Part No. 5
    Operation 1—Standard Time—1.77 min/piece
    Operation 2—Standard Time—2.51 min/piece
Sub-assembly No. 1
    Standard Time—3.80 min/unit—(Five assemblers available)

Sub-assembly No. 2
    Standard Time—3.40 min/unit—(Two assemblers available)
Final Assembly and Test
    Standard Time—8.21 min/unit—(Seven assemblers available)
Packing and Inspection
    Standard Time—1.82 min/unit—(Four packers available)

The job is divided into six (6) lots of 500 units each. In order to
simplify calculations move the pieces from one operation to another
only in lots. In moving a lot, the four-hour move factor applies.
When can this job be promised the Lambert Corporation?

15-2. a. The Davis Electric Company manufactures a complete line of
home lighting fixtures for sale through independent retail store
outlets. Several months ago, they introduced a new novelty lamp
which met with immediate sales success. This morning one of
Davis' most important customers, the Ace Electric Company, requested to know the status of their order for 8,000 of the lamps.
Upon checking it has been found that through a clerical oversight
the order has not been scheduled into the shop. To lose the Ace
account means that Davis would suffer a substantial decrease in
sales. This loss must be avoided.

Upon checking the inventory cards, it is found that there is an
adequate supply of all of the stock items needed for the order except four specially designed component parts, Part A, Part B,
Part C, and Part D.

The following are the process sheets for the four parts and the
final lamp assembly

Process Sheet Part A				*Process Sheet Part B*				*Process Sheet Part C*			
Op. No.	*Set Up*	*Prod. Rate**	*Dept.*	*Op. No.*	*Set Up*	*Prod. Rate**	*Dept.*	*Op. No.*	*Set Up*	*Prod. Rate**	*Dept.*
10	10 Hr.	50	A	10	10 Hr.	50	E	10	10 Hr.	200	A
20	20 Hr.	100	C	20	20 Hr.	200	D	20	5 Hr.	66.7	C
30	10 Hr.	66.7	B	30	10 Hr.	100	C	30	10 Hr.	66.7	B
40	15 Hr.	200	D					40	15 Hr.	200	D
								50	10 Hr.	66.7	E

Process Sheet Part D				*Process Sheet Assembly*			
Op. No.	*Set Up*	*Prod. Rate**	*Dept.*	*Op. No.*	*Set Up*	*Prod. Rate**	*Dept.*
10	10 Hr.	100	D	10	15 Hr.	200	Assy.
20	15 Hr.	100	A				

    * Production rate in units per hour.

The following is known about the plant:
a. The shop will work 8 hours of straight time plus 2 hours overtime per day during this emergency.
b. Each department can work on only one item at a time.
c. The operations for each part must be completed in sequence.
d. Although the entire 8000 lamps must be completed and shipped
at one time, the order can be split up into four lots of 2000 each
for interdepartmental movement.
e. All interdepartmental moves require one (1) day.
How soon can the Davis order of 8000 lamps be produced?

15-2. b. The following progress report on the Ace Electric Company's special order is given at the end of the 16th week.

> Part A
>> Op. No. 10—8000 units completed.
>> Op. No. 20—7500 units completed (machine broken down—15 hrs. req'd for repair).
>> Op. No. 30—4200 units completed.
>> Op. No. 40—5 hours of setup completed.
> Part B
>> Op. No. 10—8000 units completed.
> Part C
>> Op. No. 10—5000 units completed.
> Part D
>> Op. No. 10—8000 units completed.

The Ace Electric Company wishes to know when their order for 8000 units will be completed. You are to make a report as to the status of the order as of today.

After you have made your report (see above), you are to reschedule to get 4000 lamps out as soon as possible with the remaining 4000 units following at a later date. Make another report indicating delivery dates, approximate costs of the new plans, and so on.

15-3. The Clay Equipment Corporation has received a contract from the War Department for the production of an anti-tank mine. The mine is of sheet steel, approximately fifteen inches in diameter, which must be produced from an outside source requiring twenty-four (24) calendar days after placement of order. It is shipped from the factory to an arsenal for loading. A production schedule is desired for the manufacture of this mine.

Usually in production planning some time allowance should be made for necessary inspection. It will be assumed in this problem that spot checking is done by an inspector during final assembly in such a way as not to interfere with the assembling process.

Each subsequent process is to be started at such a time after the previous operation that it would be in process concurrently with other steps. No succeeding operation should begin until it can be completed without interruption.

Parts and Operations	Standard Time Required
Mine Body (Part A)	
Oper. 1 Blank	5.0 min. per 100
Oper. 2 Rough draw	20.0 min. per 100
Oper. 3 Finish draw and trim	20.0 min. per 100
Mine Top (Part B)	
Oper. 1 Blank and Pierce hole for fuze cup.	5.0 min. per 100
Oper. 2 Form flange and trim	15.0 min. per 100
Fuze Cup (Part C)	
Oper. 1 Blank	5.0 min. per 100
Oper. 2 Rough draw	15.0 min. per 100
Oper. 3 Finish draw and trim	15.0 min. per 100
Subassembly B & C	
Seam weld cup to top	10 min. per 100

Final Assembly

Oper. 1 Seam weld subassembly to body	30.0 min. per 100
Oper. 2 Notch (2)	15.0 min. per 100
Oper. 3 Gas weld notches & place on conveyor	20.0 min. per 100

Mines are hung on overhead conveyor which runs through paint dip tanks and drying oven to packing area. Conveyor speed is 20 f.p.m., hooks are 2 feet apart, 2 mines are hung on each hook, and the conveyor is 300 feet long.

The mines are packed into cartons of 25 directly from the conveyor at a rate of 1 unit per minute.

The bodies are moved between operations 1 & 2 on skids holding 500 blanks. They are moved between operations 2 & 3 and to final assembly in a cart that holds 100 pieces.

The tops and cups are moved between operations and to the subassembly in lots of 500.

Fig. QP-1.   Tube parts.

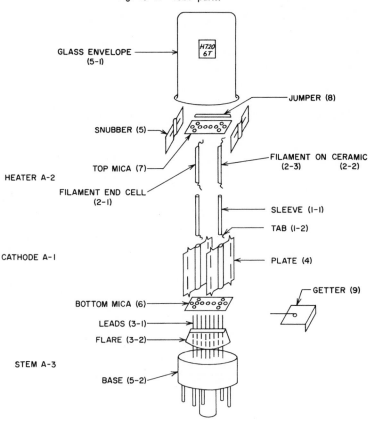

Special dies must be prepared for operation 3 on both the mine body and fuze cup, taking 20 hours and 12 hours respectively.

The subassemblies are moved to final assembly in lots of 100.

The final assembly is moved between operations in lots of 100 pieces.

The move time between each operation is 15 minutes.

The plant is on a single shift eight-hour day, five-day week.

When must this job be started in order to complete it on time?

15-4. The HiFi Radio Tube Company has received an order for 40,000 tubes as illustrated. All of the tubes are to be shipped 10 weeks from this Friday so that they may be tested and placed in production the following Wednesday at the purchaser's plant.

One full day is required to move a lot from one department to another. It would not be economical to ship more than four lots from any department for any operation—time must be allowed in each department to process the last lot received.

Information for the operation is given below. The plant is operating on an eight-hour day, five-day week basis.

Quantity Part per Tube	Part No.	Part Name	Components	Set-up and Manufacturing Time for Parts for 40,000 Tubes	Production Center
2	1-1	Cathode Sleeve		84 hours	63
2	1-2	Cathode Tab		40 hours	21
2	A-1	Cathode Tube	1-1, 1-2	80 hours	78
2	2-1	Filament End Cell			92
2	2-2	Ceramic Core			2
2	2-3	Filament			42
	A-2	Heater	2-1, 2-2, 2-3	200 hours	81
1	3-1	Stem leads (set)		Stock	2
1	3-2	Flares		Stock	2
1	A-3	Stem	3-1, 3-2	140 hours	58
2	4	Plates		Stock	2
2	5	Snubber		Stock	2
1	6	Bottom Mica		Stock	2
1	7	Top Mica		Stock	2
1	8	Filament Jumper		44 hours	16
1	9	Getter		Stock	2
1	A-4	Mount	2 of A-1 2 of A-2 1 of A-3 2 of 4 2 of 5 1 each of 6, 7, 8, 9	104	51
1	5-1	Glass Envelop		Stock	2
1	5-2	Octal Base		Stock	2
1	A-5	Completed Tube	A-4, 5-1, 5-2	72 hours	80

a. Draw a schedule chart showing the starting and completion data for each operation.

b. Draw a load chart for all production centers.

c. Four weeks before shipping the purchaser requests a report on the progress of the tubes. The following information is available Friday night:

Part 1-1 equivalent of   46 hours completed.
Part 1-2 equivalent of   24 hours completed.
Part A-1 equivalent of   10 hours completed.
Part 2-1 equivalent of 186 hours completed.
Part A-2 equivalent of   84 hours completed.
Part A-3 equivalent of   80 hours completed.

Show the progress by colored pencil. Give management a statement of how far ahead or behind we are for *each* part.

d. At this same time we are asked by the customer to set the delivery date up one week.

It is determined that production center #63 could take half ($\frac{1}{2}$) of production center #80's load—can the new schedule be met? If department #21 completes part 1-2 on schedule and can take half ($\frac{1}{2}$) department #51's load can the new schedule be met?

e. Assume you can produce for inventory, what production centers would you try to utilize and why?

Chapter 16   QUESTIONS

16-1. Explain the meaning of over-control and under-control of a productive facility.

16-2. For any given normal distribution, what do $\overline{X}$ and σ measure? J measure?

16-3. What is the usefulness of the coefficient of variation?

16-4. In scheduling what is meant by the "index" number and what is the logic behind its use?

16-5. What are the two basically different methods of manufacturing available to a concern which experiences highly seasonal sales demand? How does linear programming aid in this type of situation?

16-6. Why should the answers of any mathematical technique be analyzed subjectively before accepting them?

Chapter 16   PROBLEMS

16-1. There are six different orders to run and five machines through which the orders may run. The order requirements are as follows:

Order	Pieces Required
A	60
B	20
C	40
D	80
E	100
F	35

The standard times (in hours per piece) for each machine to do a particular order are shown below. The machine-hours available on each machine are also shown.

Order	Standard Time (hrs./pc.) for Machines				
	Mach. I	Mach. II	Mach. III	Mach. IV	Mach. V
A	3.0	0.5	1.0	1.5	—
B	—	—	—	7.0	4.0
C	2.0	3.0	6.0	4.0	9.0
D	2.0	4.0	3.5	3.0	3.2
E	—	6.0	2.0	3.0	4.0
F	—	6.0	2.0	3.0	3.0
Hours Avail.	160.0	32.0	200.0	160.0	190.0

Assuming that an order must be run on one machine, what are the minimum machine-hours required to complete the orders?

16-2. The United Transmission Corporation is a large manufacturer of communication equipment with plants in New York and North Carolina. It is the policy of this company to manufacture its products to supply eight (8) sales zones encompassing the entire U. S. United has recently designed a new type of photo-electric switch and has completed a market analysis which indicates a large demand for this switch in all eight sales zones. As a result of this survey, the following sales estimates were determined:

New England	3500 units per month
South Eastern	6000 units per month
North Central	2000 units per month
Central	8000 units per month
South Western	5000 units per month
Rocky Mountains	2500 units per month
Pacific Coast	3000 units per month
North Western	4000 units per month

The company has further determined that its existing manufacturing facilities, four plants (2 in N. Y. and 2 in N. C.), can handle the additional production requirements with the following limitations:

Plant	Max. Production in units per month	Est. Mfg. Cost
River Plant (N.Y.)	7500	$55.60 per unit
Electric City Plant (N.Y.)	14000	54.45 per unit
Old Valley Plant (N.C.)	6000	53.70 per unit
Oak Run Plant (N.C.)	6500	54.00 per unit

The following table shows the freight rate per unit to ship from the plant to the respective sales zone warehouse:

From	N.E.	S.E.	N.C.	Cen.	S.W.	R.M.	Pac.	N.W.
River Plant	0.80	1.90	2.40	2.90	4.90	5.40	5.30	5.10
Elec. City Plant	1.55	3.35	3.85	4.25	5.85	5.95	6.35	6.95
Old Valley Plant	2.60	3.90	5.10	4.90	6.30	6.90	7.40	7.90
Oak Run Plant	2.50	3.30	3.60	4.40	5.60	6.70	7.20	7.50

Wholesale Sales price at warehouse—$62.50 per unit.
Allocate production to plant, indicating the sales zone served, in order to obtain the greatest profit.

16-3. The General Radio Corporation is a large manufacturer of portable TV sets with plants in New York, New Jersey, Virginia, and North Carolina. For sales purposes, the company has established six sales territories which encompass the entire United States. Each territory has a district manager in charge of all sales activity occurring within his zone. Prior to the beginning of every month, it is the responsibility of the district manager to estimate the number of TV sets which will be sold in his territory in the subsequent month, i.e., prior to January 1st the estimated February sales is due, etc. This sales estimate is then used to establish a shipping schedule for the four plants consistent with the plant capacities.

The following sales estimates for next month have just been received at the central office.

Sales Territory	Sales Estimate
New England	5,500 units
Central Atlantic	6,000 units
South Eastern	4,000 units
Midwest	8,000 units
Rocky Mountain	5,000 units
Pacific Coast	4,300 units

The company has determined that its four manufacturing facilities must produce TV sets within the following restrictions.

	Minimum Monthly Output	Maximum Monthly Output
New York	10,000 units	14,000 units
New Jersey	3,500 units	6,500 units
Virginia	3,500 units	6,000 units
North Carolina	6,000 units	7,500 units

If the production in any plant drops below the minimum monthly output, the factory would have to be shut down because of excessive financial losses.

The estimated unit manufacturing costs based upon historical data are as follows:

New York	$31.50
New Jersey	$32.75
Virginia	$32.40
North Carolina	$32.90

The following table shows the freight rate per unit shipped from the manufacturing facilities to the respective district warehouse.

	New York	New Jersey	Virginia	North Carolina
New England	$0.80	$1.35	$1.70	$1.80
Central Atlantic	$1.20	$0.85	$1.40	$1.50
South Eastern	$1.90	$1.75	$1.00	$1.00
Midwest	$2.20	$2.55	$2.70	$2.60
Rocky Mountain	$3.40	$3.85	$3.60	$3.70
Pacific Coast	$4.90	$4.75	$4.80	$5.10

The wholesale price of the TV sets at the warehouses is $42.00 except in the Rocky Mountain and Pacific Coast areas where the TV sets are listed at $45.00.

What is the best shipping schedule in order to maximize the profits of the General Radio Corporation?

16-4. You are scheduling manager for three plants located in Atlanta, Ga. (AT), Louisville, Kentucky (LS), and New Orleans, La. (NR), which manufacture six products (A, B, C, D, E, and F). The plants supply ten sales districts located in the southern part of the United States. Each plant manufactures at a different unit cost per product. Freight rates for the products differ depending upon the plant which is supplying the sales district. The cost price per unit to the customer is the same, however, regardless of which plant is manufacturing and supplying.

The sales districts furnish you, one month in advance, their sales estimates for the coming month. Using these estimates you prepare master production and distribution plans. These plans are sent to the sales districts and to each of the three plants which, in turn, prepare their own detail schedules.

Product demand (in units) for the coming month, as taken from the sales estimates, follow below:

Sales District	A	B	C	D	E	F
1	1000	1500	1400	500	750	2000
2	250	200	300	1000	400	1600
3	700	500	200	300	500	1200
4	3000	2000	2500	100	800	1500
5	300	1200	1800	200	700	2200
6	1300	150	2200	500	900	1700
7	200	700	1300	400	550	1700
8	500	600	1500	100	600	1300
9	1100	400	1000	600	600	1800
10	1300	300	800	900	400	1400

Unit cost by plant by product:

Product	New Orleans	Atlanta	Louisville
A	$2.00	2.10	*
B	1.50	1.40	1.60
C	1.10	*	1.20
D	2.20	2.00	1.90
E	*	3.20	3.60
F	.90	1.00	.70

* Does not manufacture product.

Unit Sales Price:

Product	A	B	C	D	E	F
Sales Price	$4.00	$3.00	$2.60	$2.20	$5.00	$2.40

Plant Capacities in Units per Month:

New Orleans	35,000
Louisville	10,000
Atlanta	15,000

Special Plant Capacity Limitations on Particular Products in Units per Month:

Product	NR	AT	LS
B	—	2500	—
F	—	—	4500

Average freight charges per unit for a given plant to supply a specific sales district:

			Freight Rates by Product				
Sales Districts	A	B	C	D	E	F	Plant
1, 2, 3, 4	$0.20	.30	—	.40	.60	.50	Atlanta
5, 6, 7	.40	.50		.70	.60	.60	
8, 9	.40	.30		.80	.80	.60	
10	.30	.40		.50	.70	.50	
1, 2, 3, 4	—	.30	.30	.60	.40	.75	Louisville
5, 6, 7	—	.20	.10	.30	.40	.60	
8, 9	—	.60	.30	.50	.40	.90	
10	—	.40	.20	.40	.50	.80	
1, 2, 3, 4	.30	.30	.20	.40	—	.60	New Orleans
5, 6, 7	.40	.45	.30	.50	—	.80	
8, 9	.20	.25	.10	.20	—	.60	
10	.30	.35	.15	.30	—	.70	

From the data furnished above, you are to:
   a. Prepare a master plan by plant showing the products that they will produce for the coming month.
   b. Show what quantity of the product each plant will supply to each sales district (by each product).
   c. Determine the profit for your plan.

Chapter 17   QUESTIONS

17-1. Explain the statement, ". . . the system must be designed *first* and the commercial aid selected."

17-2. Explain the major disadvantage of accepting the free-consulting aid offered by the manufacturers of commercial scheduling devices.

17-3. Give an example of the type of situation in which a Sched-U-Graph could be effectively used. Explain the set-up of the board.

17-4. Give an example of the type of situation in which a Produc-Trol Board could be effectively used. Explain the set-up of the board.

17-5. Give an example of the type of situation in which a Boardmaster could be effectively used. Explain the set-up of the board.

17-6. Describe a situation in which an Acme Visual Control Panel could be effectively used. Explain the set-up of the board.

Chapter 18   QUESTIONS

18- 1. Explain the statement, "Without close control of the dispatching function there is little assurance that the work will be accomplished in accordance with the established plan."

18- 2. What is meant by "line prerogatives"?

18- 3. What prerogatives does the Production Control staff possess? How does it secure these prerogatives?

18- 4. Discuss the major advantages of having a staff member act in the capacity of a dispatcher.

18- 5. Discuss the advantages of having a line member act in the capacity of a dispatcher.

18- 6. If the plans established by Production Control are introduced into the line above the foreman level, who must ultimately do the dispatching? Explain.

18- 7. Discuss the proportion of time that a dispatcher should spend actually releasing work authorization.

18- 8. What kind of information pertaining to dispatching would the higher management of an organization require?

18- 9. Describe the type of relationship existing between the foreman of a department and the staff dispatcher of the same department.

18-10. Discuss the advantages and disadvantages of selecting a centralized controlled system of dispatching.

18-11. In a small informal organization would you suggest the use of centralized or decentralized dispatching? Discuss.

18-12. In a large, multiple-plant organization would you suggest the use of centralized or decentralized dispatching? Discuss.

Chapter 18   PROBLEMS

18-1. Using the procedure analysis technique described in Chapter 23, show how centralized dispatching would work in a toy manufacturing plant.

18-2. Using the procedure analysis technique described in Chapter 23, show how decentralized dispatching would work in a toy manufacing plant.

18-3. Using the procedure analysis technique described in Chapter 23, show the difference between a high level line dispatching and a staff dispatching system. Discuss the differences in terms of operational flexibility, cost, and organizational design.

Chapter 19   QUESTIONS

19- 1. Discuss the reasons why progress reporting is so important to both the Production Control and accounting groups.

19- 2. If you were the designer of a Production Control system, how would you make sure that too much information was not reported?

19- 3. How would you propose to control "illegal" files in an organization on a continuing basis?

19- 4. Discuss the advantages and disadvantages of combining forms for progress reporting.

19- 5. What would be your approach in determining how frequently a progress report should be made?

19- 6. If progress reports are to be made on the basis of work accomplishment, how would you determine the amount of accomplishment which would warrant a report?

19- 7. Discuss the use of "management by exception" in progress reporting.

19- 8. Discuss the purpose and use of "audit points."

19- 9. Explain, in your own words, the 5 elementary phases of the progress reporting cycle.

19-10. Under what type of circumstance would you propose the use of the written system of progress reporting?

19-11. Under what type of circumstance would you propose the use of the oral system of progress reporting?

19-12. Under what type of circumstance would you propose the use of the electronic system of progress reporting?

Chapter 20   QUESTIONS

20-1. Discuss the type of situation in which prewritten papers offer substantial savings.

20-2. Compare the use of pneumatic tubes with the use of TelAutograph equipment.

20-3. Under what type of circumstance would teletype equipment be justified as the means of dispatching and progress reporting?

20-4. Discuss the type of situation in which telephones could be used as the sole media for progress reporting.

20-5. In your own words describe the operation of the Telecontrol equipment. Give your evaluation of this equipment.

20-6. Discuss the use of the progress report cycle concept in selecting the commercial equipment to be used in reporting.

Chapter 20   PROBLEMS

20-1. Draw up a set of hectographic forms for a furniture manufacturer including:
   a. Assembly orders
   b. Move tickets
   c. Job tickets
   d. Material requisitions
Assume that these forms are to be made from one hectographic master.

20-2. Using the procedure analysis technique described in Chapter 23, show how the forms in Problem 1 might be used in a furniture plant.

20-3. Draw up a form which could be used on a TelAutograph telescriber to dispatch work to the various work centers in a toy manufacturing plant.

Chapter 21   QUESTIONS

21- 1. In your own words, what is meant by the binary numbering system? What is its relationship with data processing?

21- 2. Discuss the types of situations in which marginally-punched cards could be effectively used.

21- 3. Why do the marginally-punched cards have one corner cut off?

21- 4. What is a "field" in a marginally-punched card?

21- 5. Under what type of circumstance would punched cards be used for data processing?

21- 6. What is the major advantage of computers over all other systems of data processing?

21- 7. What is a "logical decision"?

21- 8. In your own words describe the storage of information on a magnetic drum.

21- 9. What is your opinion of the operation of the Navy ordnance system of data processing? Discuss.

21-10. Discuss the future use of computers in Production Control work as you see it.

Chapter 21   PROBLEMS

21-1. Set up some simple addition and subtraction problems and solve them by using the binary numbering system.

21-2. Set up some simple multiplication and division problems and solve them by using the binary numbering system.

21-3. Design a marginally-punched card that would be used by a manufacturer of kitchen appliances, i.e., toasters, coffee makers, electric frying pans and so forth. Make provision to:
   a. determine sales by item number.
   b. determine sales by salesmen.
   c. determine sales by month.
   d. determine sales by sales territory.

21-4. Design a punched card that will do the same sorting as given in Problem 21-3.

Chapter 22   QUESTIONS

22- 1. What are some examples of unanticipated events which would make corrective action necessary?

22- 2. In your opinion, would external or internal factors create the most need for corrective action? Discuss.

22- 3. What is a firm schedule? Is it firm? Discuss.

22- 4. Discuss the manner in which you would reduce the need for corrective action due to the delay in receipt of raw material.

22- 5. How would you reduce the need for corrective action due to unexpected absenteeism?

22- 6. Discuss the use of schedule flexibility in corrective action. Under what kinds of circumstances would you suggest using it?

22- 7. Discuss the pros and cons of schedule modification as a means of corrective action.

22- 8. What would be the costs associated with changing the number of standard hours available in a facility?

22- 9. Discuss advantages of changing work amount within the facility.

22-10. Under what type of situation would expediting the work through a facility be the best corrective action?

22-11. What advantage does the fathering system of expediting have over the "exception" principle?

Chapter 23   QUESTIONS

23-1. Why is it of value to the Production Control designer to use the procedure charting techniques described in Chapter 23?

23-2. Describe some other techniques of charting systems such as those used by office equipment and business forms suppliers.

23-3. What are the advantages and disadvantages of the various techniques of charting systems?

23-4. List 5 specific activities to which each of the techniques listed below would logically be applied:
   a. Work Activity Analysis Sheet
   b. Daily Clerical Utilization Chart
   c. Task List

Chapter 23 PROBLEMS

For each of the problems for Chapter 23: (a) First chart the flow of the paperwork only. (b) Then chart the flow of the paperwork showing departmental divisions on the chart and a personnel column for each department.

23-1. John Doe, a typical American male, wearily surveys the accumulation of bills for the month of January and seats himself at his desk, armed with check book and pen, and prepares to distribute his income, in the traditional manner. The desk contains a place for envelopes and a supply of stamps is kept in a convenient drawer. For each bill stacked in a corner of the desk, John picks up the sealed bill, opens it, withdraws the bill and tosses the envelope in a wastebasket. He then reads the bill, makes out a check, fills in the stub, addresses the envelope, separates the portion of the bill to be returned with the remittance, places check and portion of bill into envelope. This routine is occasionally interrupted when John calls his wife into the room and questions an expenditure of unusually large size or nature, but eventually he completes payment in the usual manner. The material is placed in the envelope, the envelope is sealed and a stamp is affixed, and the sealed envelope is placed to one side of the desk. When all bills are completed in this manner, John dons hat and coat and walks to the corner mail box and deposits the accumulated mail. Periodically the mail truck picks up the mail and carries it to the post office, where it is sorted and eventually distributed to receivers.

23-2. The following procedure describes part of the production control system in a plant manufacturing all types of industrial hydraulic presses and pumps. It is basically a job shop, manufacturing to customer's specifications in most cases, although many standard parts for subassemblies are used in the finished product. The plant is relatively small (approximately 350 employees).

The Sales Department prepares 2 typed copies of a quotation request from an order for a "special" piece of equipment. The original quotation request and one copy of the order are sent to the Engineering Department; the duplicate quotation request is sent to the Cost Section of the Production Department.

The Engineering Department determines the time to perform the Engineering work and enters this on the quotation request. A bill of material and operation sheet (sequential listing of operations) are then prepared from drawings on file or sketches prepared specially for this order. All forms are then forwarded to the production department, cost control section.

The cost section determines costs for all material, labor and burden to produce the equipment ordered, and enters them on a quotation form.

The quotation form and quotation request are returned to the production manager; all other forms are filed in the Cost Section.

The production manager sends the quotations, with delivery recommendations, to the sales department who then submits the quotation to the customer.

After approval from the customer, the sales department prepares 8 copies of a production order and submits the original to the production department with copies to engineering, inventory control, tool department, assembly department, shipping department, and billing department.

From its copy, the engineering department makes up a complete set of drawings and bill of material and forwards to Production Control. Production Control makes 4 copies of an operation sheet for all parts, including assembly operations, and forwards to cost section, stockroom, and one copy to the Shop.

23-3. The following is the written procedure for a system used in Production Control of a plant manufacturing seasonal items for stock.

ITEM PRODUCTION ORDER PROCEDURE

At the beginning of each season as soon as the items have been established in the line, the Production Scheduling Supervisor makes out the *"Master* Item Production Order" (Form No. 3). The following information is first written on a blank dittoed form: Description, item number, contents and description, materials, standard unit cost, special instructions, label, operation descriptons, cost center, machine or work center, and production rate/hour (if available). The written copy is given to the plant superintendent, production manager, and purchasing for review and corrections. Corrections are noted on the form by each reviewer and initialed. The corrected copy is returned to the scheduling supervisor. The "Master Item Production Order" is typed from the corrected written copy.

Copies are run from the spirit-duplicator master in the following quantities:

Regular items: 9 white and 3 yellow.
Assortments: 14 white and 5 yellow.

One white copy is sent to industrial engineering for production standards (production rate/hour). As soon as standards have been set, they are written on this copy and returned to the scheduling supervisor and filed with the remaining copies. If production standards have been previously set, engineering will review to determine if any changes are required or have been made and note the changes on the copy. One white copy is sent to the cost department for information purposes and determining cost standards. (See Cost Department Procedure.) The remaining copies are kept in production scheduling as reserve copies until production is scheduled on the item.

When a production run on an item is required, as determined from the schedule (See Scheduling Procedure), two white and one yellow copy of the reserve copies are pulled from the file and typed with the following information: Production order number, make quantity, start date, finish date, material quantity. The yellow copy and one white copy are sent to the plant superintendent with the required number of work cards (See Procedure for Making and Processing Work Cards). The plant superintendent gives the white copy to the stockroom supervisor as a requisition for stock. After stock has been withdrawn, the white copy is returned to the plant superintendent. The white copy is filed by item number in the plant superintendent's office for reference. The other white copy is sent to the inventory control department. Inventory control posts the scheduled quantity of the item and bulk stock to a work sheet. The white copy is then returned to production scheduling. The other white copy is filed by item number in the scheduling department as an information copy and daily production quantity is recorded on the back from the work cards.

After the job has been run, the yellow copy is returned to production scheduling showing actual quantity completed and any substitutions made. The white copy in production scheduling is marked "complete" and placed in a "completed production file" by item number. The yellow copy is sent to inventory control and quantity produced is written on the Kardex Item Inventory Record (See "Inventory Control Procedure—Item Production Analysis"). Bulk stock used is written on the "Bulk Stock Inventory Kardex Record." Inventory control sends the yellow copy to the cost department for costing of the job (See Production Cost Procedure).

23-4. Visit a local plant or any other activity selected and chart a paperwork system as it exists using the SPAS #1, #2 and #3. Using the scientific method, improve the system and chart the improved system in the same manner.

Chapter 24 QUESTIONS

24-1. Who in an organization should be assigned the responsibility of making work sampling studies? Why?

24-2. List at least 10 possible objectives or reasons why specific work sampling studies might be made.

24-3. Describe the training required for any particular person to make work sampling studies.

24-4. Assuming that you will make a work sampling study of one of the situations described in Question 24-2, describe specifically how you would proceed to inform the people affected by the study about the objectives of the study.

24-5. Describe as many reasons as you can why work sampling study results may not be valid.

Chapter 24 PROBLEMS

24-1. It is decided to make a work sampling study of the clerical operations in an inventory control section. Determine the number of observations required if:

   a. Accuracy desired is 5% of p; p is estimated to be 20%.

   b. Accuracy desired is 1%; p is estimated as 9%.

How many observations must be made each day in each situation if the study must be completed by one man in working 20 days?

24-2. In the example of problem 24-1, a maximum of one observation can be made each 10 minutes.

   A. How many observers would be needed if the study in each case must be completed in 10 working days?

   B. At the end of 5 days, p in case A was found to be closer to 16% than 20%. How many observers would be needed to get the required observations in the remaining time?

C. How many in case B if p is 12%?

24-3. Make a complete work sampling study of a selected situation for a period of one week. Obtain the assistance of others to get as many observations as possible.

Chapter 25 QUESTIONS

25-1. Why is it important to have a systematic method of evaluating a Production Control system?

25-2. Discuss the basic differences between qualitative and quantitative evaluation.

25-3. Is there any qualitative evaluation involved in a quantitative measure? Explain.

25-4. Review the six principles of a good Production Control system and develop some new ones of your own. Be sure you are not restating any of the six.

25-5. What are the advantages of discussing the Production Control system with everyone involved rather than only one or two top-ranking people?

25-6. What is the major purpose of establishing a work load analysis?

25-7. Discuss the use of the planning and performance analysis.

25-8. Discuss the use of the "cause" analysis.

Chapter 26 QUESTIONS

26-1. What are the functions that should be placed under the responsibility and authority of Production Control? Describe your reasoning in detail.

26-2. What are the important factors to consider in determining at what level in the organization Production Control should be placed?

26-3. How is the relationship of Production Control to other functions determined? Give at least two specific examples.

Chapter 26 PROBLEMS

26-1. Visit a local plant which has a Production Control organization and draw the organization chart of the plant showing the position of Production Control and its relationship with other functions. Describe fully your evaluation of the organizational structure.

26-2. Write the position description of the Production Control manager (or equivalent) as it exists. Describe fully your evaluation of the position's responsibility, authority, and relationships.

Chapter 27 QUESTIONS

27-1. Where would you start in establishing a formal Production Control function, with the organizational structure or the systems? Describe fully your reasoning.

27-2. List, in order, the steps that you would follow in setting up a formal Production Control function in a manufacturing plant.

Chapter 27 PROBLEM

27-1. For the plant in Problem 26-1, write a complete and detailed plan for organizing and implementing the Production Control function using your evaluation as a guide. Describe each step as listed in Question 27-2.

INDEX

Accounting:
 progress reporting information, 199-200
 summary punching, 326
 work authorization relationship, 67-68
Acme Visible Record, 102
Acme Visible Records Company, 191
Action phase of production control:
 control function relationship, 17
 dispatching, 18 (see also Dispatching)
 functions, 22-23
Activity levels, 29
Adjusted sales dollars, 46-47
Administration, activity classification, 29
Annual Survey of Manufacturers, 51
Assembly chart, 164, 165
Assignment of work tasks (see Scheduling)
Auditing:
 improvement, winning support for, 14-15
 quantitative evaluation, 282-284
 system benefit necessity, 6
Audit points, 205-206
Audit sheet factors, 282-284
Balance-to-date cards, 326
Basic motion time, 142
Bill of materials:
 product planning, 73
 scrap allowance, 92-94
 use, 91-94
Binary numbering system, 220-223
Block control, 37n (see also similiar-
 process control)
Boardmaster, 189-190
Brass ring system, 133
Breakeven analysis, 84-86
Breakeven point formula, 113-115
Brisley, C. L., 266
Budgetary control check, 70
Business Week, 51
Calculating, 326
Cancelled orders control, 239-240
Capacity modification, 243-244
Card punching, 318
Card-to-card transceiving, 327

Card verifying, 319
Carrying costs:
 definition, 111-112
 economic order quantity formula, 114,
 115
 quantity discount comparison, 122-123
Case study approach:
 limitations, 9-10
 work sampling, 274-275
Cause analysis, 284
Central tendency, 306
Chance control system, 8
Chip draw, 273
Choice system of control, 8, 9
Clerical activities, classifications, 28
Clock hours, 166-167
Closed-circuit television, 217
Coefficient of variation, 171-172
Collecting data (see Data processing)
Combination system of production control,
 8 (see also Production control)
Commercial data processing equipment,
 225-235
Commercial loading and scheduling de-
 vices, 183-191
Commodity organization, 298
Communications (see also Progress report-
 ing) :
 data processing basis, 23
 management coordination, 295
 progress reporting equipment, 209-219
Component parts, 72-73, 76
Computers, 230-235
Conference method, 14
Confidence limits, 58, 59, 60
Construction activities, classifying, 29
Continuous forecasting, 52-53 (see also
 Forecasting)
Continuous process production control,
 36-37
Continuous production, 29-30, 32 (see
 also Production activities)
Control, 1 (see also specific type of control)

Control charts:
  forecast aid, 61-62
  seasonality factors, 171-172
  significance, 169n
  statistical explanation, 305-307
  work sampling reliability, 274, 275
Control-limit technique:
  application, 57-62
  limitations, 64
Controlling function of management, 6-7
Control phase, 17 (see also Follow-up phase)
Coordinating function of management, 6, 294-295
Corrective action:
  causes, 238-241
  evaluation, 281
  methods, 241-244
  need, 237-238
  overview, 24
  summary, 244
  work sampling, 274
Correlation (see also Forecasting):
  data sources, 50-51
  forecasting techniques, 49-51
Cost cutting:
  index method of scheduling, 172-176
  material inventories, 109
  procedure analysis (see Procedure analysis)
  process planning improvements, 86
  two-bin material control, 104-105
  value analysis group, 76
Costs (see specific cost)
Cumulative probability distribution, 58
Current correlators, 50
Custom control system, 8-9, 64, 95-96
Custom work process (see also Production activities)
  characteristics, 33
  control, 64, 95-96 (see also Special-project control)
  examples, 32-33
Daily Clerical Utilization Chart, 249
Data processing (see also Progress reporting):
  binary numbering system, 220-223
  calculating, 326
  card punching, 318
  card-to-card transceiving, 327
  card verifying, 319
  collecting data, 23
  definition, 23
  detail printing, 324
  digital computers, 230-235
  duplicating, 319
  electronic progress reporting, 208

Data processing: (Cont.)
  end printing, 321
  exploding, 91-92
  facsimile posting, 327
  gang punching, 320
  group printing, 325
  interpreting data, 23-24, 321
  marginally punched cards, 223-225
  marked-sensed punching, 321
  matching, 324
  merging, 323
  Navy system, 233-235
  punched cards, 225-230
  reproducing, 320
  selecting, 323
  sorting, 322
  statistical work, 325
  summary, 235-236
  summary punching, 326
  tape reading, 328
  teletype feature, 215
  ticket converting, 322
  typewriter tape punching, 328
  work sampling, 273-274
Design of production control:
  assigning functions, 302
  basic principles, 296-299
  defining functions, 299-300
  evaluation, 304
  flow system, 36-37
  fundamentals v. case study, 9-10
  human relations factors, 13-15
  job-order control, 38
  loading and scheduling, 183, 184
  necessary information, 11
  personnel selection, 300-301
  planning and controlling, 303
  principles, 25-26
  procedure analysis (see Procedure analysis)
  process planning, analyzing, 88-89
  progress reporting, 207
  requirements, 17
  similar-process control, 37-38
  special-project control, 38-39
  staff size, 301
  summary, 304
  tool analysis, 136-137
  training programs, 302-303
  winning support, 13-15
  work measurement analysis, 148-149
Detail printing, 324
Digital computers, 230-235
Dimensional motion time, 142
Directing function of management, 6
Dispatching:
  centralized control, 196

Dispatching: *(Cont.)*
  decentralized control, 197-198
  definition, 22-23, 192
  duties, 192-193
  function chart, 18
  line v. staff controversy, 193-196
  Telecontrol equipment, 217-219
Dispatching control, 196-198
Dispersion, measure of, 306
Distribution, linear programming example, 311-317
Dollar usage inventory classification, 116-119
Duplicating, 319
Economic order quantity:
  balance of inventory costs, 109-110
  breakeven formula, 113-115
  carrying cost, 111-112
  modifications of formula, 116
  objective, 112
  procurement costs, 110
  quantity discount, 121-123
  restricting conditions, 119-121
  set-up costs, 110-111
  simplified calculations, 115-116
  variable usage, 123
Electrical accounting machines, 92
Electronic computers, 220-223
Electronic data processing machines, 92, 228-235
Electronic progress reporting, 208
Electronic Statistical Machine, 325
End printing, 321
Engineered standard, 141-142
Engineering activities, classification of, 29
Equipment *(see* Material management; Tool control; *or specific item)*
Equipment breakdown, 241
Estimates *(see* Forecasting)
Evaluation of production control:
  basic methods, 277-279
  qualitative, 279-281
  quantitative, 282-284
"Exception" principle, 244
Expediting:
  corrective action technique, 243-244
  definition, 24
Exploding, 91-92
Externally-generated work, 68
Facsimile posting, 327
"Farming out" corrective action, 243
"Fathering" principle, 244
Firm schedule concept, 238-239
"First come, first served" scheduling, 173-174
Fixed costs, 84
Flow charts, 256, 257, 258

Flow production control:
  characteristics, 36-37
  material control situation, 94-95
Follow-up phase:
  corrective action, 24 *(see also* Corrective action)
  function terminology, 18
  interrelationship among phases, 17
  progress reporting, 23-24 *(see also* Data processing; Progress reporting)
Forecasting:
  analyzing, 53-54
  control *(see* Forecasting control)
  correlation, 49-51
  definition, 19, 40
  deviation causes, 56
  frequency of comparisons, 55-56
  historical estimate, 43
  judgment factor, 52
  local level, 51-52
  market surveys, 48-49
  need and uses, 41-42
  revising, 56-57 *(see also* Forecasting control)
  sales force estimate, 44-45
  summary, 54
  timing, 52-53
  trend lines, 45-48
  types, 42-43
  use in process planning, 80, 89
Forecasting control:
  control-limit technique, 57-65
  summary, 64-65
  when to revise, 56-57
Foreman, dispatching function, 195-196
Formal production control *(see also* Production control):
  definition, 7
  functions, 17-19
Forms Analysis Chart, 252-253
Forms improvement and control:
  analysis, 262-263
  check list, 263
  product planning, 72
  work sampling design, 268-270
Functional organization, 298
Fundamental approach v. case study, 10
Future work, estimating, 19 *(see also* Forecasting)
Gang punching, 320
Gantt-type load chart:
  Boardmaster application, 190
  description, 159-162
  Produc-Trol board, 187-189
  time depicted, 184
  Visual Control Panels, 191
Gilbreth scientific management concept, 5

Graphic Systems Company, 189
Gross national product, 50
Group printing, 325
Hancock Industries, 217
Historical data work measurement, 146
Historical estimate, 43 (see also Forecasting)
Holden and Shallenberger, 10
Human relations:
  historical development, 12
  problems of control installation, 13-14
  techniques, 14
IBM 705 computer, 231
"In control" situation, 61 (see also Control-limit technique)
Index method of scheduling, 172-177
Informal production control (see also Production control):
  custom factors, 8-9
  definition, 7
  dispatching function, 23
Innovating function of management, 7
Intercom equipment, 215-216
Intermittent process, 31-32 (see also Job-order process; Production activities)
Internally-generated work, 68-70
Interpreting, 321
Interviewers, 300-301
Inventory:
  classifications, 91
  dollar usage classification, 116-119
  economic order quantity, 109-116
  limited quantities, 119-121
  punched card records, 228-230
  stock-out effect classification, 129-130
  variable costs, 110
Inventory control (see Material control)
Job assignment (see Loading; Scheduling)
Job-order control:
  characteristics, 38
  material control, 96
Job-order process (see also Production activities):
  characteristics, 31-32
  control (see Job-order control)
  control-limit application, 64
  examples, 31
  reporting level, 290
Kardex file, 102, 184
Kiplinger Letter, 51
Kolmogoroff formula, 58-62
Leading correlators, 49-50
Lead time in reorder point formula, 125-126
Least-squares method of linear regression:
  concept explanation, 308-310
  disadvantages, 48

Levels of activity, 29
Linear programming:
  Modi method, 311-317
  scheduling analysis, 177-182
Linear regression, least-squares method of, 308-310
Load analysis sheet, 158, 159
Load cards, 187
Load charts:
  commercial application, 184-191
  Gantt type (see Gantt-type load chart)
Loading:
  available standard hours, 165-167
  commercial devices, 183-191
  control, 37n (see also Similar-process control)
  definition, 22, 150
  establishing program, 151-152
  information required, 152-153
  operations to be controlled, 153-154
  statistical load control, 168-172
  summary, 167
  technique selection, 154
  types of systems, 154-165
  unit of measure selection, 153
Loading by schedule periods, 162-165
Long-range forecasting, 42-43 (see also Forecasting)
Magnetic disc storage system, 232, 233
Magnetic drum information storage, 232
Magnetic tapes, 231-232
Maintenance department forecast, 45
Management:
  by exception (see Management by exception)
  controls, interrelationship of, 67-68
  forecasting function (see Forecasting)
  functional relationships, 294-295
  functions, 6-7
  load evaluation, 169
  of material (see Material management)
  production control:
    position, 10-11
    reporting level, 288-291
  support for control system, 13-15
  techniques, basis of, 4-6
  work measurement uses, 139
Management by exception (see also Management):
  definition, 26
  evaluation, 281
  forecasting role, 57
  progress reporting, 205-206
Manpower (see also Personnel):
  control, 22 (see also Loading)
  shifting, 243
  staff size, 301

Manual data processing, 223-225, 227
Manufacturing:
    classifications, 28
    production control example, 27
Marginally punched cards, 223-225
Marked-sensed punching, 321
Market surveys, 48-49 (see also Forecasting)
Mass production (see also Production activities) :
    characteristics, 30
    control (see Flow production control)
    examples, 29
    intermittent process contrast, 32
    material control situation, 94-95
Master scheduling, 154-157
Matching, 324
Material control:
    bill of materials, 91-94
    concentration areas, 91
    function, 90-91
    inventory classifications, 91
    material management, 91 (see also Material management)
    overview, 21
    physical control, 97-98
    record keeping, 98-103
    summary, 106-107
    supermarket concept, 105-106
    two-bin system, 104-105
    types of situations, 94-96
Material control card, 98-102, 104
Material-in-process control, 91
Material management:
    basic problems, 91
    definition, 108
    dollar usage classification, 116-119
    economic order quantity, 109-116
    limited quantity modifications, 119-121
    obsolescence factor, 116
    quantity discount, 121-123
    reorder point determination, 123-130
    summary, 130
    supply delay control, 240
    variable usage, 123
Material nomenclature and identification, 91
Materials-handling procedures, 83
Mathematical programming:
    binary numbering system, 220-223
    military basis, $2n$-$3n$
    Modi method, 311-317
    scheduling analysis, 177-182
    value analysis, 74-76
McCaskey system, 133-135
Merging, 323

Methods improvement:
    relationship to process planning, 86
    work measurement relationship, 139-140
Methods time measurement, 142
Military basis of production control, 2-6
Modified load chart, 160
Modi method:
    basic explanation, 311-317
    linear programming, 180-182
    scheduling analysis, 177
    table, 179, 180
Monthly Labor Review, 51
Morrow, Robert, 266
Motion time analysis, 142
Multiple correlation, 50 (see also Correlation)
Navy Department progress reporting, 233-235
Nomographs in order quantity, 116
Normal curve, 306
Obsolescence factor, 116
Office layout checklist, 247
Off-standard performance (see Corrective action)
Off-the-job training, 302
On-the-job training, 302, 303
Operation methods, 29-34
Operation performance, 82
Operations research, 5
"Optimizing" in procedure analysis, 254
Order quantities, 108-130 (see also Material management)
Order scheduling technique, 161-162
Order writing, 19-20 (see also Work authorization)
Ordnance predetermined time system, 142, 144-145
Organization analysis, $2n$
Organization charts:
    procedure analysis use, 247-248
    reporting levels, 289, 290
Organization of production control:
    authority, 291-294
    basic principles, 296-299
    design (see Design of production control)
    goals, 286-287
    management relationships, 294-295
    overview, 7-9, 285-286
    range of activities, 287-288
    reporting level, 288-291
    summary, 295
Organization manual, 247-248
"Out of control" situation, 61, 64 (see also Control-limit technique)
Performance evaluation:
    analysis, 283

Performance evaluation: *(Cont.)*
  product planning requirement, 77
  work authorization relationship, 67-68
Periodic forecasting, 52-53 *(see also* Forecasting)
Perpetual loading, 158-161
Personnel:
  assigning to functions, 302
  computer record storage, 234
  control, 22 *(see also* Loading)
  evaluation *(see* Work sampling)
  process planning information, 82
  selecting, 300-301
  staff size, 301
  use in process planning, 80-81
Physical control, 97-98
Planning analysis, 283
Planning phase:
  action planning, 20-22
  forecasting *(see* Forecasting)
  function terminology, 18
  interrelationship with control functions, 17
  prior-planning, 19-20
  product planning *(see* Product planning)
  work authorization *(see* Work authorization)
Plans, definition, 1
Plant manager, dispatching duty, 194
Plant superintendent, dispatching function, 194
Pneumatic tubes, 213
Poisson Distribution, 127-128, 130
Position description, 299-300
Posted card, 98-100
Predetermined time systems, 142-143
Preparation of specifications, 20 *(see also* Product planning)
Preparation of work authorization, 19-20 *(see also* Work authorization)
Prewritten papers, 210-213
Prior planning:
  evaluation, 281
  forecasting *(see* Forecasting)
  product design *(see* Product planning)
  scope, 19-20
  work authorization *(see* Work authorization)
Probability, statistical, 267, 307
Problems, 329-359
Procedure analysis:
  definition, 245
  design tool, 245-246
  scientific method, 246
  standard procedure, 247-265
  technique summary, 265

Procedures description, 300
Procedures improvement checklist, 260-262
Procedures manual, 300
Procedure writing, 81 *(see also* Process planning)
Process planning:
  advantages, 86-87
  analyzing function, 87-89
  breakeven analysis, 84-86
  definition, 20, 79
  information needed, 20-21
  methods improvement relationship, 86
  objective, 80
  prerequisite information, 80-81
  product planning relationship, 77-78
  resulting information, 81-84
  scheduling overlap, 80 *(see also* Scheduling)
  steps, 81
  summary, 89
  tool planning *(see* Tool control)
Process sheet, 83
Procurement:
  costs, 110, 111
  custom-order industry, 96
  economic order quantity formula, 114, 115
  lead time, 125-126
  mass production industry, 95
  tools, 132-133
Product design, 20, 72-78
Production activities:
  basic processes, 29
  classification, 27-29
  custom work, 32-33
  job-order process, 31-32
  mass production:
    characteristics, 30
    examples, 29
  similar process, 30-31
  types of processes, table, 34
Production control:
  auditing *(see* Auditing)
  basic principles, 3, 4, 5
  basic systems, 35-39
  benefits, 6-7
  corrective action *(see* Corrective action)
  data processing *(see* Data processing)
  definition, 1-2, 245
  department organization *(see* Production control department)
  design *(see* Design of Production Control)
  digital computer use, 230-235
  dispatching *(see* Dispatching)
  evaluation, 277-284
  forecasting *(see* Forecasting)

Production control: *(Cont.)*
  functions, 2, 16-24, 26, 287-288
  historical development, 2-6
  human relations *(see* Human relations)
  line introduction, 193-196
  loading *(see* Loading)
  material control *(see* Material control)
  material management *(see* Material management)
  nature, summary, 11
  objectives, 6-7
  organization *(see* Organization of production control)
  phases, 17-24
  principles, 24-26
  process planning *(see* Process planning)
  product planning *(see* Product planning)
  questions and problems, 329-359
  scheduling *(see* Scheduling)
  tool control *(see* Tool control)
  use of systems, 13-15
  variable sales pattern, 177-182
Production control department:
  analysis of forecast, 53-54
  authority, 291-294
  evaluation, 304
  functions, 18
  line-staff controversy, 193-196
  organizing, 285-295
  planning and controlling, 303
  reporting level, 288-291
  staff size, 301
  work authorization coordination, 67
Product mix:
  analysis, 56
  forecasting factor, 53
Product planning:
  bill of materials, 73
  design forms, 72
  information for process planning, 80, 89
  overview, 20
  problems, 77
  sub-assemblies and component parts, 72-73
  summary, 77-78
  value analysis, 73-76
Produc-Trol board, 187-189
Progress reporting:
  abuses, 201-203
  basic information, 199-201
  commercial equipment, 209
  data processing *(see* Data processing)
  definition, 199
  fixed time intervals, 204
  frequency, 203-204
  methods, 207-208

  Navy system, 233-235
  overview, 23-24
  pneumatic tubes, 213
  prewritten papers, 210-213
  radio and television, 216-217
  TelAutograph Telescriber, 213-215
  Telecontrol equipment, 217-219
  telephone and intercom equipment, 215-216
  teletype equipment, 215
  timing, 206-207
  work accomplishment, 204-206
Project chart, 165
Punched cards:
  equipment, 102-103
  machine, 225-230
  marginal, 223-225
Purchasing function, 18 *(see also* Material management)
Qualitative evaluation:
  application, 279-281
  definition, 277
  quantitative contrast, 278-279
Quantitative evaluation:
  application, 282-284
  definition, 278
  qualitative contrast, 278-279
Quantity discount, 121-123
Questions and problems, 329-359
Radio, use in progress reporting, 216-217
Random time selection method, 271-273
Ratio Delay Study, 266 *(see also* Work sampling)
Record keeping, 98-103
Regression lines, 308-310
Remington Rand Sched-U-Graph, 184-187
Reorder point determination, 123-130
Reorder point quantity, 108-109 *(see also* Material management)
Replanning, 24 *(see also* Corrective action)
Reproducing, 320
Requisitions:
  design importance, 97
  traveling, 104
Research and development activities, classifying, 28
Route sheet, 83
Routing *(see also* Process planning):
  definition, 20, 21, 79
  dispatching overlap, 80 *(see also* Dispatching)
  resulting information, 82
Rush order analysis, 239
Safety stock:
  investment calculations, 129-130
  levels, 124, 125-127

Sales activities, classifying, 28
Sales force estimate, 44-45 (see also Forecasting)
Sched-U-Graph, 184-187
Schedule flexibility, 242
Schedule modification, 242
Scheduling:
  available standard hours, 165-167
  commercial devices, 183-191
  corrective action, 242
  definition, 150
  establishing program, 151-152
  firm concept, 238-239
  index method, 172-177
  information required, 152-153
  introducing (see Dispatching)
  linear programming method, 177-182
  operations to be controlled, 153-154
  planning, overview, 22
  product planning requirement, 77
  statistical load control, 168-172
  summary, 167
  technique selection, 154
  types of systems, 154-165
  unit of measure selection, 153
Scientific management (see also Management):
  human relations importance, 12
  item quantity, 109-110
  production control position, 10-11
  Taylor's and Gilbreth's concept, 5
Scientific method, 246
Scrap allowance, 92-94
Seasonality:
  comparison timing, 55
  determining pattern, 58-59
  economic order quantity, 124
  forecasting factor, 53
  load control, 171-172
Selecting, 323
Semi-variable budgeting, 51 (see also Correlation)
Service activities, classifying, 28
Set-up costs, 110-111
Shallenberger, Holden and, 10
Short-range forecasting, 42-43 (see also Forecasting)
Sig-Na-Lok, 102
Similiar-process control:
  characteristics, 37-38
  material situations, 95
Similar process operations (see also Production activities):
  characteristics, 30-31
  control (see Similar-process control)
  examples, 30
  material control, 95

Sorting, 322
Sorting needle, 224, 226
Space arrangement in procedure analysis, 248
SPAS (see Standard Procedure Analysis Set)
Special project control (see also Job-order control):
  characteristics, 38-39
  job-order similarity, 39
  reporting level, 290
Specifications, preparation, 20 (see also Product planning)
Spread, measure of, 306
Staff dispatching system, 195-196, 198
Staffing pattern, 51, 146 (see also Correlation)
Staff meetings, 14
Standard data concept, 147
Standard, definition of, 140
Standard deviation, 305, 306, 307
Standard hours, 166-167, 240, 243
Standardization:
  materials, 91
  tooling, 132
  work hour adjustments, 240-241
Standard operating procedure preparation, 81 (see also Process planning)
Standard Procedure Analysis Set, 255, 256-258
Statistical control:
  control-limit technique, 57-65
  limits, establishing, 170
Statistical load control, 168-172
Statistical quality control, 2n, 5
Statistical work, 325
Statistics:
  control charting, 305-307
  least-squares method of linear regression, 308-310
  load control, 168-172
  trend line analysis, 47-48
Stock-outs:
  classification by effect, 129-130
  cost, 126-127
  definition, 105, 111
  reorder point determination, 125, 126
Stop-watch method, 140-142, 267
Sub-assemblies, 72-73, 76
Subcontracting, 243
Summary punching, 326
Supermarket concept, 105-106
Supervision:
  authority principle, 297-298
  custom work skill, 33
  delegation principle, 299

Supervision: *(Cont.)*
  dispatching function, 193-196
  number of subordinates, 298-299
  process planning, 87, 88
  production processes, relative skill, 34
  reporting principle, 299
  training programs, 302-303
Survey of Current Business, 50-51
Surveys:
  designing tool, 13, 14
  workload requirements, 249-252
Systematic procedure, 1
System of chance causes, 305
Systems and procedures analysis, 245-265
Tables:
  functions of production control, 18
  Modi table, 179
  production processes, types of, 34
  random sampling numbers, 272
Tape reading, 328
Task List, 249
Taylor scientific management concept, 5
Technical estimates, 143-146
TelAutograph Telescriber, 213-215
Telecontrol equipment, 217-219
Telephone equipment, 215-216
Teletype equipment, 215
Television, use in progress reporting, 217
Testing programs, 300-301
Ticket converting, 322
Tickle file procedure, 95
Time estimates, 82-83
Time factor, 81
Time standards, stop-watch, 140-142
Tippett, L. C., 266
Tool control:
  analyzing, 136-137
  brass ring system, 133
  definition, 21-22, 131
  importance, 131-132
  McCaskey system, 133-135
  process planning information, 82
  procurement, 132-133
  summary, 137
Tool cribs, 133-135
Trade journals, 51
Training programs, 273, 302-303
Traveling requisition, 104
Trend lines, 45-48 *(see also* Forecasting)
Turnover, 303
Two-bin system, 104-105
Typewriter tape punching, 328
Usage Recap Sheet, 102
Value analysis, 73-76
Variable costs, 84
Variable usage factor, 123
Visi-Record, 102

Visual Control Panels, 191
Wage incentives, 139
Wassell Company's Produc-Trol board, 187-189
Watertown Arsenal, 5
Work Activity Analysis Sheet, 249
Work assignment *(see* Scheduling)
Work authorization:
  analyzing, 70-71
  authority forms, 66-67
  basic requirement, 67
  definition, 66
  externally-generated work, 68
  internally-generated work, 68-70
  preparing, 19-20
  process planning relationship, 80, 89
  relation to other controls, 67-68
  releasing *(see* Dispatching)
Work detail plan, 20-21
Work Distribution Chart, 249
Work factor, 142
Workload analysis, 283
Workload requirement survey, 249-252
Work measurement:
  definition, 138
  function analysis, 148-149
  historical data method, 148
  methods improvement relationship, 139-140
  predetermined time systems, 142-143
  staffing pattern, 146
  standard data, 147
  stop-watch standards, 140-142
  summary, 149
  technical estimates, 143-146
  uses, 138-139
  work sampling, 143 *(see also* Work sampling)
Work sampling:
  application, 266-267
  case study, 274-275
  categorizing, 268
  corrective action, 274
  data processing, 273-274
  defining objectives, 268
  form design, 268-270
  informing observed, 273
  number of observations, 271
  origins, 266
  procedure analysis, 251
  random time selection method, 271-273
  summary and references, 276
  theory, 267
  timing, 270-271
  training observers, 273
  work measurement technique, 143
Work simplification, 258